THE GOVERNMENT OF THINGS

The Government of Things

Foucault and the New Materialisms

Thomas Lemke

NEW YORK UNIVERSITY PRESS
New York

NEW YORK UNIVERSITY PRESS
New York
www.nyupress.org

© 2021 by New York University
All rights reserved

References to Internet websites (URLs) were accurate at the time of writing. Neither the author nor New York University Press is responsible for URLs that may have expired or changed since the manuscript was prepared.

Library of Congress Cataloging-in-Publication Data
Names: Lemke, Thomas, author.
Title: The government of things : Foucault and the new materialisms / Thomas Lemke.
Description: New York, N.Y. : NYU Press, 2021. | Includes bibliographical references and index.
Identifiers: LCCN 2021003107 | ISBN 9781479808816 (hardback) |
ISBN 9781479829934 (paperback) | ISBN 9781479810536 (ebook) |
ISBN 9781479890712 (ebook other)
Subjects: LCSH: Foucault, Michel, 1926–1984—Political and social views. | Political science—Philosophy. | Philosophy and education.
Classification: LCC JC261.F68 L46 2021 | DDC 320.01—dc23
LC record available at https://lccn.loc.gov/2021003107

Chapters 1 and 2 are based on previously published material: "Materialism Without Matter: The Recurrence of Subjectivism in Object-Oriented Ontology," *Distinktion: Journal of Social Theory* 18(2), 2017: 133–152; and "An Alternative Model of Politics? Prospects and Problems of Jane Bennett's Vital Materialism," *Theory, Culture & Society* 35(6), 2018: 31–54. Reprinted by the permission of the publishers.

New York University Press books are printed on acid-free paper, and their binding materials are chosen for strength and durability. We strive to use environmentally responsible suppliers and materials to the greatest extent possible in publishing our books.

Manufactured in the United States of America

10 9 8 7 6 5 4 3 2 1

Also available as an ebook

There is thus a tendency for any materialism, at any point in its history, to find itself stuck with its own recent generalizations, and in defence of these to mistake its own character: to suppose that it is a system like others, of a presumptive explanatory kind, or that it is reasonable to set up contrasts with other (categorical) systems, at the level not of procedures but of its own past "findings" or "laws." What then happens is obvious. The results of new material investigations are interpreted as having outdated "materialism."
(Williams 1980, 103)

I think that you are completely free to do what you like with what I am saying. These are suggestions for research, ideas, schemata, outlines, instruments; do what you like with them. [. . .] I could tell you that these things are trails to be followed, that it didn't matter where they led, or even that the one thing that did matter was that they didn't lead anywhere, or at least not in some predetermined direction. I could say they were like an outline for something. It's up to you to go on with them or to go off on a tangent [. . .].
(Foucault 2003, 2, 4)

CONTENTS

Introduction	1
PART I: VARIETIES OF MATERIALISM	19
1. Immaterialism: Graham Harman and the Weirdness of Objects	21
2. Vital Materialism: Jane Bennett and the Vibrancy of Things	40
3. Diffractive Materialism: Karen Barad and the Performativity of Phenomena	57
PART II: ELEMENTS OF A MORE-THAN-HUMAN ANALYTICS OF GOVERNMENT	79
4. Material-Discursive Entanglements: Grasping the Concept of the Dispositive	81
5. More-Than-Social Configurations: Expanding the Understanding of Technology	103
6. Beyond Anthropocentric Framings: Circulating the Idea of the Milieu	121
PART III: TOWARD A RELATIONAL MATERIALISM	141
7. Aligning Science and Technology Studies and an Analytics of Government	144
8. Environmentality: Mapping Contemporary Political Topographies	168
Conclusion: Multiple Materialisms	191
Acknowledgments	203
Notes	207
Bibliography	249
Index	289
About the Author	301

Introduction

Materialism is a rich philosophical tradition that goes back to antiquity. It started with the works of Democritus and Lucretius, was taken up and rearticulated in modern philosophy in the writings of Hobbes, Spinoza, and many others, and flourished in the nineteenth and twentieth centuries, especially due to the achievements of the natural sciences and the rise of Marxism (see, e.g., Braun 1982; Lange 2010). While materialist thought always had an important critical role in contesting different versions of idealism and spiritualism, its impact went well beyond academic disputes and intellectual debates. It not only denoted a position in a philosophical controversy but also figured prominently in popular discourse. Interestingly, "materialists" suffered from a bad reputation both in the world of theory and in the view of common sense. For centuries they were regarded as people of questionable character who did not believe in God, adhered to dubious morals, and expressed dangerous thoughts: an "evil sect" ("schlimme Sekte"), as an important German encyclopedia put it in the eighteenth century (Zedler 1739, 2026; see also Post and Schmidt 1975, 7).[1]

Things have changed today. At least in academia, materialism has become something respectable, serious, and even fashionable. And "things" have played a decisive role in this transformation: materials, artifacts, and objects are increasingly attracting scientific interest and are being freshly conceptualized. The past two decades have seen a remarkable development in the social sciences and the humanities: the rise of new materialisms (see, e.g., Hird 2004; Coole and Frost 2010a; Dolphijn and van der Tuin 2012).[2] Theoretical perspectives and empirical studies that focus on the diverse and plural forms of materiality are complementing or replacing research on social constructions, cultural practices, and discursive processes. New materialist scholarship shares the conviction that the "linguistic turn" or primarily textual accounts are insufficient for an adequate understanding of the complex and dynamic interplay of meaning and matter. It claims that the "hegemony of cultural, discursive, and textual methodologies" (Kirby 2017, 9) not only leads to impoverished theoretical accounts and conceptual flaws; the "perceived *neglect* or *diminishment* of matter" (Gamble et al. 2019, 111; emphases in original) also results in serious ethical quandaries and political problems, as it fails to address central challenges facing contemporary societies, especially economic change and the environmental crisis.[3]

The new materialisms are the result of a double historical and theoretical conjuncture. The 1970s and 1980s were marked by the decline of once popular materialist approaches, especially Marxism, and the rise of poststructuralist and cultural theories. While the latter rendered problematic any direct reference to matter as naïvely representational or naturalistic, new materialists endorse a novel concept of matter. In contrast to traditional forms of materialism, the new "theoretical paradigm" (Rekret 2018, 49–50) refers to the idea that matter itself is to be conceived as active, dynamic, and plural rather than passive, inert, and unitary (Bennett 2004, 348–49; Alaimo and Hekman 2008; Colebrook 2008; Coole and Frost 2010b, 3–4). New materialist scholarship criticizes the idea of the natural world and technical artifacts as a mere resource or raw material for technological progress, economic production, or social construction. The "material turn" (Bennett and Joyce 2010) aims at a new understanding of ontology, epistemology, ethics, and politics to overcome anthropocentrism and discursive idealism. It proposes to reconceptualize central dualisms of modern thought including the split between nature and culture, matter and mind, the human and the nonhuman (Dolphijn and van der Tuin 2012; Connolly 2013; Wilson 2015).

As new materialisms are still a vivid and dynamic field, it is difficult to chart the terrain, to specify its frontiers and foundations, and to establish what is distinctive about it. In the following I will briefly present a preliminary mapping of the new materialisms before explaining the argument and the structure of the book.

Situating New Materialisms

New materialisms need to be distinguished from three alternative conceptualizations of matter. First, material culture studies in anthropology investigate "the social life of things" (Appadurai 1998; see also Henare et al. 2007). They focus on how human subjects engage with material objects to generate or maintain social relations (see Miller 2005; 2008). In contrast to this approach, new materialists problematize the (hierarchical) conceptual distinction between a material and a non-material world. Starting out from a more comprehensive understanding of materiality, they insist that the differentiation between the human and the nonhuman is itself precarious and mobile (Anderson 2011a, 393; Tischleder 2014, 23–7). Secondly, new materialisms also differ from a "materialist essentialism" (Castree 2003, 8) that ascribes fixed and stable ontological features to things that result in unequivocal moral positions. Most new materialists rather conceive of "plastic materialities" (Hawkins 2010; Kirby 2017, 11; see also Bhandar and Goldberg-Hiller 2015), understanding matter as flexible and dynamic instead of rigid and solid—and so undermining principal and general moral judgments. Thirdly, new materialisms are directed against a "reductionist materialism" that has dominated science for a long time. It promotes a concept of matter as composed of discrete, simple, and passive elements. In biology, for example, it is characterized by the tendency to assert that "the apparently distinctive attributes of organisms arise from the properties of their component parts—cellular and, ultimately, molecular" (Benton 1991, 14; Gamble et al. 2019, 116; see also Stengers 2011).

New materialist scholarship does not represent a homogeneous style of thought, but rather encompasses a heterogeneous group of different approaches and theoretical orientations (Coole 2013, 452). Some take their inspiration from the phenomenological tradition (Harman 2005; Bogost 2012), while others turn to modern vitalism (Bennett 2010a). New materialists include Derrideans (Kirby 2011) as well as Deleuzeans (Braidotti 2002); there are works that rely on theorems from quantum mechanics (Barad 2007), perspectives of complexity theory (de Landa 2000), principles from evolutionary theory (Grosz 2008), and insights from neuroscientific research (Wilson 2015). New materialists endorse a "transversally new intellectual orientation" (Dolphijn and van der Tuin 2012, 86) that cuts across traditional academic profiles and established intellectual borders. The disciplinary spectrum extends from feminist theory (Braidotti 2002; Alaimo and Hekman 2008; van der Tuin 2011a) via theory of art (Bolt and Barrett 2013), political theory (Bennett 2010a), international relations (Lundborg and Vaughan-Williams 2015) and philosophy (Meillassoux 2008; Bryant et al. 2011; Harman 2011a), to media studies (Fuller 2005), geography (Wiley 2005), sociology (Fox and Alldred 2016), legal studies (Kang and Kendall 2020), archaeology (Witmore 2014), and literary studies (Tischleder 2014).

While the commitment to take matter more seriously is a thread common to all new materialisms, they differ substantially in the way matter is conceptualized and in how they conceive of the relation between ontology and epistemology. In fact, the terrain of new materialisms covers "partly incompatible trajectories" (Gamble et al. 2019, 111) and even includes "contradictory" identities (Kirby 2017, 8). It might therefore be premature or misplaced to seek to establish "criteria for inclusion and exclusion" (Devellennes and Dillet 2018, 6), as any clear-cut definition is open to contestation and dispute. Still, it is possible to identify a number of broadly shared interests and concerns. I propose four distinct but interlinked themes or topics to chart the terrain of the new materialisms: ontology, epistemology, politics, and ethics.[4]

New materialists first emphasize the need to reconsider *ontological* questions by taking into account the productivity and dynamism of matter. They propose to take a critical distance from the Cartesian-Newtonian understanding of ontology and to reconceptualize agency beyond the human subject. This "ontological reorientation" (Coole and Frost 2010b, 6–7) promises to transcend the modernist dualism of nature and culture, affirming the inventiveness and indeterminacy of matter. The second distinctive aspect of new materialist scholarship concerns *epistemology*, as it takes up or is even based on developments in the natural sciences. New materialists call for a stronger engagement of the social sciences and humanities with knowledge production in the natural sciences (Hird 2009; see also Alaimo and Heckman 2008; Wilson 2015; Devellennes and Dillet 2018). This scholarship invites us to question established disciplinary borders, in order to understand biology and nature as historical and contingent rather than governed by eternal and deterministic laws (Grosz 2008; Kirby 2011; Wilson 2015). The third aspect connects the rethinking of materiality to the matter of *politics*, seeking to develop a new form of analyzing power relations beyond the sphere of the human (Barad 2007, 35). New materialist scholarship envisions bringing together an interest in political economy with environmental issues and questions of social justice (Bennett 2010a; Connolly

2013). It proposes a "consideration of a raft of biopolitical and bioethical issues concerning the status of life and the human" and "a critical and non-dogmatic reengagement with political economy" (Coole and Frost 2010b, 7). Conceptual propositions such as "vibrant matter" (Bennett 2010a) or "vital matter" (Braidotti 2018) challenge existing concepts of subjectivity and agency and re-chart the scope and the terrain of the political (Rekret 2018, 50–1; Lundborg and Vaughan-Williams 2015, 4). The interest in political matters in new materialist work is often complemented by a reconsideration of *ethical* concerns. Instead of taking the autonomous individual or the moral subject as the point of reference, these concepts of ethics are based on the complex encounters between human and nonhuman entities and their constitutive relations of mutual dependence and exchange (Braidotti 2006; Haraway 2008).

Taken together, these ontological, epistemological, political, and ethical propositions promise to broaden and deepen empirical investigations and critical thought. The new accent on the dynamism and the vitality of matter contests the concept of an eternal and unchanging nature that is intimately connected to sexist, racist, and capitalist practices, thereby revising and expanding feminist, postcolonial, and Marxist critique. In this vein, new materialist scholarship seeks to open up the critical arsenal for inspection and innovation. It helps to problematize notions like "reification" or "naturalization," as their use might contribute to reinforcing the processes they critically address, operating with an undercomplex and inadequate understanding of nature that conceives of matter as passive, inert, and solid.

It was this promise of a radical rethinking of the political and a re-mapping of the terrain upon which politics and political contestation take place that captured my interest in new materialisms in the first place some years ago. However, while I shared many concerns and commitments with new materialist scholarship, it was difficult not to sense a certain uneasiness. There are two main reasons for this. First, representatives of new materialisms often seek to distinguish themselves sharply from what now has come to be labeled "old materialisms" (Cudworth and Hobden 2015; Bennett 2015a, 237, note 10; Kirby 2017, 15), especially the rich tradition of Marxism and important strands of materialist feminism (see, e.g., Hennessy and Ingraham 1997). Many self-declared new materialists employ to excess the rhetoric of novelty, breakthroughs, and originality, disregarding important materialist lines of thought and ignoring possible affinities and alliances (Devellennes and Dillet 2018). They often give the impression of a marketing brochure repeating one and the same message, over and over again. We are assured new materialisms are a "revolution in thought" (Dolphjin and van der Tuin 2012, 85) or that we are witnessing "the bliss of a new thinking" (Morton 2011a, 163). However, permanently rehearsing how "revolutionary" and "radical" new materialisms are tends to obscure the fact that materialism was always engaged in renegotiating and updating its agenda in confronting its counterpart, whether it was idealism, spiritualism, or something else. In this perspective, materialism as a "revolution in thought" is not breaking news but business as usual.[5]

The second source of my unease is closely related to the first and concerns the question of critique. New materialist scholarship puts forward an understanding of critique as a somehow outdated or particularly ill-conceived mode of engaging with the present. Instead of invigorating and extending critical investigations, it stresses the limits of critique and conceives of it as an essentially destructive, dismissive, and negative enterprise. In this perspective, critique no longer provides "an acceptable stopping point of analysis" because it relies on narratives of "separateness and exteriority" (Barad 2012a, 14; see also 2012b, 49). It appears to be a futile and fruitless endeavor, as "a critical stance re-affirms what is critiqued" (Dolphjin and van der Tuin 2012, 138). New materialists often refer implicitly or explicitly to Bruno Latour's influential diagnosis that "critique has run out of steam" (2004a).[6] Latour famously argues that critique has become irrelevant to contemporary political realities, as it was too much focused on revealing (human) false consciousness and "the discovery of a true world of realities lying behind a veil of appearances" (2010, 474–5; 2004a). According to this reasoning, critique is intimately connected to the practice of demystification—a highly problematic undertaking, as it deflects attention from the dynamism of matter and tends to "reduce *political* agency to *human* agency" (Bennett 2010a, xv; emphases in original). As it concentrates on epistemological conditions of human knowledge production, critique reaffirms humanist hubris rather than undermining it. What is needed, then, is "an alternative to *critique*" (Latour 2010, 474; emphasis in original) that engages with the real and makes it possible to experience the world beyond the limitations of anthropocentrism.

This call to "suspend [. . .] the critical gesture" (ibid., 476) relies on a surprisingly poor and static understanding of critique that does not do justice to the richness and dynamism of critical theory and practice, and that also precludes the possibility of revising and transforming critical impulses and trajectories. The focus on "ideology critique" endorses a very selective concept of critique. It neither disqualifies the necessity of critical inquiries in general, nor does it rule out the need to develop a more complex idea of critique (see, e.g., Lemke 2011a). The new materialist narrative operates with a dualistic logic of either-or, as the negativity of critique is neatly distinguished from affirmative and creative engagements with the present that seek to design "positive, even utopian alternatives" (Bennett 2010a, xv; see also Latour 2010, 474–7). Ironically, the understanding of critique as an essentially epistemological and anthropocentric project contrasts significantly with the more general emphasis on fluidity and flexibility in new materialisms; moreover, it ignores shared concerns and potential coalitions beyond the materialist imagination.[7] There is a certain paradox to be observed in new materialist endeavors to say farewell to critique. In pointing to the diversity of mechanisms that critical practices employ in overlooking or sidelining nonhuman agencies, new materialists often mobilize the very grammar of critical revelation they claim to have superseded. Instead of resisting the "relentless approach toward demystification" (Bennett 2010a, xv), they actively engage in the project of critique by claiming to present the true picture of the real beyond humanist distortions.[8]

The rhetoric of branding and the rejection of critique seriously curbed my enthusiasm for new materialisms. However, I remained convinced that new materialists were raising important questions. But what are the answers they provide? How exactly do they conceive of matter, and how does the reformulation of matter translate into a different understanding of politics? The next step of my engagement was to move beyond the initial sense of uneasiness and ambivalence to follow more closely the different trajectories within the domain of the new materialisms. Given the heterogeneity of the field, I decided to engage in this book with three exemplary positions that represent main strands or directions of new materialist scholarship: Graham Harman's project of an object-oriented ontology, Jane Bennett's account of a vibrancy of things, and Karen Barad's proposal of agential realism. While object-oriented ontology focuses on discrete and bounded "objects," isolated and separated from human subjects, vital materialism and agential realism are concerned with "things" and "phenomena" respectively. In contrast to Harman, Bennett and Barad are both interested in processes of "becoming" rather than states of "being." They focus on hybrid assemblages and relational entanglements in which the subject is "already part of the substances, systems, and becomings of the world" (Alaimo 2014, 14; Taylor 2016, 202). However, as we will see, vital materialism and agential realism differ considerably in how they conceptualize these interactional patterns and collective practices.

Objects, things, and phenomena—these three important signposts mark the landscape of new materialisms, covering a spectrum from non-relational to radically relational positions. These materialisms range from an explicit essentialism in object-oriented ontology via an unresolved theoretical tension between relationalism and foundationalism in vital materialism to the performative ontology of agential realism.[9] In the first part of this book I will engage in depth with each of the three positions, presenting their central ideas and distinctive concepts in advancing a posthumanist account and a re-appreciation of matter. The discussion will also expose some analytic inconsistencies and conceptual ambiguities. I argue that due to these theoretical problems the political purchase of the new materialisms under review is often limited or even ambiguous—sometimes in stark contrast to the self-declared and rather bold claim of providing "a more radical theory of democracy" (Bennett 2005, 142) or a "new materialist understanding of power" (Barad 2007, 35, 224).[10] In fact, instead of further politicizing ontological questions, new materialisms might paradoxically contribute to depoliticizing political matters as they tend to displace political questions by ethical and aesthetic concerns and ignore how matter and nonhuman nature is often mobilized rather than suppressed in contemporary governmental practices.

By presenting this provisionary balance sheet of the new materialisms, I do not mean to dismiss their concerns and commitments. Quite on the contrary, I endorse unambiguously the new materialist call for a critical reconsideration of matter and materiality. However, instead of dismissing the materialist tradition and saying farewell to critique, it seems more pertinent to relate the "material turn" to the concerns of earlier materialist thought and to investigate its potential for rethinking and broadening critical theory. In this light, new materialist ontology needs to be more strongly connected to an analytics of power that draws on the tradition of critical theory and is informed by a political agenda for change (Coole 2013; Cudworth and Hobden 2015). This book argues that such a "tool-box"[11] can be found in revisiting and revising the work of Michel Foucault, especially in exploring the concept of a "government of things" (2007a, 97).

Revis(it)ing Foucault's Tool-Box

In new materialist scholarship, Foucault's work plays an ambiguous role. While his genealogies are often mentioned as an influential source and inspiration for problematizing any stable concept of the "human" or the "subject", he is also perceived as a crucial proponent of discourse theory and the "cultural turn," which appears to dispute or negate the relevance of matter. In particular, Foucault's concept of the body and his insistence on the productivity of power relations serve as positive references in the new materialisms (see, e.g., Coole and Frost 2010b, 32–3; Barad 2003, 809). His work stresses the materiality of the physical body and focuses on the technologies of power that constitute disciplined and docile subjects. Foucault thus helps to undermine "corporeal fetishism" (Haraway 1997, 143), which takes it for granted that bodies are self-identical, fixed and closed entities; he analyzes the interplay of history and biology by demonstrating how the body in its materiality is affected and modified by power relations.[12]

While many new materialists praise Foucault's writings for the important insights they offer, they conceive of his account of the body and power as only partly convincing and in the end unsatisfactory. Even though these scholars rarely explicitly engage with his work, there seems to be a general consensus that Foucault has to be subsumed under the category of social constructivism and anthropocentrism (see, e.g., Meillassoux in Dolphijn and van der Tuin 2012, 77; Dolphijn and van der Tuin 2012, 167). A highly influential version of the critique has been put forward by Barad, one of the most important representatives of new materialist thought. According to this reading, Foucault's work remains within the "traditional humanist orbit" (2007, 235), confining agency to human subjects without taking into consideration the agential properties of nonhuman forces.

Barad identifies several problems with Foucault's account of matter. First, she argues that Foucault restricts the productivity of power to the domain of the social (ibid., 145; Barad 2003, 820, note 25). Accordingly, he "honor[s] the nature-culture binary [. . .], thereby deferring a thoroughgoing genealogy of its production" (Barad 2007, 146). By privileging the "social" Foucault, in Barad's view, cannot understand the complex dynamics between human and nonhuman actors. The second criticism is closely connected to the first. Barad contends that Foucault's analysis remains one-sided and limited as it "focuses on the production of human bodies, to the exclusion of nonhuman bodies whose constitution he takes for granted" (ibid., 169). The third concern addresses what Barad considers Foucault's flawed account of the "precise nature of the relationship between discursive practices and material phenomena" (ibid., 200, 146; Barad 2003, 809–10). As Foucault, according to Barad, regards the boundaries between nature and culture, human and nonhuman as self-evident and given, he also fails to give an adequate account of the complex and dynamic relations between meaning and matter. She argues that in Foucault's work matter serves as a passive background to or instrument for social power relations.

This book offers a reconsideration of the three critical charges Barad levels against Foucault: (1) a privileging of the social, (2) persistent anthropocentrism, and (3) an under-theorized relation between discursive practices and material phenomena. I will show that contrary to this widely accepted and rather dismissive assessment, elements of a "more-than-human"[13] approach can be found in Foucault's work. It is expressed in the idea of a government of things, which he introduces in his lectures at the Collège de France in 1977–78. By stressing the "intrication of men and things" (Foucault 2007a, 97), this theoretical project makes it hard to read Foucault's work as unequivocally anthropocentric. Using material that has only partly been translated into English, I argue that Foucault's account of government goes beyond a concern for ethics and forms of subjectivation to address the entanglements of humans and nonhumans. My theoretical claim is that the conceptual proposal of a government of things makes it possible to arrive at a relational and strategic understanding of agency and ontology that productively engages with some of the issues the new materialisms raise. This account builds on elements in Foucault's writings that he himself never coherently discussed or further developed. I will focus on three concepts that respond to Barad's critical points.

First, against Barad's charge that Foucault's description of the relationship between discursive practices and material phenomena remains unsatisfactory, I show that Foucault's notion of the *dispositif* is informed by a material-discursive understanding of government as "arranging things." I highlight Foucault's distinctive use of "dispositive" that assembles discursive and non-discursive elements and spell out its ontological, technological, and strategic dimensions as well as its analytical and critical value. Secondly, while Barad claims that Foucault's analysis limits the productivity of power to the sphere of the social, I argue on the contrary that it extends the meaning of *technology*. Instead of reserving the term for manipulating and mobilizing things in a literal sense, Foucault's vocabulary also applies it to "human affairs"—or rather it transgresses the divide between the social and the more-than-social. Third, against Barad's charge that Foucault's work exclusively attends to the production of human bodies, I emphasize the crucial importance of his notion of the *milieu*. This constitutes a strategic element in a liberal governmentality that seeks to govern the interface between the human and the nonhuman and sets the stage for a more conclusive framing of biopolitics.

To be sure, Foucault never directly inquired into the nature of matter or investigated the specifics of human-nonhuman relations. However, as Krithika Srinivasan has noted, Foucault explicitly considered his writings as "trails to be followed" (Foucault 2003, 4) rather than "theoretical frameworks to be adopted or rejected as a whole" (Srinivasan 2014, 505). This "conceptual generosity on Foucault's part" (ibid.) has successfully invited scholars to selectively take up, adapt, and transform his ideas and concepts in approaching issues and questions that Foucault himself did not address or that remained marginal to his historical and philosophical agenda. From gender studies (Sawicki 1991; Butler 1990) to postcolonial theory (Stoler 1995; Mbembe 2003) and the environmental sciences (Youatt 2008; Holloway and Morris 2012), to name just a few, Foucauldian concepts have been adopted for theoretical and empirical work—despite Foucault's own blindness to or his lack of interest in these research fields. The purpose of the following experiment is therefore what Brian Massumi once termed "working from Foucault after Foucault" (2009, 158) to revise and update his "tool-box" for addressing contemporary problems.

My thesis is that putting forward the notion of the dispositive, a comprehensive understanding of technology, and a complex reading of the milieu provides elements for a thoroughly relational materialism. It not only significantly differs from some theoretical inconsistencies and blind spots in the new materialisms but opens up an avenue for a more material account of politics. The analytic grid of a government of things rearticulates a range of new materialist concerns and commitments within a Foucauldian conceptual framing to explore the political and historical dimension of ontologies.[14] In Foucault's conception, the government of things "tries to work within reality, by getting the components of reality to work in relation to each other" (2007a, 47). To further develop this "*relational materialism*" (Law 1992, 389; emphasis in original; Mol 2013, 381), I propose to align Foucault's analytics of government with insights from science and technology studies (STS), especially actor-network theory (ANT) and feminist and postcolonial technoscience. My argument builds on John Law's observation of crucial "similarities" between STS work and Foucault's history of the present, as both lines of research "attend to material and linguistic heterogeneities, and how these generate effects including asymmetries and dualisms" (Law 2017, no paging; see also 1994; Law and Singleton 2013, 494, note 20). Both STS and Foucault's analytics of government approach ontological questions in terms of contingency, openness, and malleability instead of necessity, determinism, and stability. However, they differ in their modes of investigation and research interests. Their alignment allows us to join the diachronic sensibility of STS with Foucault's interest in the synchronic (see Law 1994), combining the analytical and critical strengths of the two accounts to investigate the mobile trajectories of "ontological politics" (Mol 1999).[15]

To address and avoid two possible misunderstandings: The idea of a government of things remains an underdeveloped theme in Foucault's work. His writings did not so much systematically pursue this conceptual move as offer promising elements for it. I am not suggesting that there is a fully-fledged, more-than-human account of government to be found in Foucault's work. Rather, I will remain faithful to Foucault by betraying some of his original concepts. I will draw on analytical tools and methodological suggestions Foucault put forward to further develop and sometimes distort them, in order to make them useful for contemporary intellectual debates and political struggles. In doing so, I am not primarily interested in extending Foucault's analysis into research areas he did not deal with in his own work. I will not focus on absences and deficiencies. Rather, I am trying to identify ways in which the concepts already present in Foucault's writings might offer new, underexplored, or unexpected insights into governmental practices.

Also, I do not intend to formulate a "theory of the government of things"; rather, I use the term as a productive conceptual proposition to investigate its potential—but also its limitations. It is an "epistemic thing" (Rheinberger 1997) designed to open up new spaces for thinking, to make improbable encounters possible, to establish alternative conceptual ties, and to envision avenues for empirical research. The analytic frame of a government of things is an experimental device that refrains from narrow definitions and pre-established criteria. It serves as a provocation and a promise, and its lack of coherence and conceptual rigor is a strength rather than a weakness. This "outline for something" (Foucault 2003, 4) not yet known will be helpful as a way of directing research interests to focus on some aspects rather than others, but it also remains open enough for unexpected turns and surprises. I hope the prospects of this concept as well as its problems will materialize in future work.[16]

Structure of the Book

The Government of Things: Foucault and the New Materialisms follows a threefold objective. First, it aims to bring some clarity to the picture of new materialisms, as the debate is often confusing and positions within the field sometimes contradictory. The book will systematically discuss and critically evaluate the innovative potential and explanatory perspectives of the material turn as well as its political prospects. It offers a critical review of a range of different streams of new materialism: Graham Harman's project of an object-oriented ontology (OOO), Jane Bennett's account of a vibrancy of things, and Karen Barad's proposal of an agential realism.

Secondly, the book identifies elements in Foucault's work that make it possible to take up some of the concerns and issues new materialists raise. The concepts of the dispositive, of technology, and of the milieu give substance to the idea of a government of things Foucault proposes in his lectures at the Collège de France. The book suggests new ways of engaging with Foucault's work by proposing a more-than-human understanding of government that further develops and extends important analytical and critical dimensions of his work.

Thirdly, the book explores the theoretical potential and empirical prospects of a relational materialism based on Foucault's notion of a government of things. I argue that combining an analytics of government with STS work makes it possible to arrive at a more convincing conceptual apparatus for addressing material practices and a better understanding of political matters. This theoretical synthesis also provides a way of spelling out Foucault's concept of "environmentality" to analyze contemporary forms of government that seek to design and modulate socio-techno-ecological milieus.

The first part (*Varieties of Materialism*) discusses three main currents of new materialist scholarship. Chapter 1 focuses on the work of Graham Harman, the most important proponent of "object-oriented ontology" (OOO). This seeks to unravel the true existence of objects, emphasizing their unpredictability, weirdness, darkness, and inconceivability. OOO's refusal to distinguish between human and nonhuman objects might be considered a particular conceptual strength that makes it possible to transgress "subjective" or "human-centered" categories. However, OOO does not offer any theoretical orientation on how differences between objects are established and how they become meaningful. Nor does it account for how some human objects and their power disproportionally affect humans and nonhumans. As I show, attending to isolated objects and their inner obscurity in OOO finally translates into an extreme form of subjectivism and a serious lack of conceptual clarity.

Jane Bennett's vital materialism, analyzed in Chapter 2, differs significantly from OOO's understanding of objects. It puts forward the idea of a comprehensive vitality that undermines traditional ontological and normative divisions and runs through both human and nonhuman matter. Bennett's account of "thing-power" (2010a, xvii) provides important elements for designing a posthumanist political theory and goes beyond many conceptual limitations of OOO. However, her "positive ontology" (ibid., x) fails to account for the negative processes and destructive patterns that obstruct and hinder the progressive politics she envisions. If the primary task of this vital materialism is to "alter or derail the machine so as to minimize its harms and distribute more equally its costs and benefits" (Bennett 2015b, 85), Bennett proves unable to provide the analytic and conceptual tools needed to achieve this objective. She tends to displace political questions by the appeal for a new ethical sensibility. Unfortunately, Bennett offers no convincing argument as to how the "energetics of ethics" (Bennett 2001, 132) is coupled with political dynamics or how the vital politics she advocates makes possible a radical change in the contemporary structures of production and consumption.

Chapter 3 focuses on Karen Barad's agential realism. While the term seems contradictory at first sight, it epitomizes Barad's interest in conceiving of materiality without subscribing to the idea of matter as a stable and solid fundament. I first present the different components of agential realism (epistemology, methodology, ontology, ethics) in sequence—acknowledging Barad's claim that they cannot be separated or understood independently of one another. My intention is not to identify enclosed conceptual bricks or blocks that add up or complement one another, but rather to attend carefully to the distinctive theoretical transgressions and movements Barad performs. The second part of this chapter discusses the claim that agential realism contributes to a "new materialist understanding of power" (Barad 2007, 35, 224). It focuses on Barad's productive account of the "apparatus" that makes it possible to investigate how materializations are entangled with forms of exclusion, and analyzes how temporalities, spatialities, and materialities are mutually constituted. I will highlight the distinctive theoretical and analytical strengths of agential realism (especially in comparison to OOO and vital materialism). However, the analysis also highlights some inconsistencies and problems that relate to the foundational role of quantum mechanics in Barad's account and to the comprehensive concept of ethics she endorses.

The second part of the book (*Elements of a More-Than-Human Analytics of Government*) turns to Foucault's work. It explores important conceptual tools for a non-anthropocentric and relational-materialist analytics of government. Chapter 4 elucidates the emergence of the idea of a government of things in Foucault's lectures at the Collège de France from 1978 on. I argue that this more-than-human understanding of government informs Foucault's concept of the *dispositif* (dispositive), which focuses on steering and regulating agentic forces. The chapter discusses the diverse meanings of the French notion of *dispositif* and contrasts them with the current usages of "apparatus" (*appareil*) on the one hand, and "assemblage" (*agencement*) on the other in new materialist scholarship and elsewhere. It spells out the ontological, technological, and strategic dimensions of the concept of the dispositive as well as its analytical and critical value in attending to the political dimensions of ontologies.

Chapter 5 begins by presenting Foucault's understanding of government as a technological invention. I will distinguish his notion of technology from social constructivism on the one hand and technological determinism on the other (and from Marxist and humanist accounts). The next section analyzes two technological metaphors and models for imagining political structures and processes that display distinctive rationalities of governing: the clock and the steam-engine governor. The last sections of this chapter engage with Foucault's concept of technologies of security as a specific feature of liberalism, and expose important dimensions of their role as feedback devices. The refinement and increasing uptake of these technologies in many practical domains and fields of knowledge led to the rise of cybernetic forms of communication, command, and control in the twentieth century.

Building on this more-than-social account of government, Chapter 6 puts forward an understanding of biopolitics that no longer exclusively addresses human individuals and populations but attends to the complex associations of humans and nonhumans. The chapter starts by reviewing Foucault's writings on genetics and heredity, suggesting that they advance a material-semiotic understanding of life and incorporate insights from contemporary genetics and molecular biology. The next sections focus on the notion of the milieu in Foucault's work. After revisiting the brief genealogy of the term in his lectures on governmentality at the Collège de France, I argue that the milieu occupies a central role in liberal governmentality as it seeks to control and canalize "free" circulations across the human-nonhuman divide. Since it attends to the co-constitution of humans and nonhumans, the milieu also allows for a non-anthropocentric framing of biopolitics.

The third part of the book (*Toward a Relational Materialism*) argues for an alignment of Foucault's analytics of government with work in STS to better account for contemporary topologies and political trajectories. I suggest that this theoretical synthesis makes possible a relational materialism that goes beyond the shortcomings and limitations of the important streams of new materialist scholarship outlined in the first part of this book. Chapter 7 starts by proposing a material-semiotic understanding of governing processes that provides a better way of grasping the ontological dimension of politics. STS research and Foucault's analytics of government both shift the emphasis from epistemological or theoretical questions to practical issues, multiplying the term "ontology." Or rather: the two lines of research start from "nontology" in order to investigate how distinctive modes of existence emerge. I then argue that we should replace the notion of agency, which is quite prominent in new materialist scholarship, with modes of "doing" (Mol 2002), which focuses on performative and praxeological aspects instead of properties and capacities. The next section elucidates how this more-than-human account of government differs substantially from the endorsement of posthumanism in many strands of new materialism and elsewhere. It proposes to develop novel approaches, vocabularies, and concepts to meet the dual challenge of questioning and decentering human privilege and power, while still acknowledging the asymmetrically destructive and oppressive power of (some) humans. The last part of this chapter invites us to envision a different—experimental—style of analysis and critique that gives credit to the heterogeneous, dynamic, and mobile character of dispositives. It calls for alternative political imaginaries and projects, possibly leading to forms of "counter-conduct" (Foucault 2007a, 201).

Chapter 8 diagnoses an important transformation in contemporary modes of government. Instead of targeting directly human individuals or populations, they seek to modulate and control the social, ecological, and technological conditions of life. I will start by presenting Foucault's concept of environmentality as a way of addressing this new constellation of power. The remaining parts of the chapter discuss distinctive elements of environmental modes of government. I will first analyze the rise of the resilience discourse and a neo-cybernetic regime of control that problematize conventional notions of stability to exploit and foster differences and deviances. The next part is devoted to new—"probiotic"

(Lorimer 2017)—modes of intervention that seek to govern through 'nature' rather than against it and the emergence of "vital systems security" (Collier and Lakoff 2015). Like classical biopolitics, the latter concept seeks to foster the welfare and the health of populations, but it does so by addressing a new object: material infrastructures, functions, and services deemed to be indispensable for collective life in contemporary societies. Revisiting and extending Foucault's notion of pastoral power, the last part of the chapter engages with the idea of "panarchy" (Holling et al. 2002). This concept seeks to capture the dynamics of creation and destruction in adaptation cycles of complex systems, enacting a normative grammar that is informed by logics of resilience and converts ethical responsibility into a technical responsiveness to future catastrophic events.

The Conclusion summarizes the key findings of the book. The conceptual proposal of a government of things takes up important insights and theoretical achievements of new materialist scholarship. It shares the interest in reconceptualizing matter and the focus on the interplay of epistemological, ontological, political, and ethical issues, and insists on the limits of anthropocentric modes of thought. However, in synthesizing Foucault's analytics of government with the STS-inspired work, the concept of a government of things also goes beyond new materialist scholarship. It puts forward a relational and performative account of materialities that more closely attends to the historical and political dimensions of ontogenesis.

PART I

Varieties of Materialism

1

Immaterialism

Graham Harman and the Weirdness of Objects

What is real in the cosmos are forms wrapped inside forms, not durable specks of material that reduce everything else to derivative status. If this is "materialism," *then it is the first materialism in history to deny the existence of matter.* (Harman 2002, 293; emphasis in original)

The root error [. . .] is the presumption that the world somehow already comes naturally composed of discrete objects. (Rouse 2002, 313)

The term object-oriented ontology (OOO) was coined by Graham Harman, and it defines a theoretical commitment to thinking the real beyond human experience. It seeks to uncover the true existence of things, favoring concepts of stability, essence, solidity, and permanence over notions of flux, relationality, process, and contingency. In addition to Harman's books, which include *Tool-Being* (2002), *Guerilla Metaphysics* (2005), *Prince of Networks* (2009), *The Quadruple Object* (2011a) and *Immaterialism* (2016a), Levi Bryant's *The Democracy of Objects* (2011a), Ian Bogost's *Alien Phenomenology* (2012) and Timothy Morton's *Hyperobjects* (2013a) are important contributions to OOO. These works seek to revise realism while criticizing recent tendencies in the humanities and social sciences, especially the linguistic turn in its structuralist and poststructuralist versions, but also some variants of the new materialism that endorse vitalist and performative concepts of matter. OOO's ideas have proliferated to various non-academic forms of publishing and blogging, attracting a lot of interest among younger scholars. While OOO has had repercussions mostly in philosophy, archeology, architecture, and art, works by Harman and his associates have also influenced debates in many other disciplines, for example in educational science (Oral 2014; 2015), geography (Meehan et al. 2013), and sociology (Pierides and Woodman 2012).

This chapter critically assesses OOO's ambition to explore "the heart of the things themselves" (Harman 2010a, 95). The argument focuses on Harman's work, but will additionally refer to other protagonists of OOO. I will first situate OOO within a broader philosophical perspective known by the name of speculative realism.[1] The second section analyzes the distinctive understanding of objects that OOO puts forward when it claims to attend to their obscurity and autonomy. Advocates of OOO also shift the philosophical accent from epistemology to ontology. They do not investigate the performative effects or the material dimensions of epistemologies but conceive of them as modes of representation and epiphenomena unable to account for the inner core of things. I will then discuss the critique of relational accounts that proponents of OOO develop and their fierce rejection of process philosophies and praxeological accounts. The fourth part critically engages with OOO's appeal to aesthetics and the displacement of ethical and political concerns it enacts. The final section demonstrates that OOO's promise to break once and for all with subject-object dualism results in a revived form of subjectivism. In claiming to go "back" to the object to explore its inner depth, protagonists of OOO cherish a naïve encounter with objects that reintroduces the most subjective values and arbitrary criteria.

Speculative Realism and the Critique of Correlationism

The label "speculative realism" originally derives from the title of a one-day workshop that took place at Goldsmiths College, University of London in 2007, in which the philosophers Ray Brassier, Iain Hamilton Grant, Graham Harman, and Quentin Meillassoux participated (Brassier et al. 2007). It soon became the name of a "loose philosophical movement" (Harman 2012, 184). While its theoretical references are quite heterogeneous and include such diverse figures as Whitehead, Latour, Heidegger, Nietzsche, Levinas, Badiou, and Schelling, speculative realists have a common adversary: correlationism.

The term was first introduced by Quentin Meillassoux in his book *After Finitude: An Essay on the Necessity of Contingency* (2008 [2006]). It means that "we only ever have access to the correlation between thinking and being, and never to either term considered apart from the other" (2008, 5). According to Meillassoux, all post-Kantian philosophies from Marxism and phenomenology to poststructuralism and deconstruction suffer from some version of correlationism. The "Kantian 'catastrophe'" (ibid., 124) consists of focusing on transcendental conditions for human knowledge while at the same time denying any genuine access to the real. Meillassoux conceives of the external world as an absolute reality, as what he calls "the *great outdoors* [. . .]: that outside which was not relative to us, [. . .] existing in itself regardless of whether we are thinking of it or not" (ibid., 7; emphasis in original). For him and speculative realism in general, the effect of correlationism has been to dramatically limit the range of theoretical speculation to things that fall within human knowledge systems.[2]

Contra the "mortal enemy" (Harman 2012, 184) of correlationism, speculative realists insist that the world itself cannot be reduced to the question of human access to it. They claim to be able to cure modern philosophy of its perceived obsession with mediation, which relates everything back to human knowledge, and instead to discover the true being of things outside of thought. While Kant warned us "never to venture with speculative reason beyond the boundaries of experience" (Kant 1998, B xxiv), speculative realists invite us to strive toward the unmediated and irreducible thingness of things. They share the realist conviction that there is a world that exists independently of human access, but they seek to go beyond the frontiers of traditional realism by encouraging speculation about what exists without limiting being to categories of (human) thought and knowledge. They "do not wish to establish a commonsense middle-aged realism of objective atoms and billiard balls located outside the human mind. Instead, the speculative realists have all pursued a model of reality as something far *weirder* than realists had ever guessed" (Harman 2012, 184, emphasis in original; see also Harman 2010a, 2; Zahavi 2016, 293–97).

Speculative realists seek to break with the epistemological privilege accorded to human knowledge to engage with the real nature of objects. However, to criticize correlationism for its focus on the (human) subject does not mean to embrace scientific objectivism. According to Harman, taking objects seriously confines the ambitions of scientific investigations and epistemological claims. Rather than providing an alternative to correlationism, "scientific naturalism" (Harman 2011a, 30) is perceived as one of its forms, as it reduces the concrete richness of objects to abstract categories and the narrow grid of human consciousness: "[T]he thing as portrayed by the natural sciences is the thing made dependent on our knowledge, and not the thing in its untamed, subterranean reality" (ibid., 54). In this perspective, things in themselves are inaccessible by human knowledge as it only grasps phenomena but not the true being of things.

Some speculative realists have a much more positive view of science, and not all would subscribe to Harman's skepticism. Indeed, it is about the consequences of correlationism that the partisans of OOO tend to disagree. Meillassoux, for instance, still endorses a kind of Cartesian rationalism, stressing the importance of scientific statements in rejecting correlationist accounts. While for Meillassoux mathematics provides knowledge in a non-anthropocentric world and shares the epistemic virtues of speculative realism, Harman denies that positive knowledge is possible at all.[3]

Phenomena, Objects, and Hyperobjects

Harman displaces questions of consciousness, subjectivity, rationality, and human access to an external world in order to affirm the mysterious depths and the "marvelous plurality of concrete objects" (2009, 156). This account is informed by an original reading of Edmund Husserl and Martin Heidegger. Harman productively engages with the phenomenological call for a return "back to the 'things themselves'" (Husserl 2001 [1900/1901], 168) to elucidate what he considers to be "a new kind of philosophy" (Harman 2011a, 50). While Husserl refrained from analyzing the real world in favor of an accurate description of how it appears to consciousness, focusing on *phenomena* instead of *noumena*, Harman nevertheless credits him with a crucial discovery. Husserl might be regarded as an idealist for confining his interest to the intentional realm, but Harman still considers him to be an "*object-oriented* idealist" (ibid., 20; emphasis in original) as he accounts for sensual objects and the essential difference between these objects and their qualities (ibid., 20–34). In Harman's reading, Heidegger's focus on the absence or impenetrability of things complements and further develops Husserl's emphasis on the presence of things for human consciousness. He praises Heidegger for attending to "what lies behind all phenomena" (ibid., 36), adding "real objects" to Husserl's "sensual objects." According to Heidegger, objects have two basic modes of being. They are either "present-at-hand" (*vorhanden*) in consciousness or "ready-to-hand" (*zuhanden*) when we use them. However, as Heidegger stresses, most things we encounter are never completely present to the mind; rather, they "withdraw into a shadowy subterranean realm that supports our conscious activity while seldom erupting into view" (ibid., 37, 35–50). To explore this dark and obscure dimension of objects, Harman engages with the famous broken tool episode in *Being and Time* (Heidegger 1962 [1927]).[4]

Heidegger shows that a simple malfunction of a hammer can suddenly disturb the taken-for-grantedness of the things we use. Harman insists that there is much more at stake than breaking tools and failing "equipment":

> When using a hammer, for instance, I am focused on the building project currently underway, and I am probably taking the hammer for granted. Unless the hammer is too heavy or too slippery, or unless it breaks, I tend not to notice it at all. The fact that the hammer can break proves it is deeper than my understanding of it. [. . .]Object-oriented philosophy pushes this another step further by saying that objects distort one another even in sheer causal interaction. (Harman 2012, 186–87)–87)

OOO's "step further" extends the analysis beyond Heidegger's anthropocentric approach to subject-object relations to include relations between objects, as humans are conceived of as just another kind of object. Proponents of OOO insist that things are always more fundamental than perceptions, theories, or uses of them. Things recede into an ontological obscurity characterized by a sub-phenomenal level that can never be completely accessed by humans (or other objects).[5] Instead of exploring how objects appear to human consciousness and how they could be known, object-oriented ontologists evoke their inner truth and essence. According to Harman, "[n]o theory of numbers, birds, chemicals, or Stone Age societies will ever be able to exhaust the reality of these topics" (Harman 2007a, 23). In this view, OOO puts forward a "'flat ontology'" (Morton 2011a, 165) that does not distinguish between different ontological layers or hierarchically ordered spheres. The term "object" applies to real or unreal, natural or artificial, living or nonliving, human or nonhuman entities. Accordingly, humans no longer figure as subjects that observe or oppose the realm of objects but are an integral part of it. Harman proposes to use "object" "in the broadest possible sense to designate *anything with some sort of unitary reality*. "Object" can refer to trees, atoms, and songs, and also to armies, banks, sport franchises, and fictional characters" (Harman 2010a, 147; emphasis TL).[6] Thus, technological artifacts, artworks, animals, or fantasies are equally objects—though they might not be the same kind of objects.[7] For Harman, the world is populated by innumerable objects both withdrawn and manifest, isolated and reaching out to one another.

Harman combines this radicalized reading of Husserl and Heidegger with the interest in objects found in actor-network theory (ANT). He regards ANT as "the most important philosophical method" (Harman 2016a, 1) since the emergence of phenomenology in the early twentieth century and credits it, and especially the work of Latour, with an inclusive account of objects as it acknowledges the agency of things.[8] However, ANT does not go far enough according to Harman, as it reduces objects to their actions. While this theoretical perspective successfully overcomes anthropocentric limitations by taking seriously the agency of nonhumans, it tends to "overlook the question what objects are when *not* acting" (ibid., 7; emphasis in original). Ultimately, Harman finds ANT guilty of putting too much emphasis on action and relationality as it ignores that "a thing acts because it exists rather than existing because it acts" (ibid., 7).

Harman positions OOO as a theoretical hybrid that synthesizes the phenomenological tradition with contemporary ANT against three "basic forms of knowledge"(ibid., 12) that neglect or negate the autonomy of things: undermining, overmining, and duomining. The first strategy ("undermining") seeks to explain objects in terms of their smaller constituents and is prevalent in the natural sciences. Harman insists that no object is reducible to or fully explicable in terms of a "downward reduction" (2016a, 8; see also 2011b, 2011d) to some underlying substratum, as its existence always exceeds the parts it is composed of. The second strategy ("overmining") is especially potent in the humanities and social sciences, and ANT and Latour are its most important advocates. It reduces an object to what it is doing as it is nothing more than "its relations or discernible actions" (2016a, 10). Harman cautions against this understanding, as it "allows objects no surplus of reality beyond whatever they modify, transform, perturb, or create" (ibid., 10; see also 2011b). The third approach ("duomining") consists of combining the two previous strategies, instigating a double reductionist move. Harman sees "duomining" as a general tendency of alternative approaches to thingness and materiality, especially prevalent in material feminism and material semiotics (ibid., 26; 2016b).

Against these distinctive forms of (deficient) knowledge, proponents of OOO argue for a realism that acknowledges the inner obscurity of things, their autonomy, unavailability, and inaccessibility. They follow the "principle of irreducibility" (Morton 2011a, 177), seeking to reaffirm or reinstate the "thing-in-itself": "[T]he point is that each object in *this* world is a thing-in-itself, since it cannot be translated without energy loss into any sort of knowledge, practice, or causal relation" (Harman 2016a, 32–33; emphasis in original).[9] According to OOO, the essence of objects is withdrawn and they impact and encounter each other on their own terms, which are not accessible to human knowledge. Thus, the "essential butteriness" of butter remains "radically unavailable" to us (Morton 2011a, 177). But there is something more to this principle of withdrawal. Harman and his associates are convinced that objects are characterized by an "uncanny essence" (ibid., 165). They persistently stress the "irreducible dark side" (ibid., 165) of things, which incessantly surprises and permanently escapes any attempt to know or use them.

Timothy Morton pushes this account further by identifying a particular category of objects: "hyperobjects." He reserves this term for objects that exist on such temporal and spatial scales that they challenge any idea of understanding or capturing them by human knowledge. Hyperobjects necessitate "new ways of thinking about objects, and revise our ideas about the subjects that think about them" (ibid., 167). Morton argues that hyperobjects are never fully accessible or conceivable in their entirety. He lists a number of human-made and naturally occurring hyperobjects—nuclear radiation, oil spills, tectonic plates, Hurricane Katrina, and so forth—in order to contemplate the implications of human co-existence with these entities in the Anthropocene. Global warming is one example Morton often refers to. While it is "withdrawn," as it is impossible to directly see or touch it, this hyperobject affects all weather on Earth—without being reducible to its particular manifestations, e.g., rain or drought (see ibid., 167; Morton 2013a; 2013b).[10]

According to Morton, the very existence of these hyperobjects renders untenable concepts such as "world," "nature" or "environment," as they are based on the idea of a single and stable grounding. He claims that contemporary Marxist, postmodernist, and environmental thought have failed to account for the affective presence and inescapable potency of hyperobjects that swirl around and stick to us. By stressing their "strange strangeness" (2013a, 91), Morton seeks to rule out any reference to a harmonious nature or environmental romanticism (see ibid., 129–30). Hyperobjects remain beyond human understanding, and the accumulation of knowledge on objects will only nourish the insight concerning their incessant withdrawal: "The more data we have about hyperobjects the less we know about them—the more we realize we can *never* truly know them" (ibid., 180; emphasis in original). Instead Morton turns to the aesthetic, arguing that only painting, music, art installations, poetry, and film are able to take "the hyperobject as its form" (ibid., 180).

The Turn to Essentialism

OOO's basic idea that objects possess intrinsic qualities that mark their individual character and enable them to interact with other objects informs a new version of "essentialism" (Harman 2016a, 17). It seeks to reevaluate the theoretical and normative imaginary by inviting us to recalibrate the "political reflexes associated with terms such as essence ('bad') and reciprocal interplay ('good')" (Harman 2007b, 22). Thus, OOO fights against the "prejudices" (Harman 2012, 188) that it claims are still dominant in the humanities and social sciences and that privilege performances and events over substances and essences, malleability over stability, and networks over objects (see ibid., 188). It strives to give a positive meaning to terms like "essence" or "the real" by opposing the idea that things are generated in practices and that change and contingency are central explanatory references (see Harman 2016a, 14–16).

By contrast, proponents of OOO endorse a "non-relational metaphysics" (ibid., 17) that argues forcefully against "relationism"—the idea "that a thing is defined solely by its effects and alliances rather than by a lonely inner kernel of essence" (Harman 2009, 75; see alsoMorton 2011a, 184). According to Harman and his associates, "relationism" is guilty of negating the existence of "the thing itself" by reducing objects to the relations they engage in, thereby ignoring the question of "what preciselyis interconnected with what" (Morton 2011a, 185). The issue of relationality also occupies a pivotal place in Harman's analysis of Latour's version of ANT. Harman claims that actors must be accorded "a reality beyond all relationality"; each actor (or object) must be "in and of itself [. . .] *apart from any relations*" (Harman 2009, 187; emphasis in original). In this view, Latour's work is unable to account for the concrete individuality of objects and how they can change over time: "[I]f a thing is entirely relational, then there would be no reason for it to change. The thing would be fully deployed or exhausted in its reality here and now, and the same would be true of all of the things with which it relates. Why, then, would the universe ever change? [. . .] Unless the thing holds something in reserve behind its current relations, nothing would ever change" (ibid., 187).

Thus, advocates of OOO operate with a conceptual opposition and hierarchy of objects and relations: "The term 'object' as I use it means anything that exists. The term 'relation' means any interaction between these objects. I hold that such interaction is always a kind of translation or distortion, even at the level of inanimate things" (Harman 2010b, 2). While this understanding suggests that relations are always ontologically secondary, as they miss or even caricature the inner complexity and impenetrable depth of objects, Harman does allow for a somewhat "weak" relationism in his writings. Rather than being inferior or additional to objects, he suggests that new objects are formed out of the relations between already existing objects (Harman 2011a). Thus, relations are themselves conceived of as "new compound objects" (Harman 2016a, 17) that emerge when objects encounter one another.

Harman proposes to "narrow down" a comprehensive and broad understanding of relationality, in order to focus on "a special type of relation that changes the reality of one of its *relata*" (ibid., 49). He introduces the biological concept of symbiosis to mediate between OOO's preference for ontological stability and the theoretical interest in accounting for temporal change (ibid., 42–51). According to Harman, treating every relation as significant for its relata runs the risk of ending up with a "'gradualist' ontology in which every moment is just as important as every other" (ibid., 45). He seeks to circumvent the problematic alternatives of positing eternal objects immune to transformation on the one hand and "a nominalistic flux of 'performative' identities that shift and flicker with the flow of time itself" (ibid., 47) on the other. Instead, Harman proposes a new understanding of "symbiosis as the key to unlocking a finite number of distinct phases in the life of the same object" (ibid., 49–50).[11]

While these formulations indicate theoretical tensions in Harman's work on how to account for relationality, it is in any case defined by a deep hostility towards process philosophies and performative or praxeological accounts.[12] Harman claims that OOO's essentialism has nothing to do with the ideological constructs that helped to mask forms of domination in the past; quite on the contrary, it operates as a tool for liberation and contestation as it promotes "a *weird* realism in which real individual objects resist all forms of causal or cognitive mastery" (Harman 2012, 188; emphasis in original):

> This deeply *non-relational* conception of the reality of things is the heart of object-oriented philosophy. To some readers it will immediately sound deeply reactionary. After all, most recent advances in the humanities pride themselves on having abandoned the notion of stale autonomous substances or individual human subjects in favor of networks, negotiations, relations, interactions, and dynamic fluctuations. This has been the guiding theme of our time. But the wager of object-oriented philosophy is that this programmatic movement towards holistic interaction is an idea once but no longer liberating, and that the real discoveries now lie on the other side of the yard. (Ibid., 187–88; emphasis in original)[13]

Beyond the critical engagement with "relationism," the second "special opponent" (Harman 2011a, 13) of OOO is materialism. Harman insists that OOO is "a resolutely anti-materialist theory" (Harman 2016a, 95–96; see also Harman 2010c).[14] As there is nothing more fundamental than objects, it rejects the idea that objects are "composed of some intrinsic, essential stuff" (Morton 2011a, 177). Therefore, Harman and his associates dispute the idea of "any kind of substrate" (ibid., 179) below or behind objects, regardless of whether the material basis or background is conceptualized as discrete atoms or a fluid substance: "Materialism lopes along hampered by a Newtonian-Cartesian atomistic mechanism on the one hand and the formless goo of Spinoza on the other" (ibid., 179). For OOO philosophers, the problem with matter is that it is unable to account for the existence of objects as it either reduces them to a substratum that allegedly bears them or dissipates them into their historical contexts or epistemological conditions of emergence.

OOO endorses instead an "immaterialism" (Harman 2016a) that operates as a programmatic antidote, on the one hand, to the different approaches in the natural sciences that refer to materiality, and, on the other, to those in the social sciences and humanities that engage in a critical reappraisal of matter. Interestingly, the rejection of materialism leads many object-oriented ontologists to affirm "a new sort of 'formalism'" (Harman 2002, 293). Harman illustrates this move with the example of the Ferris wheel, a rotating upright wheel with multiple passenger-carrying components in an amusement park. He invites us to imagine the Ferris wheel broken down into "numerous bolts, beams, and gears in its mechanism" (ibid., 293). While these pieces and parts once composed the Ferris wheel, they are now something different and can be used for diverse purposes and new arrangements. However, this does not mean that we are dealing with some unspecific matter, Harman insists:

> Far from it! Above all else, the "parts" in question here are *form*, not matter. When taking apart the ferris [sic] wheel in my mind, I do not immediately posit a set of inert iron granules from which all pieces of the wheel are molded. I begin more proximately with bold-machine and engine-machine. In turn, each of these pieces is composed of formal parts: bolts and screws are never terminal points of reality, but always composite relational systems. (Ibid., 293; emphasis in original)

However, given OOO's general hostility to any substantial notion of relationality, the matterless formalism it advocates risks revitalizing an old philosophical debate that opposes form and matter to each other. By stressing the persistence of forms, OOO postulates a stable object that is immune to any genuine transformation of "the thing itself," with change limited to the object's surface without penetrating its inner depths.[15]

Aesthetics as First Philosophy

In commenting on Emmanuel Levinas' philosophy and his concept of ethics, Harman concludes that aesthetics rather than ethics has to be considered as "first philosophy" (Harman 2007b). He regards ethics as a particular form of aesthetics that transcends the sphere of human intersubjectivity as it accounts for the interactions of objects, given their withdrawal from one another:

> [T]he real problem of metaphysics is not how beings interact in a system: instead, the problem is how they withdraw from that system as independent realities while somehow communicating through the proximity, the touching without touching, that has been termed allusion or allure. If we identify this event with "aesthetics" in the broadest sense of the term, it becomes clear why first philosophy is aesthetics, not ethics. The ethical relation to other humans is merely a special case of substances communicating without touching. Aesthetics is first philosophy, because the key problem of metaphysics has turned out to be as follows: how do individual substances interact in their proximity to one another? (Ibid., 30)

OOO's appeal to aesthetics is much broader than simply redefining ethical concerns as a subset of more fundamental aesthetic questions. Rather, it is informed by a genuine interest in aesthetics, seeing it as more sensitive to the insight that "things reside in infinite depths, and all things erupt into enjoyment along the shallowest façades of the world" (ibid., 21; see also Morton 2012). As we have seen, Harman cautions against existing forms of knowledge as they are unable to grasp the complexity and depth of objects. However, OOO does not problematize the limits or shortcomings of contemporary epistemologies by providing a better or more comprehensive knowledge, but rather affirms "the ultimate unknowability and autonomy of things" (Harman 2016a, 12–13). Thus the ambition of knowledge production to enlighten the obscure depths of objects always already fails to take seriously their "irreducible dark side" (Morton 2011a, 165). While it is certainly right to state that OOO and speculative realism "do not really offer much in terms of a theory of knowledge that could justify their metaphysical claims" (Zahavi 2016, 304), this critique misses an essential point: proponents of OOO do not seek to provide a correct or adequate knowledge but rather aim at establishing a "*counter*method" (Harman 2012, 200; emphasis in original) that resists all endeavors to dissolve the object into its composite parts or to situate it in the relations it engages in. It mobilizes knowledge to illustrate the limits of knowledge and to criticize its ambition to control and dominate things.

Thus, aesthetics not only displaces ethical concerns but also disqualifies epistemological questions—insofar as these ignore or negate the inconceivable richness and ungraspable depth of objects. The focus on aesthetics also informs the rhetorical strategy Harman employs. His arguments mostly rely on discussing singular cases to illustrate the extreme diversity or the real essence of objects, without displaying any theoretical interest in examining their general conditions of emergence or the specific contexts that determine their dynamics and their patterns of evolving and changing. This attitude, of refraining from causal explanations and conceptual explorations of the structures and trajectories of objects, might be due to the impact of Heidegger's philosophy. It rejects science-led concerns as the result of an old Western tradition that inaugurated a totalizing technological rationality and nurtures a will to power over objects: "Harman's thinking has this much in common with depth-ontology in the *echt*-Heideggerian mode: that it finds no room for anything like what a scientist (or science-led philosopher of science) would count as a contribution to knowledge or a claim worth serious evaluation in point of truth-content or validity" (Norris 2013, 195; emphasis in original).[16]

Given the aesthetic imperative, OOO tends to cherish objects in their astounding richness, irrevocable individuality, and inner depth, instead of making an effort to analyze or explain their properties and dynamics. Or rather, any attempt to know the object under investigation would risk impeding its singularity by reverting to forms of causal explanation that reside in general ideas of regularity and repetition. As Harman and his associates constantly emphasize, objects are "strange" (Morton 2011a,

165), "alien" (Bogost 2012), or "weird" (Harman 2012, 188). They cannot be known, and this is why they will never cease to surprise other objects (including humans). Thus, for OOO advocates the best philosophical option is to refrain from "the will to knowledge" and to be attentive to how things are by appreciating their uniqueness and potentiality.

OOO, and also speculative realism more generally, are characterized by a strong commitment to aesthetic questions. The meaning of the aesthetic is redefined and extended, as it defines the primary mode of interacting between objects and restricts the scope of epistemological and ethical concerns:

> [T]he force of speculative philosophy lies in its *specifically aesthetic* appeal. To put this another way, the reason I might want to encounter Harman's OOO or Brassier's bracing nihilism, or any of the other variants of SR [speculative realism], is their exhilarating aesthetic work: they allow or engender an encounter with the absolute that is beautiful or sublime—or voluptuous, fecund, terrifying, horrifying, intense, or any of the other aesthetic effects the various options in SR might generate. [. . .] To be sure, scientific knowledge claims show up all over the place in SR debates, but I get the sense that their rightness or wrongness, what it is that they suggest we know [. . .] won't actually change anybody's mind. (Richmond 2015, 400; emphasis in original)[17]

Speculation is set in motion here as a counterforce to knowledge and rationality. It serves as a means to uncover transhistorical stability and persistence, and claims to grasp the concrete texture of "the thing itself." As Carol A. Taylor notes, speculation in OOO denotes "a practice of enweirding, in which the weird other is approached, named, storied, fictionalized, and turned into words (or 'other' things), but never known, because it remains the autonomous real in itself. In other words, we 'know' about things—or think we do—because we (humans) make up stories, fictions, and narratives about them" (Taylor 2016, 210).[18]

This aesthetic and essentialist understanding of "the universe of things" (see Shaviro 2014) has dire consequences for the empirical analysis of objects. OOO's call to appreciate the richness of objects results in a highly subjective and impoverished account. The problems are apparent when we look more closely at one of the examples Harman provides:

the *Dutch East India Company* (Vereenigde Oostindische Compagnie: VOC), which existed in the seventeenth and eighteenth centuries. Surprisingly, Harman approaches this object "by looking for outstanding individual humans in the history of the VOC" (2016a, 55). Not only does Harman revitalize old and outdated traditions of historiography that focused on "great men" in accounting for historical developments and events, he also fails to give an explanation of what qualifies a particular human being (or object) as "outstanding." The approach remains ultimately arbitrary—especially given the fact that it is presented as a new "research program" (Harman 2007b, 22). Another example of this lack of analytic clarity is the political geography of the state proposed by Katharine Meehan and colleagues (2013), which seeks to understand how objects like wiretaps, cameras, and standardized tests affect state power.[19] These authors explore the implications of OOO for an analysis of the state as an object, but their account is in the end unconvincing, since it tends to switch between a reified concept of the state as a homogeneous though retreating actor and a more relational understanding (see Schmidt 2014).

Subjectivism Reloaded

To question the epistemological and ontological privilege accorded to human subjectivity, OOO starts from the idea of bounded, absolute, and discrete objects—treating humans as just one kind of object. However, this egalitarian or "democratic" impulse (see Bryant 2011a) with an emphasis on aesthetics and a straightforward rejection of any analytic ambition has some serious side effects. It tends to throw out the baby with the bathwater, as this decentering move "shies away from any interrogation of the effect of the human object (let's not even say 'subject') on other objects. [It] easily mistake[s] a consideration of *human* agency and ontology for the old-fashioned, outdated investment in subjectivity that is correlationism" (Johns-Putra 2013, 128; emphasis in original). It is quite unfortunate, but OOO has—in spite of advancing the notion of hyperobjects—nothing to offer as a way of accounting for the environmental crisis or indeed any form of asymmetrical "communication" (Harman 2007b, 30) between objects. Beyond the general topic of human mastery and the critique of anthropocentrism, OOO remains silent on how objects (human and nonhuman) affect one another differently; its main interest is in speculating about their irreducible strangeness and surprising weirdness.

It is weird, though, that cherishing the strangeness of objects finally reaffirms the quite familiar narrative of human subjectivism. Advocates of OOO endorse "a startling theory of subjectivity" (Morton 2011a, 168), which paradoxically combines the idea of an asubjective nature with a crude subjectivist stance.[20] Strange as it might sound, "[t]he move to the object is [. . .] not a move away from but rather a renewed move towards the Subject (with a capital S)" (Åsberg et al. 2015, 164). As Carol Taylor (2016) shows, OOO's appraisal of objects ends up resurrecting a blunt form of subjectivism that not only remains within the humanist frame but is also distinctively gendered. Taylor discusses this explicitly in commenting on Bogost's claim that an "alien phenomenologist's carpentry" "offers a rendering satisfactory enough to allow the artifact's operator to gain some insight into an alien thing's experience" (Bogost 2012, 100). She wonders

> who [. . .] is doing the "rendering"? By whose criteria is this rendering deemed to be "satisfactory"? And, again, who is doing the "speculating"? Undoubtedly Bogost himself, alone or in collusion with other—male?—philosopher carpenters. [. . .] Bogost (2012) complains that posthumanism is not post-human enough. If so, that is also the case for object-oriented ontology but in this case, the human who is reinstalled as recorder of traces is indubitably male, embodying an opaque set of values, and judging from a distance. (Taylor 2016, 210)[21]

Surprisingly, the genuine unavailability and withdrawal of objects has not prevented OOO philosophers making various claims about their inner truth and their essential properties. In fact, the proclaimed "new thinking" (Morton 2011a, 163) turns out to be nothing more than a revitalization or reenactment of a very old philosophical play where subject and object are opposed (Åsberg et al. 2015, 162). While protagonists of OOO express the wish to refute all subjectivism and the problem of correlationism, the reader is puzzled to see that the desire to account for the object independently of and beyond the human subject ironically tends to be rendered in anthropomorphic terms. It is quite debatable if the ambition to develop a theoretical vocabulary "applicable to the primitive psyches of rocks and electrons as well as to humans" (Harman 2011a, 103) takes seriously the very alterity and "strangeness" that, according to this line of thought, characterizes (nonhuman) objects. How can we reconcile what appears to be the simple extension of qualities formerly reserved to humans to nonhuman objects (e.g., communication, thought, or experience) with the essential insight that objects are "weird" and "withdrawn"? So far, OOO theorists have offered no convincing argument about how to address this considerable tension and how to escape the charge of anthropomorphism (Booth 2015).[22]

As OOO protagonists insist that there is an ontological difference in degree but not in kind between objects, this theoretical stance allows in principle for an empirical analysis of human and nonhuman entities and their relations. However, no examples of such an investigation can be found in the OOO literature. This disturbing dearth of empirical investigations is all the more irritating as it is accompanied by a striking absence of conceptual precision. While the focus on objects might be a helpful way of circumventing any pre-analytic distinction between living and non-living entities in order to open it up for critical inquiry, Harman reintroduces this dichotomy by discussing the life and death of objects and their "lifespan" (Harman 2016a, 47). He thereby not only repeats conceptual ambiguities in determining life (see Helmreich 2011), but also reaffirms the ontological and (bio)political distinction between life and non-life (see Povinelli 2016).

OOO and speculative realism both tend to overstate their originality by selectively reading the philosophical tradition and by ignoring alternative versions of posthumanist thought. Concerning the first point, commentators have remarked that the richness and diversity of the phenomenological tradition is not taken into account by Harman and his associates. Dan Zahavi argues that they not only misinterpret classical texts but also fail to address important differences within phenomenological thought (see Zahavi 2016). The same observation applies to the concept of matter that is evoked in OOO, representing a very limited and caricatured understanding of matter and the materialist tradition. As Jane Bennett has rightly noted, OOO recognizes matter only as "a flat, fixed, or law-like substrate" (Bennett 2015a, 233). As we will see in the next chapters, none of the new materialists Harman and other OOO theorists criticize for "duomining" shares this "idealist" notion of matter that Harman attributes to them (and neither do most "old" materialists). It is quite obvious that OOO has not made any effort to seriously engage with the concept of matter—a shortcoming that is advertised as a theoretical advantage under the label of "immaterialism" (Harman 2016a).

The disregard of feminist materialism is especially troubling. Not only are the "founding fathers" (Taylor 2016, 205)—Graham Harman, Ian Bogost, Levi Bryant, and Timothy Morton—all male; they also engage with an exclusively male philosophical lineage in which Kant, Husserl, and Heidegger are the most prominent figures. While feminist materialists have for a very long time developed posthumanist accounts and flat ontologies (see, e.g., Haraway 1991; Suchman 2007), this literature is only marginally taken into account (see, e.g., Harman 2016a, 14). As Rebekah Sheldon remarks, "OOO has been so provocative for feminist theorists because of its cannily unknowing usurpation of the energies of feminist thought and its relegation of that history to footnotes within its own autobiography" (2015, 204; Alaimo 2014; van der Tuin 2014, 231).

This willful theoretical ignorance has serious consequences. While it might indeed be instructive to point out problems with current accounts of relationality and matter, the conceptual proposals OOO delivers are ultimately unconvincing. The rejection of "relationism" relies on the idea of a "phantasmatic transparency" that conceives the other as "transparent only in isolation, absent any distorting relations or textual mediation" (Zalloua 2015, 406).[23] But maybe the most problematic aspect of OOO (and speculative realism) is not its naïve encounter with objects but rather its attack on the critical tradition in philosophy and social theory. By limiting critique to the Kantian tradition of correlationism,[24] OOO effectively rules out any critical account of the real (Åsberg et al. 2015, 162–63). The speculative turn implicitly rejects the critical tradition as it theoretically privileges stability and inaccessibility; its focus on individual objects tends to ignore structural hierarchies and asymmetries, and it makes no attempt to link these to human objects and their differential power to affect human and nonhuman objects. Ironically, "at the very moment when humans have caused a state shift in the earth's biosphere and are presiding over a mass extinction, we are witness to the ascendency of a social theory that massively redistributes agency to the nonhuman and promotes withdrawal as the primary mode of being" (Campbell et al. 2019, 129–130). Thus, the critique of anthropocentrism and the ontological egalitarianism OOO endorses tends to flatten distinctions among objects (instead of exploring them) and obscures the de facto privileged role and the planetary power of humans. This is a weird realism indeed.

2

Vital Materialism

Jane Bennett and the Vibrancy of Things

> This does not mean that vitalism, which put so many images in circulation and perpetuated so many myths, is true. [. . .] But it does mean that it had and no doubt still has an essential role as an "indicator" in the history of biology. And in two ways: as a theoretical indicator of problems to be solved [. . .]; and as a critical indicator of the reductions to be avoided [. . .]. (Foucault 1998a, 474)

While object-oriented ontology praises itself for its "coolness" (see, e.g., Morton 2011a, 163) by focusing on bounded and fixed objects and their withdrawal, another important strand of the new materialisms stresses the "vitality" or "vibrancy" of things. It replaces the focus on abstract concepts, cool speculation, and the incessant search for the inner depth of isolated objects with an interest in the always unstable interconnections and contingent associations of human and nonhuman bodies.

The theoretical profile and the political perspectives of this "vital materialism" (Bennett 2010a, x) are probably best spelled out in the writings of Jane Bennett, who has extensively and over a long period elaborated the implications of the material turn for politics and political theory. In *The Enchantment of Modern Life: Attachments, Crossings and Ethics* (2001), *Vibrant Matter: A Political Ecology of Things* (2010a) and in her work in general, Bennett starts from the assumption that matter must be addressed as an active part of political processes that have so far been seen as dominated by human subjectivity. She coins the term "thing-power" (2010a, xvii) to account for the ability of inanimate entities to produce effects by operating in conjunction with other material bodies. Bennett's work seeks to rethink the traditional divides between matter and life, inorganic and organic, passive object and active subject in order to establish that agency is not an exclusive property of human beings.

This chapter discusses the basic arguments and important achievements of this "enchanted materialism" (Bennett 2001, 156–58; 2005, 135) as well as some problems and limitations of this theoretical perspective. I first analyze the ontological underpinnings of Bennett's vitalist account and examine the concept of "thing-power." The second part presents two examples she uses to illustrate the "force of things" (Bennett 2004): the breakdown of the electric grid in North America in 2003, and the nutritional effects of a particular kind of fatty acid, omega-3. The next section engages with Bennett's posthumanist political theory that successfully problematizes anthropocentric conceptions of politics and liberal accounts of agency. I will then address central conceptual problems of Bennett's variant of materialism, arguing that the idea of an all-encompassing "vitality of matter" and an originary "force of things" still endorses an essentialist account of matter as such, independent of the "dense network of relations" (Bennett 2010a, 13) that constitutes them. The concept of thingness is also empirically limited, as it provides only a selective understanding of material agency. These conceptual shortcomings and analytical problems affect the political perspectives of Bennett's vital materialism, which are discussed in the final part of this chapter. I diagnose a tendency in her work to displace political considerations by invoking new ethical sensibilities. Thus, being attentive to the vibrancy of things tends to ignore the inequalities, asymmetries, and hierarchies enacted in vital materializations.

Imagining a Different Onto-Story: Exploring Thing-Power

Bennett's work breaks with the idea, so dominant in the Western political tradition, that nature is ruled by deterministic laws while human societies are governed by free will. It invites us to imagine a different "onto-story" (2001, 15; 2005, 136) foregrounding the "vitality *of* matter" (2010a, vii; emphasis in original), a concept that disturbs conventional understandings of agency as it acknowledges the force of nonhuman entities: "By 'vitality' I mean the capacity of things—edibles, commodities, storms, metals—not only to impede or block the will and designs of humans but also to act as quasi-agents or forces with trajectories, propensities, or tendencies of their own" (ibid., viii).

This "neo-animist ontology" (Bennett 2011a, 120) synthesizes a heterogeneous bunch of theoretical concepts and ideas from Lucretius, Spinoza, Adorno, Latour, Thoreau, Bergson, Dewey, and Deleuze and Guattari to arrive at a different concept of agency. Bennett argues that agency needs to be "distributed across a wider range of ontological types" (2010a, 9) that cuts across the human/nonhuman divide, so that things like food and minerals can be reconceptualized as having the ability to produce effects. Furthermore, she moves beyond the focus on individual bodies and their borders to propose a concept of action that is based on certain configurations of human and nonhuman forces she calls—following Deleuze and Guattari—assemblages. In her reading, assemblages are "ad hoc groupings of diverse elements, of vibrant materials of all sorts" (ibid., 23; Bennett in Khan 2009, 92). This fluid and open concept of things is the main point of disagreement with the proponents of object-oriented ontology I discussed in the last chapter. According to Harman, Bennett's insistence on vibrant matter ultimately leads to the rejection of "fully formed individuals" and their replacement by a "pampered layer of ultimate particles" (Harman 2016a, 20, 96; 2011e).

Bennett's ambition is not limited to expressing a new ontological narrative that embraces the "force of things." Her project is as much a philosophical exercise as it is a political endeavor. The philosophical aim is to rethink and refute the idea of matter as dead or passive stuff, reworking and revising the notion of "vibrant matter" that has traditionally been marginalized in philosophical thought but played a central role in the materialist tradition from Spinoza to Nietzsche. Bennett claims that being attentive to matter as "vibrant, vital, energetic, lively, quivering, vibratory, evanescent, and efflorescent" (2010a, 112) will encourage "the emergence of more ecological and more materially sustainable modes of production and consumption," promoting "greener forms of human culture" (ibid., ix–x). According to Bennett, the idea of a "shared materiality of all things" (ibid., 13) will lead to a "positive ontology" (ibid., x) that enables a fundamental political transformation as it inspires a new environmental sensibility and informs a radical restructuring of economic relations. Thus, the "guiding question" of her work is: "How would political responses to public problems change were we to take seriously the vitality of (nonhuman) bodies?" (Ibid., viii; see also Bennett and Loenhart, 2011)[1]

Bennett's vital materialism needs to be distinguished from older versions of vitalism on the one hand and more traditional forms of materialism on the other. It differs from the former as it does not suggest a universal or isolated life force, an "élan vital" (Bergson 1998) or "entelechy" (Driesch 1908a; 1908b) that is shared by and animates organisms. While Hans Driesch and Henri Bergson were anti-materialists and unable to imagine a materiality that informed the vital processes they discerned in nature, Bennett follows Spinoza's lead by stressing the creative and dynamic role of affectivity in constituting matter (Bennett 2010a, xiii; 62–81; Bennett in Khan 2009, 93–95; see also Gamble et al. 2019, 119–20).[2] Bennett's "enchanted materialism" also claims to go beyond existing forms of materialism and critical inquiries. She points to the limits of strategies of "demystification" prominent in the materialist tradition, which often buy into the anthropocentric imaginary according to which human agency "has illicitly been projected into things" (2010a, xiv; 2001, 111–130; Bennett in Khan 2009, 93–95). For Bennett it is not sufficient to "expose social hegemonies" (2010a, xiii) as historical materialists did, as we also need to address "nonhuman, thingly power" (ibid., xiii). Thus, the negative critique of existing institutions must be corrected and complemented by designing "positive, even utopian alternatives" (ibid., xv).[3]

The Force of Things: Two Examples

Like Harman, Bennett takes up theoretical propositions originally formulated within actor-network theory, especially the suggestion to replace the (human) actor with the figure of the actant, which can be human or nonhuman. An actant is defined by its ability to produce effects and alter situations rather than by a capacity for action (Bennett 2004, 355; 2005, 133–35). While Bennett states that there are undeniable "affinities" (2010a, 98) between humans and nonhumans, she cautions that they do not exhibit the "same kind of agency" (ibid., 98). However, she still holds that not only humans but also comparatively simple organisms like earthworms respond to changing contexts and situations. Their actions display a "certain 'freedom of choice'" (ibid., 97) that cannot be grasped by inscribing them into a divine plan or reducing them to a mechanical instinct. Bennett introduces the concept of "things" in order to exceed the traditional opposition between "objects" and "subjects." Against the proponents of object-oriented ontology, she argues that we should differentiate between the terms "thing" (or body) on the one hand and "object" on the other, as only the former makes it possible to "disrupt the political parsing that yields only active (manly, American) subjects and passive objects" (Bennett 2015a, 234; Bennett in Watson 2013, 156–57). Instead of imagining isolated objects, she insists that "matter has an inclination to make connections and form networks of relations with varying degrees of stability" (2004, 354).

Bennett's vital materialism extends beyond nonhuman animals or plants. She argues that even inorganic matter, such as litter or minerals, exhibit "powers of life, resistance, and even a kind of will" (ibid., 360). In the following, I discuss two of her examples that prompt us to rethink the notion of agency for political analysis. The first case Bennett invokes to illustrate her concept of "*distributive* agency" (2010a, 21; emphasis in original) is the famous power blackout in North America in 2003. It affected 50 million people in the US and Canada for approximately twenty-four hours and resulted in the shutdown of over one hundred power plants, including twenty-two nuclear reactors. Bennett proposes that we see the electric grid as an "agentic assemblage" (ibid., 21), consisting of a list of heterogeneous actants that in one way or another contributed to the blackout. The electric power grid assemblage comprises human actors that build and manage the power sites, maintain the networks, supervise the operations, consume the electricity, and pass the regulatory laws, but it also contains a different species of actants: electrons, trees, wind, fire, coal, sweat, computer programs, plastic, wire, wood, and electromagnetic fields (ibid., 24–25). Bennett reconstructs the events of August 2003 in the United States and Canada as follows:

> several initially unrelated generator withdrawals in Ohio and Michigan caused the electron flow pattern to change over the transmission lines, which led, after a series of events including one brush fire that burnt a transmission line and then several wire-tree encounters, to a successive overloading of other lines and a vortex of disconnects. One generating plant after another separated from the grid, placing more and more stress on the remaining participants. In a one-minute period, "twenty generators (loaded to 2174 MW) tripped off line along Lake Erie." (Ibid., 25)

According to Bennett the blackout was the "end point of a cascade" (ibid., 25), a dynamic interplay of human and nonhuman actants. The participants not only included human decisions, motives, and omissions such as profit interests and insufficient maintenance programs, but also trees, computer programs, regulatory environments such as the neoliberal organization of the energy market and solid infrastructural networks, and also singular incidents and spontaneous events like a fire. Pointing to the hybridity of human and nonhuman agencies, Bennett successfully disturbs linear concepts of causality. She not only cautions against attempts to predict and control the dynamics of action, but also makes it more difficult to engage in a moral "blame game" (ibid., 37). Given the complexity of the event and the multitude of actants involved, Bennett suggests there is no simple answer to questions of responsibility and accountability.

The second example that Bennett explores, in a paper entitled "Edible Matter" (2007) and again in a chapter of *Vibrant Matter* (2010a, 39–51), concerns the agency of a particular kind of fatty acid that is prevalent in some wild fish: omega-3. Bennett points to several scientific studies that, she argues, have proven the positive effects of omega-3 on the human body (ibid., 39–40). These studies seem to suggest that omega-3 significantly modulates and enhances human behavior, affective states, and cognitive abilities, as it "can make prisoners less prone to violent acts, inattentive schoolchildren better able to focus, and bipolar persons less depressed" (ibid., 41). Bennett considers omega-3 to be a potent example of the agency of nutrients being more than the milieu or the resource for human agency. She conceives of bodies of food as interacting with human bodies, forming assemblages that Bennett proposes to call "American consumption" or the "crisis of obesity" (ibid., 39).

For Bennett, the agentive capacities of omega-3 call for a revision of how dietary problems are addressed by scientific and political authorities and for a re-evaluation of the proposed solutions. This suggests, first, the need to shift the focus from the liberal concept of the free will of individual humans to the more heterogeneous and complex agentive assemblages that they are part of. Second, the analysis would not only include social, cultural, or economic factors that shape and govern dietary routines and nutritional regimes, but also the agentive capacities of fatty acids and other nutrients that regulate human affective states or well-being. Thus, the vital materialist account provides a different perspective on obesity: "The problem of obesity would thus have to index not only the large humans and their economic-cultural prostheses (agribusiness, snack-food vending machines, insulin injections, bariatric surgery, serving sizes, systems of food marketing and distribution, microwave ovens) but also the strivings and trajectories of fats as they weaken or enhance the power of human wills, habits, and ideas" (ibid., 42–43).[4]

Combining Latour and Rancière: Designing a Posthumanist Political Theory

The ontological recognition of a "force of things" is intimately linked to a normative project. Bennett aims to reinvent political theory by questioning (and ultimately superseding) its anthropocentric underpinnings. While contemporary democratic theory is still dominated by the figure of the human subject, she suggests a more complex concept of the "demos" (2010a, 30) whose composition is not restricted to human beings. This theoretical move makes it possible, according to Bennett, to draft a "more radical theory of democracy" (2005, 142) that attends to a dynamic and vital force shared by all materialities.

Bennett's argument relies on the assumption of "a *structural parallel* between formations in nature and cultural formations" (ibid., 137; emphasis in original). To explore the democratic potential of this version of complexity theory, she proposes a posthumanist reading of Jacques Rancière's political theory. Bennett takes up Rancière's claim that democracy is defined by the eruption of those who protest against their exclusion from politics (Rancière 1999). Rather than being marked by the search for consensus, this concept of politics is characterized by the lack of agreement on the constitution of the demos, on who counts as a political subject. Democracy then exposes the contingency of the dominant "partition of the sensible" (ibid.) that provides an ontological framing, rendering some groups and acts visible while keeping others "out of the picture" (see Bennett 2005, 138).

However, as Bennett points out Rancière's political theory is still shaped by an anthropocentric imaginary that relies on a radical divide between humans and nonhumans. She seeks to expand this limited conception of democracy by referring to the self-organizing power that assemblages exhibit. Bennett takes up Latour's proposal to imagine new forms of exchange and communication between humans and nonhumans that might ultimately give rise to a "parliament of things" (Bennett 2010a, 103–4; Latour 1993, 142–45).

The posthumanist political theory Bennett proposes successfully breaks with the longstanding tradition of humanism and anthropocentrism in this discipline. It problematizes the idea of an ontological hierarchy that places humans at the top, proposing instead a flattened conception that regards humans "merely" as "a *particularly rich and complex* collection of materials" (ibid., 11; emphasis in original). Furthermore, by favoring assemblages and associations Bennett's vital materialism is helpful in displacing liberal accounts of individual self-determination on the one hand and OOO's focus on the autonomy of isolated objects on the other.

This posthumanist account is combined with an innovative concept of "life" that cuts across traditional ontological and normative divisions. Indeed, the very notion of a vitality of matter combines two elements that have often been regarded as oppositional or even contradictory, namely matter and life. This semantic hybrid makes it possible to question the modern concept of life that relies on a strict division between organic and inorganic, organism and machine. Bennett does not reserve the notion of life for organisms but extends certain qualities and properties that define the modern concept of life (e.g., self-organization, reproduction) to inorganic matter, claiming for example that minerals, too, reproduce and self-organize (see 2010a, 11). She also reminds us of the fact that in specific historical epochs and cultural contexts nonhumans have been considered to be legal and moral entities.[5] Bennett argues that this historical reminder of pre-modern (and non-modern) ways of recognizing nonhuman "doings" helps us to imagine alternative concepts of agency that no longer restrict it to humans alone.

Finally, Bennett's vital materialism also goes beyond conventional (environmental) politics in two important ways (see Chandler 2011, 303). First, the move from "the environment" to "vibrant matter" breaks with ideas of human stewardship or any exterior relation between humans and the environmental conditions of life; rather, materiality is shared by humans and nonhumans alike, and human agency is always already part of more complex assemblages that fuse and hybridize human and nonhuman forces. Second, Bennett's concept of a vitality of things extends traditional concepts of embodiment. Following Donna Haraway (2008; 2016), Bennett points out: "In a world of vibrant matter, it is thus not enough to say we are 'embodied.' We are, rather, *an array of bodies*, many different kinds of them in a nested set of biomes" (2010a, 112–13; emphasis in original). In this perspective, human agency depends on the existence of nonhumans.

A More-Than-Relational Account of Agency

Bennett's work has contributed significantly to a more complex understanding of politics, problematizing the idea that it is restricted to the realm of the social and to human beings. However, the idea of vibrant matter has also given rise to criticism and caution. Unfortunately, there are some conceptual and analytical problems associated with Bennett's vital materialism that ultimately undermine her claim to provide an "alternative approach to democracy" (2005, 136).

Bennett's central concept of a vitality of things, in particular, has provoked widespread criticism. The critical comments pertain to both elements of the concept, "vitality" and "thingness." Concerning the former, some commentators have dismissed the diagnosis that matter is reduced to dead and passive substance in contemporary political thought and practice, leading to the claim that matter needs to be liberated "from its long history of attachment to automatism or mechanism" (Bennett 2010a, 3). Ben Anderson has stressed that in the wake of processes of digitalization and molecularization and within contemporary discourses of resilience and preparedness, matter is being conceived of more and more as informational and malleable, as well as "in terms of circulation, connectivity and complexity" (Anderson 2011a, 395; see also Braun 2011). Similarily, Andrew Barry (2013), in his empirical study on the construction of the Baku-Tbilisi-Ceyhan (BTC) pipeline and the controversies and disputes surrounding the project, has shown that assessing the properties and performances of materials and artifacts increasingly intersects with the generation and circulation of information.[6]

Secondly, Bennett tends to endorse a romanticized and one-sided picture of the "vibrancy of things." Her materialist account cherishes vitality and enchantment as "a positive resource" (Bennett 2001, 15) while neglecting or ignoring destructive aspects. Material surprises may not always be sources of delight and pleasure but include dangerous and possibly fatal consequences such as those exhibited in natural disasters, climate change, or materials like asbestos (Gregson et al. 2010, 1080–82; Jensen 2015, 19). Even more importantly, Bennett's appraisal of "generosity" and "joy" (Bennett 2001, 174 and 12–13) tends to disregard or discard more negative sentiments such as hostility, boredom, disappointment, or rejection. However, these affective energies might play a fundamental role in bringing about (political) change (see Anderson 2004; Wilson 2015).

The third problem concerns Bennett's comprehensive understanding of life. As we have seen, Bennett goes beyond organic concepts of life and claims that "everything is, in a sense, alive" (2010a, 117). However, in its generality this position is only partly convincing. While it is certainly right to conceive of life not as a property that pertains to specific bodies but as a process or rather the outcome of certain materializations, it might be more accurate to distinguish between differently composed materialities and various complexities of conjunctions between bodies—in which the distinction between animate and inanimate bodies may play a crucial role. As Bruce Braun and Sarah Whatmore put it: "Is more gained from a closer attention to the *specificity* of the matter at hand, as opposed to a generic analogy to 'life' that could be described as a metaphysics?" (Braun and Whatmore 2010, xxix–xxx, emphasis in original; Braun 2008, 675–77)[7]

Similar reservations apply to the concept of "things." When Bennett conceives of things as "the ordinary stuff around us that we possess and use, and are possessed and used by" (Bennett and Loenhart 2011, 6), this definition captures only a very restricted and superficial understanding of thingness. As the many examples from litter to worms she uses in her work indicate, Bennett's treatment of matter privileges visual contact and bodily presence (Gregson 2011; Princen 2011). The ontological spectrum of things she presents is limited to natural entities and technological objects (see Bennett 2005, 134–5)—thereby excluding a large variety of other "things" beyond these categories or below the threshold of visibility and physicality. We are left to "wonder about how to encounter vibrant matters that unsettle distinctions between near and far, presence and absence, and might take place with properties and capacities closer to a liquid or gas, water or air. How might we understand encounters with forces such as flows of finance or systems such as climate that exceed the intimate bodily presence of some of Bennett's examples?" (Anderson 2011a, 394)

Unfortunately, Bennett's idea of things is not only empirically limited. While it might be possible to extend and enlarge the category of "thingness," vital materialism is also characterized by fundamental conceptual problems. Steve Hinchliffe (2011) notes a significant ambiguity in Bennett's concept of "things." While she claims that things are to be understood as relational as they are not stable and solid entities but participate in dynamic and processual assemblages, she simultaneously regards them as "things in themselves" that have persistency and activity that extends beyond and prior to their relationality. In fact, the very idea of a "force of things" amounts to a "naive realism" that allows things to have a "more-than-relational character" (Hinchliffe 2011, 398; see also Cudworth and Hobden 2015). Bennett is partly aware of these conceptual problems, but merely refers to them as "disadvantages": thing-power, she acknowledges, "tends to overstate the thinginess or fixed stability of materiality" (2010a, 20). A second self-admitted problem is the "latent individualism" involved, as in fact "an actant never really acts alone" (ibid., 20–21). Nonetheless, she still endorses the idea of an original thing power, which is at odds with the general focus on assemblages in her work.

In fact, Bennett's ambivalent account of agency, alternating between a relational and a "more-than-relational" concept, informs her understanding of assemblages and their composition. She claims that an assemblage consists of "member[s] and proto-member[s]" possessing "a certain vital force" (ibid., 24) before they enter into an association. In other words, there is a vital force that pertains to the assembled individual entities regardless of their constitutive relations. Bennett is certainly right in proposing a distributed concept of agency that recognizes a more complex and extensive range of entities and processes in order to account for "making a difference" in the world. The theoretical commitment to reject abstract concepts and analytic presuppositions about actants and "their" capacities, and to explore how agents emerge in particular fields of forces, is a move in the right direction.

However, it is not convincing to simply extend the category of the actor beyond humans to include formerly excluded entities, affirming the agentive capacities of things. This theoretical proposition still buys into the liberal concept of agency that sees it as a property of individual entities, focusing on will, freedom, and choice. While this extension empirically broadens the range of those included in the collective of (political) actants, it leaves intact the liberal imaginary and the conceptual divide between causality and agency, external forces and inner will. It would be more pertinent to abandon the notion of agency altogether and to put the emphasis on "modes of doing," which better brings out how materialities work together (Abrahamsson et al. 2015, 13–15).

While Bennett repeatedly states that agency is not a capacity or a property, but rather the outcome of an assemblage, she still resorts to the vitalist idea that agency is a quality that pertains to material beings. This ambivalence is also present in Bennett's definition of vital matter as the "capacity of things" to "act as quasi-agents or forces with trajectories, propensities, or tendencies of their own" (2010a, viii). Here again, the focus on the performative nature of relations is displaced by the idea of pre-established identities.[8]

From Politics to Ethics

The analytical problems and conceptual ambiguities associated with the idea of a vitality of things seriously limit the political purchase of Bennett's work. In fact, her analysis stops halfway. While it is certainly necessary to address the composition of the collective and open up the demos for more-than-human encounters, this theoretical gesture is not sufficient to account for the political issues at stake. It still remains to be seen how exactly forces come to be determined in one way rather than another. It is not enough to embrace the vitality of things as a way of focusing on how the collective is assembled; we need to attend to how vital forces are mobilized and enacted, and to analyze what comes to matter (and what does not): "Perhaps politics does not name the formation of publics as such, but rather the *determination* of incipient events, the ongoing and ever renewed work of turning contingency into necessity. [...] Politics, then, is perhaps not *equivalent* to the vitality of matter, rather it consists in *rejoining* this vitality, in contributing to its ongoing and ever-renewed determination" (Braun 2011, 392; emphases in original).

To clarify this point, it is helpful to go back to the two examples Bennett uses to illustrate her idea of a vital materialism. Her analysis of the breakdown of the electric power grid in the US attends to the multitude of different agencies involved in the event and their complex relationships. However, in the end she agrees with the directors of the FirstEnergy corporation in declaring that no one was to blame for the event (2010a, 37), claiming that the complex interconnections between humans and things render obsolete any idea of a "strong responsibility" (ibid., 37). While the causal issue is indeed difficult to resolve, the final conclusion is clearly unsatisfactory and makes one wonder what exactly the prospects of this "weak" understanding of responsibility might be. As one commentator succinctly put it: "Overall, I worry that Bennett's laudable ethical stance becomes an avoidance of politics. Bennett's response to this catastrophe, certainly one of the anxieties precipitated by a nonregulated, profit-driven, globalized reality where sovereignty has been undermined, is an individual, rather than a political one" (Marso 2011, 426). It is not enough to see "individuals as simply incapable of bearing *full* responsibility for their effects" (Bennett 2010a, 37; emphasis in original); rather, it is necessary to develop a different understanding of responsibility that dispenses with the liberal idea of "full" or "partial" responsibilities altogether.

Bennett seeks to avoid reducing political processes to moral questions, in order to get away from the "blame game." She fears that a politics that engages in moral condemnation will be unable to address the "cultivated discernment of the web of agentic capacities" (Bennett in Khan 2009, 93; Bennett 2010a, 38). While it is certainly right to question singling out individuals and groups as a way of making scapegoats of them, Bennett's alternative route leads to a dead end. The proposed way of overcoming a moralized politics results in an ethics grounded in personal choice: "Perhaps the ethical responsibility of an individual human now resides in one's response to the assemblages in which one finds oneself participating: Do I attempt to extricate myself from assemblages whose trajectory is likely to do harm? Do I enter into the proximity of assemblages whose conglomerate effectivity tends toward the enactment of nobler ends?" (Bennett 2010a, 37–38)

As agency is—according to Bennett—based in assemblages that make up who and what "we" are, it is difficult to imagine how it is supposed to be even possible to "extricate myself" from "harmful assemblages." Bennett's call to move away from a politics of moral condemnation leaves us with nothing more than a voluntaristic ethics. It seems that the only remaining option is to focus on the self, as there are no others to be held accountable. As Bonnie Washick and Elizabeth Wingrove argue:

> [I]nsofar as our assemblaged agency leaves us never "fully responsible" for its effects, holding oneself accountable seems to telescope the ethical terrain to entanglements from which one could unilaterally extricate oneself or into which one could unilaterally insert oneself. Not only does such an ethical call conjure relations between self and world (its human and nonhuman constituents) that appear problematically voluntarist, but it in addition limits the possibilities of "action-in-concert" to individuals' ethical calibration: notions of harm and "noble ends" are apparently what we apply when assessing entry to or exit from collective processes, rather than concerns and values shaped within and through collective contestation and world-making practice. (2015, 75)

Let us now turn to the second example. Bennett's claim about the agentive powers of omega-3 has been scrutinized in an article by Sebastian Abrahamsson and his colleagues (2015), and I rely on their arguments in the following. These authors analyze closely the argument in one of the scientific articles to which Bennett refers in order to bolster her claim that empirical studies "support the idea that lipids have the power not just to increase human flesh but also to induce human moods, modes of sociality and states of mind" (Bennett 2007, 137). Abrahamsson et al. show that the article in question (Gesch et al. 2002) does not provide any conclusive evidence for the potency of omega-3 to affect positively the human body. The study only diagnoses an improvement, without claiming to have unraveled the causal chain that led to it. Thus, the observed changes in mood and behavior could result from omega-3, but also from other fatty acids or different vitamins and minerals that were given to the participants in the study. Equally, it might have been effected by a particular combination of some substances. Whatever the case may be, as Abrahamsson et al. (2015) show on the basis of a careful reading of the study, it is impossible to make an argument for the isolated agency of omega-3. Rather, the evidence suggests that "omega-3 is not matter *itself* all by itself, but rather matter *in context*. It is engaged in many relations" (Abrahamsson et al. 2015, 5; emphases in original).

While Bennett only discusses whether or not omega-3 "acts" to change "human moods and cognitive dispositions" (2010a, 40), Abrahamsson et al. (2015) also inquire where the edible omega-3 comes from. They do not focus on health-related questions but show that the production of this fatty acid is part of a global political economy of food. The authors propose a "shift in attention away from the human moods affected by omega-3 and onto the worlds from which omega-3 is procured" (Abrahamsson et al. 2015, 11). Referring to an article in an epidemiological journal (Brunner et al. 2009), Abrahamsson et al. point to important social and ecological effects of the materiality of omega-3 that are left out of Bennett's analysis. Most of the omega-3 fatty acids sold to health-conscious individuals in the Global North come from fish caught in the Global South, contributing to the depletion of the oceans and scarcity of food in these regions. Thus, the well-being of human beings in rich countries is entangled with the worsening of the living conditions of those inhabiting the seashores of the poorer parts of the world. Abrahamsson and colleagues conclude their analysis of omega-3 by pointing out what is obviously missing in Bennett's account of politics: "[R]ather than getting enthusiastic about the liveliness of 'matter itself', it might be more relevant to face the complexities, frictions, intractabilities, and conundrums of 'matter in relation'. For it is in their relations that matters become political, whether those politics are loudly contested or silently endured" (Abrahamsson et al. 2015, 13).

Examining the two examples Bennett presents to illustrate her idea of enchanted materialism shows that her work suffers from an undercomplex, insufficient, or "undercooked" (Gregson 2011, 403) account of politics (see also Lettow 2017, 109–10; Gamble et al. 2019, 120).[9] Ironically, the general call to acknowledge "thing power" results in a political theory without any specific analysis of power relations. Rather than cherishing a vitality of things per se, it is necessary to investigate how material human and nonhuman forces come to be determined in one way or another. Instead of celebrating the move from dead and passive to vibrant and active matter, we need to analyze how matter is differentially set in motion and to what ends.[10] Therefore, we need to "account not only for the force of life, of the vibrancy of matter, but the force of the negative as well, the forces that demarcate the field of becoming into the possible and impossible, determining what matter can come to matter" (van Wyk 2012, 135).

Bennett has responded to this criticism by suggesting that vital materialism consists exactly in questioning and overcoming traditional modes of political analysis and critique: "I have found it to be more helpful for my particular political problem (which I would characterize as this: 'How to transform our unsustainable, unjust, and earth-destroying modes of consumption and production into their opposites?') to forego the category of the political as such . . ." (Bennett 2015b, 84). She distances her project (and new materialisms more generally) from an idea of politics considered as an institutional ensemble or a separate sphere characterized by distinct boundaries and norms (see ibid., 83–84). While this is an important clarification, it does not answer the question of how exactly politics and ethics intersect and how to attend to "the economic and social factors that condition ethical relations" (Bennett 2001, 132).

As a consequence, the political prospects of Bennett's vital materialism are clearly limited. Her call "to make the political more poetic" (Bennett in Watson 2013, 158) does not take into account the material preconditions and contexts necessary for political change. Rather than endorsing a different politics, it envisions a new "ethical sensibility" (Bennett 2001, 12; Bennett in Watson 2013, 151; see Coole 2013, 462). While this is not a bad thing as such, serious problems arise from the lack of a critical analysis of the political and social conditions that resist or constrain the material change to a sustainable economy and to a more democratic politics Bennett calls for.[11]

Bennett seeks to promote a political theory that goes beyond a focus on (human) power to assemble the political collective more inclusively. However, the vital materialism she proposes undermines the analysis of how the two are linked. The enlarged concept of the demos she advocates is coupled with a very limited and selective account of politics that presents a positive image of vital powers while ignoring or at least downplaying how they control and channel what comes to matter. While the need for a move to a posthumanist politics might well be more urgent than ever, Bennett's concept of vital matter is, in the end, unable to address the ontological and political questions at stake. It avoids any substantial analysis and critique of power to focus on a new ethical sensibility, declaring that the "newfound attentiveness to matter and its powers will not solve the problem of human exploitation or oppression, but it can inspire a greater sense of the extent to which all bodies are kin" (Bennett 2010a, 13). In attacking a humanist account of politics, Bennett not only exposes the limits of humanism but also gets rid of politics. Hence, rather than providing an "alternative model of politics" (Bennett 2011b, 406), vital materialism quite surprisingly results in an alternative to politics.

3

Diffractive Materialism

Karen Barad and the Performativity of Phenomena

The third strand of new materialism discussed in this book is Karen Barad's agential realism. This differs significantly from OOO's focus on the inner depth of objects on the one hand and vital materialism's fascination with the vibrancy of things on the other. In contrast to OOO, agential realism rejects the idea of isolated objects but is interested in complex entanglements with bodies of different kinds. However, to account for the "extraordinary liveliness" (Barad 2007, 91) of the world, agential realism does not follow the vitalist imagination of an originary "force of things," promising instead to provide a thoroughly relational ontology.

Barad's understanding of agential realism is systematically presented in her book *Meeting the Universe Halfway: Quantum Physics and the Entanglement of Matter and Meaning* (2007) and has been further developed since its publication in many articles and interviews. Barad was originally trained in theoretical physics, and she draws on a diversity of theoretical sources encompassing physical as well as social theories, including quantum physics, science studies, feminist theory, critical race theory, postcolonial theory, and (post-)Marxist theory (see ibid., 25–28). Most importantly, she combines insights from the physicist Niels Bohr, one of the most prominent figures in quantum mechanics, with elements of poststructuralist theory and feminist technoscience studies. Barad seeks to break with the concept of matter as a passive substance that exists independently of epistemic practices, putting forward the idea that "matter plays an agentive role in its iterative materialization" (ibid., 177). Starting from this proposition, she reconceptualizes the interrelations between humans and nonhumans and rethinks the categories of subjectivity, agency, and causality. In sum, she claims to have developed "an epistemological-ontological-ethical framework that provides an understanding of the role of human *and* non-human, material *and* discursive, and natural *and* cultural factors in scientific and other social-material practices, thereby moving such considerations beyond the well-worn debates that pit constructivism against realism, agency against structure, and idealism against materialism" (ibid., 26; emphases in original).

The following discussion of agential realism starts with the epistemological issues raised by Barad's uptake of Bohr's quantum mechanics. I also address questions of method, focusing on how Barad extends Donna Haraway's concept of diffraction. The second section turns to ontological and ethical concerns, analyzing how Barad partially revises the lessons of quantum mechanics in order to connect them to ontological considerations. In doing so, Barad especially engages with Judith Butler's work and her concept of gender performativity, as well as with Foucault's analytics of power. She also refers to Levinas' understanding of ethics, redefining the concept of responsibility he proposes as a way of overcoming its anthropocentric implications. The third part presents Barad's concept of the apparatus, which differs from both a technical and a social understanding of the term. It examines the claim that agential realism contributes to a new understanding of power relations. The next two sections aim at an overall assessment of agential realism, showing its strengths as well as some limitations. I critically discuss the foundational role of quantum physics in Barad's account as well as her inflated concept of ethics, which risks marginalizing the political dimensions of ontological matters.

From Reflection to Diffraction: Epistemological and Methodological Revisions

One longstanding concern of feminist theory has been taking seriously the fundamental historicity and situatedness of knowledge without giving up or undermining epistemological claims (Haraway 1991, 183–201; Harding 2004). Inscribing her work within this tradition, Barad proposes a radical re-reading of Bohr's quantum mechanics. She argues that the central lesson to be learned from Bohr is that *"we are part of that nature that we seek to understand"* (2007, 26; emphasis in original). Barad claims that Bohr's "philosophy-physics" (ibid., 121) opened up epistemological questions to include ontological concerns. She refers to the famous scientific debate between Heisenberg and Bohr in the first half of the twentieth century on the measurement of the position of an electron (ibid., 3–25, 115–18). Heisenberg's initial formulation of the "uncertainty principle" suggests that it is impossible to *know* both the momentum and the position of a particle at the same time. As the measurement process inevitably distorts the properties of the particle, (human) observers cannot access the "real" properties. While Heisenberg focuses on epistemological uncertainty, Bohr makes an ontological argument. In this understanding, particles do not *have* any determinate properties "behind" or "before" the observing apparatus that allows for measuring and assessing them. According to Bohr, their distinctive ontological qualities are contingent on the specific experimental configuration designed to observe them. Thus, in contrast to Heisenberg, "Bohr is making a point about the nature of reality, not merely our knowledge of it" (ibid., 19; 402–3; see also Barad 2010, 258–59).

Bohr's "proto-performative" (Barad 2007, 31) account is the basis for Barad's critical engagement with "representationalism," by which she understands the idea that "beings exist as individuals with inherent attributes, anterior to their representation" (ibid., 46; Barad 2003, 804). According to Barad, the narrative of representationalism is firmly rooted in (Western) philosophy and thought and comes in different guises, e.g., as positing an opposition between words and things, nature and society, represented and representation. She refers especially to two contemporary epistemological positions that seemingly endorse opposing paradigms while in fact both share the representationalist framing: social constructivism on the one hand and scientific positivism—"traditional realism" (Barad 2007, 225; see also 2007, 41) in her understanding—on the other. While the former takes society as a self-evident and given entity, the latter is grounded in the belief of a fixed and stable nature. Both are, in Barad's view, untenable epistemological options as they share a common "metaphysical substrate": the "belief in the power of words to mirror preexisting phenomena" (ibid., 133). She suggests that representationalism subscribes to a correspondence theory of truth, endorsing the phantasm of how to correctly reflect already existing "things" (see ibid., 56).[1] Barad seeks to counter the dominant narrative of representationalism that positions the observer outside or above the world. As a theoretical alternative she proposes a performative account, insisting that "practices of knowing and being are not isolable; they are mutually implicated" (ibid., 185).[2]

Barad's critique of representationalism cannot be separated from the method of "diffractive reading" (Barad 2012c, 218; 2012a, 13; 2014; van der Tuin 2011b; 2014). The notion of diffraction plays an important role in physics, but it also emerges in Haraway's work in the 1990s (Haraway 1992, 300; 1997, 16, 272–7). Diffraction—or interference[3]—as a physical phenomenon occurs when waves (light, water, sound) encounter an obstruction (see Barad 2007, 74–85). This process results in specific patterns that can be observed when waves overlap, which makes it possible to explain their dynamics. Haraway uses the term as a strong metaphor, proposing an optics of relationality that opposes "diffraction" to "reflection." Barad takes up the notion, but suggests some modifications to Haraway's understanding of the term. For both authors reflection is bound to "the themes of mirroring and sameness" (Barad 2007, 71), while diffraction establishes deviance and difference. However, while Barad sees Haraway using the term primarily as a metaphor and a "semiotic category" (ibid., 416, note 2), she proposes diffraction as both a physical phenomenon and a methodological perspective.

Still, Barad is more interested in diffraction as a specific mode of investigation than as an object of study (see ibid., 73). To her, it serves as a "productive model for thinking about nonrepresentationalist methodological approaches" (ibid., 88). To explore this methodology, Barad points to the central role of optical metaphors in addressing matters of epistemology and the intimate link between knowledge and vision in the Western tradition. The metaphor of reflexivity specifies the criterion for the "correctness" of the scientific investigation and informs the idea of a mirror image between representations and the represented. Moreover, reflexivity also provides a critical yardstick for the objectivity of scientific knowledge. It presupposes the idea of a sovereign and self-identical subject by allowing for a "self-referential glance back on oneself" (ibid., 88) as a self-reflection on the role of the investigating subject in the production of knowledge. According to Barad this epistemological account is flawed, as the principle of reflexivity relies on a fixed and asymmetrical opposition between the subject of knowledge and the object of scientific inquiry. In contrast, diffraction does not presuppose a prior identification of "subject" and "object"; rather, it explores how they and the boundaries between them are produced as part of the process under investigation (ibid., 93; 418, note 17; 89–90).[4]

There are two important dimensions to this methodological commitment. First, in focusing on mutual entanglements and patterns of difference, diffraction enacts a strategic move from mediation to relationalism. In Barad's account there is no need for an intermediary between subject and object, between knowledge and the world, culture and nature, the material and the discursive. None of the usual suspects—consciousness, theory, words—is needed to bridge the gap or to offer a means of communication between the representations and the represented (ibid., 409, note 9). Barad introduces the neologism "intra-action" to capture a relational dynamics defined by processes of co-constitution and mutual emergence. While "interaction" implies that two already given subjects encounter one another, intra-action does not start with the assumption of preexisting entities. Quite on the contrary, intra-action stresses that "things" as such do not exist as they only materialize in a dynamic and ongoing agentic process (see ibid., 33, 140, 178; see also Neimanis 2014, 16–18).[5]

Secondly, diffraction for Barad is also "an ethico-onto-epistemological matter" (2007, 381). It is not only a tool to break with representationalism in its different guises, but it informs her practice of doing theory responsibly. While diffraction theoretically exposes the problems and limitations of a representationalist epistemology with its focus on reflection, it also articulates an ethical commitment as it "attends to the relational nature of difference" (ibid., 72). Instead of conceiving of different theoretical perspectives and positions as fixed and closed in order to contrast them with one another, agential realism conceives of them as principally open and "in *dynamic* relationality to the other" (ibid., 93; emphasis in original). This ethical gesture is not restricted to how we deal with theories and texts; diffraction also points to how to relate to a world that is always already "our world," reminding us that "we" are "part of the intra-active ongoing articulation of the world in its differential mattering" (ibid., 381).

From Things to Phenomena: Ontological Questions and Ethical Concerns

Against the philosophy of knowledge, but also contrary to most streams of feminist theory, Barad insists that it is necessary to attend to the "ontological dimensions of scientific practice" (2007, 42). This interest in the ontological does not mean leaving out or bracketing epistemological questions; rather, Barad questions a particular epistemology (and ontology) that conceives of the two domains as independent and separate. Agential realism serves as a "new 'ontoepistemological' framework" (ibid., 43) that seeks to overcome the limitations of conventional forms of realism that focus on individual entities with inherent properties and boundaries by a performative account that stresses that "*agencies are only distinct in relation to their mutual entanglement*" (ibid. 33; emphasis in original).

This "relational ontology" (ibid., 93) is informed by a critical reading of Butler's notion of gender performativity and Foucault's understanding of discursive practices. Barad credits both theorists with developing an account of power relations that focuses on productivity and performativity. In this perspective, "power is not an external force that acts on a subject; there is only a reiterated acting that is power in its stabilizing and sedimenting effects" (ibid., 235). However, as I pointed out in the introduction to this book, Barad argues that while this approach makes it possible in principle to investigate the materialization of bodies, Foucault restricts the productivity of power "to the limited domain of the 'social'" (Barad 2003, 810). The conceptual privilege Foucault attributes to the social precludes—according to this reading—engaging with matter in a substantive way, since he regards "matter merelyas an end product rather than an active factor in further materializations" (ibid., 810; see also Barad 2007, 235). Barad claims that this approach restages matter's passivity and is unable to acknowledge the contribution of non-social factors in materialization processes.

In a similar vein, Barad undertakes a diffractive reading of Butler's writings. She stresses the importance of Butler's work, which successfully links the formation of the subject to the production of the body's materiality, thereby criticizing narrow conceptions of social constructivism that circulate within and outside of feminist theory. Accordingly, matter should be understood as "*a process of materialization that stabi-*

lizes over time to produce the effect of boundary, fixity, and surface" (Butler 1993, 9; emphasis in original). Barad credits Butler for her proposal to "'return to the notion of matter'" (2007, 61), thereby exposing the flawed idea of "gender as a cultural inscription on the naturally sexed body" (ibid., 60). However, she also claims that this account of materialization is limited in important ways as Butler only attends to the materialization of human bodies (see ibid., 209).[6] Barad argues that "for both Butler and Foucault, agency belongs solely to the human domain, and neither addresses the nature of technoscientific practices and their profoundly productive effects on human bodies, as well as the ways in which these practices are deeply implicated in what constitutes the human" (ibid., 145–46). What is needed, in Barad's eyes, is a concept of performativity that accounts for the materialization of all bodies and finally allows for an investigation of the practices through which the boundaries between the categories of human and nonhuman emerge and are stabilized. In the light of these criticisms, Barad suggests "a reworking of Butler's notion of performativity from iterative citationality to iterative intra-activity" (ibid., 208).[7]

According to Barad, this theoretical move to "posthumanist performativity" (Barad 2003) makes it possible to eliminate the anthropocentric bias of Butler's (and Foucault's) concept of materialization. For Barad, both humanism and anti-humanism fail to account for the "boundary-making practices by which the 'human' and its others are differentially delineated and defined" (Barad 2007, 136). While humanism supports human exceptionalism, anti-humanism as employed by poststructuralists to question and subvert the humanist conviction nevertheless takes "the boundary between nature and culture, the human and the nonhuman, to be a given" (ibid., 428, note 6).[8] By contrast, posthumanism as she conceives of it "is about taking issue with human exceptionalism while being accountable for the role we play in the differential constitution and differential positioning of the human among other creatures (both living and nonliving)" (ibid., 136).

Informed by this idea of posthumanist performativity, agential realism fundamentally challenges traditional understandings of realism and ontological dualisms between nature and culture, subject and object, the material and the discursive.[9] It invites us to dispense with the representationalist idea of individual and independent things and to conceive of the world as inhabited by "phenomena." Barad here takes up Bohr's understanding of phenomena, which has nothing to do with the classical philosophical distinction between *phenomena* and *noumena* but revises the question of referentiality and the concept of the real: "the referent is not an observation-independent object but a phenomenon" (ibid., 198). Thus, phenomena underline the "epistemological inseparability of observer and observed" (ibid., 33, 139, 308). In this perspective, phenomena are not distinct entities with inherent boundaries and isolated properties but "ontologically primitive relations—relations without preexisting relata" (ibid., 139). Agential realism takes a distance from the idea of a vibrancy of things in vital materialism on the one hand and the concept of stable objects in OOO on the other: "Reality is composed not of things-in-themselves or things-behind-phenomena but of things-in-phenomena" (ibid., 140).[10]

The ontological shift from "things" to "phenomena" is informed by an innovative understanding of the apparatus. Barad reconceptualizes and appropriates this notion by diffractively reading Bohr's concept of the apparatus together with Foucault's account of discursive practices. She argues that Bohr's understanding of the apparatus is limited to a laboratory setup, ignoring the whole range of material-discursive practices[11] that enable experimental configurations to work. While this account of the apparatus still rests on a humanist concept of the observing subject, agential realism seeks to provide a "posthumanist understanding of the role of the apparatus and of the human and the relationship between them" (ibid., 145). It is characterized by three main principles.

First, Barad's reading of the apparatus invites us to rethink ontological boundaries. It proposes a move from a static and stable to a performative and dynamic concept of the apparatus that accounts for its boundary-making practices, where what is inside and what is outside is intrinsically indeterminate and can only be understood by the workings of the apparatus itself (see ibid., 170). Thus, agential realism breaks with more conventional usages of the notion. For Barad, an apparatus is neither a technical instrument nor a scientific device, nor is it a cultural structure or a social system. It does not resemble the "inscription devices," "ordering devices," or "market devices" (Latour and Woolgar 1979; Suchman 2007; Callon et al. 2007, respectively) described in STS, but neither can it be reduced to "repressive" or "ideological state apparatuses" (Althusser 1971; 2014) or a "cultural apparatus" (Mills 1972; see also Sawchuk 2001).[12]

Secondly, Barad's notion of the apparatus entails a revision of scientific objectivity. Apparatuses do not operate in a distinctive ontological realm (e.g., the social or the technical) hierarchically separated from the one that it determines and shapes; nor are they neutral devices that passively record phenomena existing independently of them. Within the agential-realist framework we find a systematic refusal to accept any pre-established idea of exteriority and separateness, in order to endorse entanglements and intra-actions leading to what might initially seem to be a surprising conclusion: "apparatuses are themselves phenomena" (ibid., 146, 170), i.e., they produce phenomena and are part of (more comprehensive) phenomena (see ibid., 148). How is this seemingly paradoxical formula to be understood? Following Barad, the crucial point is to conceive of apparatuses not as definite objects or fixed structures but rather as "open-ended practices" (ibid., 170) that change according to the agential determinations they enact: "Different agential cuts produce different phenomena. Crucially, then, the apparatus is both causally significant (providing the conditions for enacting a local causal structure) *and* the condition for the possibility of the objective description of material phenomena" (ibid., 175; emphasis in original).

The third feature of the apparatus put forward in agential realism concerns the problematization of conventional understandings of causality. What is at stake is not a different causality that replaces or shifts the accent from one set of causes to another (e.g., from the social to the natural or vice versa) but an alternative concept of causality. By following Bohr's lead, Barad seeks to go beyond the "usual dualist thinking about causality" that opposes freedom and determinism to envision "a third possibility" (ibid., 198). This means that agential realism not only remains at a distance from determinist accounts; it is also critical of the opposite (but mirroring idea) of free will or complete arbitrariness (see ibid., 170–71). While in both conceptions causality is conceived of as a relation between separate and independent entities, agential realism claims that there are no originary forces that pre-exist their intra-action. Thus, this idea of causality differs from the "common choices of absolute exteriority and absolute interiority and of determinism and free will" (ibid., 176). Apparatuses are world-making practices that determine what counts as a "cause" or an "effect."

To sum up, agential realism rejects the idea of a pre-existing world characterized by fixed causal schemes and pre-established patterns of time, space, and matter. In this reading, apparatuses do not merely evolve in time; they are not situated in space and do not mobilize matter, but should be understood as "specific material reconfigurings of the world that [. . .] iteratively reconfigure spacetimematter as part of the ongoing dynamism of becoming" (ibid., 142). They are not preexisting and invariant structures or setups, but dynamic material-discursive practices that are "perpetually open to rearrangements, rearticulations, and other reworkings" (ibid., 170). This does not mean, however, that everything is possible or that the trajectory is an arbitrary one. Quite on the contrary, "*apparatuses are the material conditions of possibility and impossibility of mattering*; they enact what matters and what is excluded from mattering" (ibid., 148; emphasis in original). While there is no determining agent or force exerting an external power on someone or something, nevertheless the semantic and ontic indeterminacy within the phenomenon is resolved by the particular arrangements and distinctive properties of apparatuses that enact what matters—and what does not matter.

Barad considers ontological questions as deeply entangled with ethical concerns, since we are part of the particular materializations that exist, by their very existence excluding others and creating new possibilities for future re-configurations of matter. This material ethics is about taking responsibility for knowledge making and the normative preferences we enact (see ibid., 382). It is not restricted to acknowledging that exclusions are inevitable to mattering; the important point in this concept of ethicality is to accept responsibility for the specific material intra-actions and to permanently review and rework their boundaries (see ibid., 205). As Barad claims that we participate in the making of the phenomena we seek to understand, she calls for "an appreciation of the intertwining of ethics, knowing, and being," an endeavor she terms "*ethico-onto-epistem-ology*" (ibid., 185; emphasis in original).

For this comprehensive understanding of ethics, Barad turns to the work of Emmanuel Levinas and his concept of responsibility. Levinas rejects the metaphysics of the self that provides the foundation for conventional approaches to ethics. In his view, ethics grounds human experience (not the other way around). According to Levinas, ethics is characterized by the encounter with the "face of the other" (see Levinas 1969, 279). Responsibility is not a commitment that a subject chooses or a relation between two subjects but is rather based in the ability to respond to the Other. It is an embodied relation that precedes the intentionality of the subject (see Barad 2007, 391; see also Critchley 2008, 69–76). However, Barad argues that Levinas' understanding of responsibility is limited in the sense that it "ignore[s] the full set of possibilities of alterity—that 'having-the-other-in-one's-skin' includes a spectrum of possibilities, including the 'other than human' as well as the 'human'" (Barad 2007, 392). While Levinas restricts his concept of responsibility to human encounters, Barad calls for a "posthumanist ethics, an ethics of worlding" (ibid., 392) that goes beyond anthropocentric concerns to open up the field of ethics.[13]

As a result the meaning (and the matter) of ethics are transformed. It accounts for the materializations we are part of in bringing them into being (while producing "us" in the very same process). This posthumanist ethics relates to a material concept of responsibility that goes beyond providing different responses to changing configurations and constellations (Barad 2007, 393). What is at stake in this "ethics of entanglement" (Barad 2012d, 47; Åsberg 2013) is not something additional or subsequent that comes after the facts are established, evaluating and possibly revising them. In agential realism there is no neat line that distinguishes facts and values; rather, facts are already value-laden, they embody normative preferences that give rise to some material configurations rather than others (see Barad 2012a, 15). Thus, Barad is interested in "how values matter and [are] materialized" (ibid., 15). This comprehensive and encompassing understanding of ethics as "part of the fabric of the world" (Barad 2007, 182) is to be distinguished from two rather limited ethical engagements. First, from a moralizing discourse about the right normative answer that correctly reflects and addresses the ethical problem at hand, and secondly from a bioethical approach that seeks to come up with regulations and normative guidelines after scientific facts are produced in order to evaluate them and restrict or prohibit illegitimate usages (see Barad 2012a, 15).

Barad makes it clear that this "ethics of mattering" (2007, 36) does not contradict or conflict with scientific objectivity. Rather, there is no objectivity without ethics, the two cannot be conceived of as separate. Knowing is not a passive observing practice but a material engagement with the real that takes into account what matters and what does not matter. As "different material intra-actions produce different materializations of the world" and indeed different worlds, excluding alternatives by the very practice of world-making, Barad states that "it matters to the world how the world comes to matter" (ibid., 380).[14]

A New Materialist Understanding of Power

Conceptualizing intra-actions as "nonarbitrary, nondeterministic causal enactments" (ibid., 179), agential realism proposes a different concept of power. It puts forward the idea that power relations do not merely take place *in* time, *in* space, or *on* matter. Rather than conceiving of time as linear, space as a container and matter as stable, agential realism draws attention to the intra-active dynamics *"through which temporality and spatiality are produced and iteratively reconfigured in the materialization of phenomena"* (ibid., 179; emphasis in original). To be clear: the argument is not that time, space and matter are relative instead of absolute parameters determined by the dynamics of power that shapes and forms them. While this account still nurtures the idea of a temporality and spatiality separate and exterior to material practices, Barad puts forward a more ambitious claim: temporality, spatiality, and matter are themselves made/marked by intra-actions (see ibid., 179), a dynamics of mutual constitution and re-configuration that is captured by the term "spacetimematter" (ibid., 142).

Barad elaborates this "new materialist understanding of power" (ibid., 35) by discussing two case studies. She first comments on Leela Fernandes's study of relations of production in a Calcutta jute mill (Fernandes 1997), which brings together Marxist theory and poststructuralist thought (Barad 2007, 226–46). Barad credits *Producing Workers: The Politics of Gender, Class, and Culture in the Calcutta Jute Mills* with providing a concrete and careful analysis of how the spatiality of capitalism is iteratively (re)produced through the politics of gender, caste, and class without neglecting how structural relations of power are permanently contested on the shop floor. She sees another important merit of the book in how it attends to the reworking of temporalities by showing how "premodern," "modern" and "postmodern" modes of production materialize through one another instead of contradicting each other. According to Barad, Fernandes successfully disentangles her study from the idea of "a single deterministic trajectory of power" and takes into account the "important role that multiple intra-actions, exclusions, and agencies play in the dynamics of power" (ibid., 236).

While Barad claims that Fernandes' book convincingly shows how issues of political economy and cultural identity are inseparably intertwined, she also points to what she perceives as important shortcomings of *Producing Workers*. In her view, Fernandes' study remains confined to the boundaries of the factory, thereby neglecting the topological intra-actions and reworkings of the "global" and the "local." Barad claims that Fernandes presents an incomplete and selective account of the multiple material-discursive practices that come to matter in her analysis. Furthermore, Barad argues that *Producing Workers* fails to acknowledge how meanings, bodies, and boundaries are co-constituted in iterative materializations. She emphasizes the need to bring in issues of responsibility in order to account for "the changing nature of the dynamics itself" (see ibid., 242). Barad proposes to take up insights from agential realism to rectify the two problems she perceives: "Hence, using the framework of agential realism, the jute mill can be understood as an intra-acting multiplicity of material-discursive apparatuses of bodily production that are themselves phenomena materializing through iterative intra-actions among workers, management, machines, and other materials and beings which are enfolded into these apparatuses" (ibid., 237).[15]

The second case study draws on an article by Monica Casper on fetal surgery that critically addresses the debate on nonhuman agency within science studies (Casper 1994; see also Casper 1998). Casper notes that many theorists have ignored how the very notion of nonhuman agency in this debate is premised on "a dichotomous ontological positioning in which [nonhuman] is opposed to human" (ibid., 840). To illustrate this point, she turns to prenatal diagnosis and warns that "constructions of active fetal agency may render pregnant women invisible as human actors and reduce them to technomaternal environments for fetal patients" (ibid., 844).

Barad shares Casper's conviction that we need to rethink general attributions of agency to nonhumans and to open up the boundaries between the human and the nonhuman for critical inquiry. However, she disagrees with Casper's conclusion that it is necessary to deny agency to fetuses in order to protect the self-determination and agency of pregnant persons. For Barad, it is "the presumed alignment of agency and subjectivity" (Barad 2007, 217) that is problematic rather than the idea of fetal agency per se. She argues that it is the attribution of subjectivity, not agency, that has played such a crucial role in public debates on abortion and reproductive choice in the last few decades (ibid., 216–17; see also Barad 2012a, 16–17).

In Barad's account the fetus is not a subject with inherent properties, nor is it a stable object. The fact that there are fetal enactments (e.g., the fetus "kicking back") does not entail the concession of fetal agency per se as something fetuses possess (to varying degrees) (see Barad 2012b). Rather, it is "a phenomenon that is constituted and reconstituted out of historically and culturally specific iterative intra-actions of material-discursive apparatuses of bodily production" (Barad 2007, 217). To support this claim Barad turns to ultrasound technology. The piezo-electric transducer does not merely visualize the fetus; nor does it obscure or obstruct the vision of the real. Instead, it contributes to produce the body it "pictures." The sonogram images "refer to a *phenomenon* that is constituted in the intra-action of the 'object' (commonly referred to as the 'fetus') and the 'agencies of observation'" (ibid., 202; emphasis in original). Thus, there is no pre-existing object that the piezo-electric transducer helps to make visible. The apparatus in question is not a simple observing instrument or a technological device; rather, it encompasses a complex variety of material-discursive practices: "medical needs; design constraints (including legal, economic, biomedical, physics, and engineering ones); market factors; political issues; [. . .] and the nature of training of technicians and physicians who use the technology" (ibid., 203–4). The piezo-electric transducer contributes to and materializes in these intra-actions, and it is itself "the interface between the objectification of the fetus and subjectivation of the technician, physician, engineer, andscientist" (ibid., 204).[16] The visualization of something called "the fetus" that serves as an objective referent for risk analysis and decision-making is enacted by specific apparatuses.[17]

The notion of the apparatus is at the heart of the agential-realist account of power, as it offers several analytic advantages. First, the understanding of the apparatus as a material-discursive practice (instead of a social structure or a technical device) allows for a more comprehensive analysis of power, rejecting pre-established dualisms and divisions ("the social" vs. "the technical," etc.) and investigating them as part and product of processes of differential materializations. This concept provides a more convincing picture of how to account for exclusions and asymmetries. The intra-active reworking of causality makes it possible to treat them not as external "effects" or "consequences" of material practices (conceived of as separate and saturated) but as an integral part of specific materializations (ibid., 237–38).[18]

Secondly, the agential-realist concept of the apparatus enables a more concrete analysis of the making/marking of human and nonhuman bodies that goes beyond investigating the (re)configuration of the boundaries between them. Intra-actions not only shape the contours or surface of the body but also its very materiality, accounting even for "the very atoms that make up the biological body" (Barad 1998, 106). This is an important achievement, as it does more than simply shift the accent from the exterior to the interior of the body. It also makes it possible to address the very distinction between the interior and the exterior, and to question the "exteriorization of the interior" in political analysis. As Mariam Fraser has pointed out, this move extends the analysis of how power affects bodies beyond the limitations of a politics of identity that relies on persistent and visible features (2002, 617).[19] The comprehensive concept of the apparatus of bodily production overcomes narrow understandings of body politics that focus on ideological processes and/or on the surface of the body without investigating how the materiality of bodies, their meaning, and their physicalities are produced through the workings of power.

Thirdly, the concept of the apparatus incorporates ethical considerations and concerns. These are not something that comes after or lies beyond processes of power. Ethics is part and parcel of the workings of material-discursive apparatuses. The example of obstetrical ultrasound and new reproductive technologies in general shows how intimately issues of responsibility and accountability are linked with technological and political questions. The construction of the fetus as a subject by a material-discursive apparatus makes it possible to hold the pregnant person (mostly addressed as "the mother") accountable for fetal well-being; this imaginary control generally includes many social and biological factors beyond individual choice or control. This material-discursive apparatus not only affirms fetal subjectivity and constructs the accountability of the subject called "the mother"; the workings of the apparatus simultaneously exclude from the picture a different catalogue of responsibilities (what Barad terms "the real questions"): "consequences of inadequate health care and nutrition apparatuses in their differential effects on particular pregnant women; accountability for the consequences of global neocolonialism, including the uneven distribution of wealth and poverty; and many other factors" (2007, 218).

Fourthly, agential realism offers a reconsideration and reevaluation of how subversive or oppositional practices operate. As Barad makes clear, claiming to go beyond Butler on this point, acts of subversion include "but are not limited to, changes in the specific material reconfigurations of apparatuses through the enfolding of particular resignifications" (ibid., 219). Rather, the "hegemonic apparatuses" (ibid., 219) are simultaneously (re)constituted and challenged by the very material-discursive intra-actions they enact. Thus, "subversion" is always already present as the boundaries of the apparatus are indefinite and contingent, but these forms of contestation and change are also local and limited—at least compared to the imaginary of "revolutionary breaks" (ibid., 452, note 29) that radically overturn the totality of the existing power structures. They will never reach a point of saturation or closure, since new possible worlds open up as others are excluded in the "ongoing dynamism of becoming" (ibid., 142).[20]

Material Foundationalism

Barad's work offers valuable insights and important perspectives for social theory, gender studies and STS. It is interesting to note that most attempts to take up or translate agential realism for empirical research have focused on technoscientific practices (Schwennesen and Koch 2009; Aradau 2010; Fitsch and Engelmann 2013; Fitsch 2014). However, the fact that agential realism works well in accounting for these constellations does not necessarily indicate a systematic limitation of this theoretical framework (Meißner 2013). Since apparatuses in Barad's conception are conceived of as open-ended and extendible, transcending the boundaries of the laboratory, the empirical and explanatory potential of agential realism is not necessarily restricted to experimental or technological set-ups (see, e.g., Højgaard and Søndergaard 2011). Still, Barad's work is characterized by some unresolved conceptual problems and theoretical tensions that limit the analytic prospects of agential realism—at least in its current form. I will focus on two problems that subvert its critical potency.

The first concerns Barad's uptake of quantum physics and its quasi-foundational role in agential realism. It has been noted that the exclusive focus on Bohr's insights tends to marginalize or exclude other important contributors to quantum mechanics (see Schweber 2008, 881). However, the main issue is not whether or not Barad's account is historically accurate but rather how she invokes physics. Trevor Pinch has observed that Barad subscribes to a "form of scientism" (2011, 440; see also Lettow 2017, 110) that ignores important lessons from the tradition of STS.[21] Rather than deconstructing or situating the experiments in quantum physics she discusses, she employs them "as the obvious grounding for a new ontology in science studies" (ibid., 434). In this light, the emphasis on intra-actions, diffraction patterns, and a relational ontology come at a price. It is built on the idea of a straightforward transfer of insights from science that are seen to provide a solid foundation for science studies and feminist theory. Pinch, being a physicist himself, claims that it is only the absence of any contextualization of quantum mechanics that allows Barad to use Bohr as a theoretical grounding for agential realism.[22] He argues that Barad adheres to an obvious paradox:

> I find it deeply puzzling that Barad can call for a more situated account of science and at the same time fail to situate the very part of science she is talking about, while drawing in a realist mode upon experiments to support her position. Perhaps this is where we do need Bohr after all—there does seem to be something mutually exclusive in Barad's way of doing the science and writing about the science *as a scientist*, and doing and writing about it *as a science studies practitioner*. (Ibid., 439; emphases in original)

In her response to Pinch's review, Barad addresses what she takes to be "misunderstandings" (2011, 443) of agential realism. Contrary to Pinch, for Barad there is no mutual exclusivity between science and science studies; she sees Pinch as stuck with epistemological concerns and adhering to the "uncertainty principle" (ibid., 444). While Barad correctly detects a polemical sub-current in Pinch's argument, he is still right in sensing inconsistencies or at least unresolved tensions in Barad's attempt to stress relationality and intra-activity on the basis of a rather fixed and stable onto-epistemological framework.[23]

The main point of the controversy concerns the foundational role of quantum physics in agential realism. It is indeed troubling that Barad seems to suggest that (her reading of) Bohr's quantum mechanics is the final analytic key to opening up any and every question or problem. In fact, she states that quantum mechanics is "the correct theory of nature that applies to all scales" (Barad 2007, 85)—a claim quite difficult to sustain. While quantum physics certainly offers useful explanations for an impressive range of physical processes, it is debatable if it is able to account for all phenomena from the microscopic to the macroscopic (Schweber 2008, 881). It is by no means a "general theory," and one "may well ask what differences are entailed when her [Barad's] analyses move from elementary particles to organisms, humans, and cultures" (Hayles 2017, 69). This reminder to refrain from general theorizing is consistent with agential realism, as Barad not only points to the exclusions required for certain materializations to emerge but also cautions that scales are not independent variables but produced in agential cuts. A diffractive reading of agential realism seems necessary, especially given the often (too) bold claims Barad advances in promising insights into "the nature of nature" (2007, 247). A more modest and situated understanding of the matter and the meaning of quantum physics would certainly help to further clarify and demonstrate the usefulness and productivity of this framework by exposing its limits and the exclusions it enacts.[24]

The epistemological privilege accorded to Bohr's version of quantum mechanics is coupled with a conceptual ambivalence that resembles Bennett's position of a more-than-relational ontology. On the one hand, Barad claims that matter is not a stable and given property but rather the fluid and contingent effect of practices, asserting that "matter does not refer to a fixed substance; rather, *matter is substance in its intra-active becoming—not a thing but a doing, a congealing of agency*" (Barad 2003, 822; emphasis in original). On the other hand, this strict relational account is sometimes complemented by a "material foundationalism"

(Bruining 2013; see also Hoppe and Lemke 2015; Hoppe 2017a, 43). Dennis Bruining proposes this term to describe a certain tendency in new materialisms, "how 'materiality,' the very 'matter' of bodies, nature and things is (perhaps inadvertently) posited [. . .] as *a priori* and as, allegedly, beyond culture, despite an awareness of the untenability of such claims" (Bruining 2013, 151). In Barad's case this foundational gesture materializes in the idea that there is something like "matter's dynamism" (Barad 2007, 135), a basic agentic force that informs the "marrow of being" (ibid., 396).[25]

This conceptual ambivalence is more than an unsolved tension or a residual fundamentalism in Barad's work, as it fulfills an important strategic function. The simultaneous focus on radical relationality and stress on a quasi-fundamental role of matter give Barad's critique of social constructivism and poststructuralism its particular strength. Sara Ahmed criticizes Barad for endorsing a simplified and one-sided account of feminist theory and the linguistic turn. She argues that Barad's account results in a "caricature of poststructuralism as matter-phobic" (Ahmed 2008, 34) that does not credit the complexity and the richness of this theoretical tradition. While this critique tends to commit the very mistake it diagnoses in Barad's work, namely simplifying and caricaturing it by ignoring its conceptual achievements and innovations, it nevertheless captures a paradox in agential realism. To the extent that relations and intra-actions are situated at the center of the theoretical engagement, conceptual bonds and entanglements tend to fade into the background. There is no place for alternative options or interpretations, and theoretical preferences and selections are not sufficiently accounted for.[26]

Apart from the selective account of poststructuralism, Barad sometimes also seems to revert to what might be called a representational critique of representationalism. She tends to discuss representationalism as a flawed, misguided, and under-complex form of knowledge that needs to be replaced by something more attuned to difference, change and heterogeneity: diffraction.[27] However, the matter (and the meaning) of representationalism depends on constellations and circumstances, on the historical and cultural materializations it helps to enact. To put it succinctly: Barad's "representation" of representationalism is itself underdeveloped, as it tends to reproduce a non-relational and isolationist account that Barad convincingly criticizes throughout her work.[28]

The Call for Ethics

Since Barad considers all forms of matter as agentive and responsibility an integral part of intra-actions, every single intra-action becomes a relation of responsibility (2007, 178–79, 184–85). Thus, "particular possibilities for (intra-)acting exist at every moment, and these changing possibilities entail an ethical obligation to intra-act responsibly in the world's becoming" (ibid., 178, 396). However, placing ethics everywhere, seeing it interwoven with the "fabric of the world" (ibid., 182), risks situating it nowhere specific. What gets lost in this totalizing conception is a sense of particular responses to alternative normative values articulated in materializations. While the notion of responsibility suggests a normative horizon, it remains unclear how intra-actions differ in their ethical value (see also Braunmühl 2018).[29] This provokes the question of what criteria to draw on to discriminate intra-actions that are "fuller" or "more just" than others, what materializations are to be preferred over others.[30] Or in Barad's own words: how do we account for "ways of responsibly imagining and intervening in the configurations of power" (2007, 246)?

Barad provides an answer, albeit an unsatisfactory one, to this question. She states that the normative criterion for this ethics of responsibility is to be "responsive to the possibilities that might help us to flourish" (ibid., 396). However, apart from the uncertainty about what exactly "flourishing" might mean in this context, it is also open whom the "us" includes. While Barad claims that agential realism does not diminish or deflect "human accountability" but rather "requires [. . .] much more attentiveness to existing power asymmetries" (ibid., 219; see also Barad 2012b), she often resorts to an undefined "we." But how is this "we" constituted, and whom does it include? How does this ethical responsibility translate into political options, given that responsibility is itself a differential resource based on the existing power asymmetries and structural exclusions? So far, Barad has not provided a convincing answer to these questions (Garske 2014, 122–24).

In fact, it seems necessary to supplement the "ethics of mattering" (Barad 2007, 36) by a material understanding of politics. According to Barad an analysis of power relations necessitates "an understanding of the nature of power in the fullness of its materiality" (2003, 810).

However, as Bonnie Washick and Elizabeth Wingrove (2015) argue, this account remains limited as the focus on a radically open future (see, e.g., Barad 2007, 235) and the never-ending flow of agentic possibilities tends to downplay the significance of more durable structural patterns and hegemonic positions. These authors draw particular attention to Barad's comments on Fernandes' "structural-discursive" account of power relations (ibid., 226). They note an important shift in the analytic spotlight in Barad's reading of Fernandes' study that privileges the shop floor level while shifting attention away from more systematic patters of oppression and exploitation: "The distance between this vision of a plentiful futurity and Fernandes' persistent attention to how *different* workers are *differentially* constrained is profound: where the latter sees horizons of possibility delimited not deterministically and for all time but rather *systematically*, the former sees the never exhausted, because never delimitable *in advance*, scope of intra-acting agencies" (Washick and Wingrove 2015, 69; emphases in original).

The stress on radical contingency and the "ever-changing relations of power" (Barad 2007, 237) in agential realism is surprisingly paired with a systematic omission of contestation and conflict. To put it differently, Barad's account lacks an understanding of the political dimension of "worldly configurations" (ibid., 91), their controversial and disputed nature (Hoppe and Lemke 2015). Conceiving of the multiple possibilities of "worlding" (Barad 2007, 181) as potentially conflicting or competing alternatives, is necessary for an investigation of power relations. This dimension is already present in Barad's work, especially in her concept of the apparatus and her reconceptualization and appropriation of Foucault's analytics of power. However, she does not systematically explore this path.

In the next part of this book, I will turn to Foucault's work on governmentality as a contribution that has been ignored or too swiftly discarded by Barad and many other new materialist scholars. I engage in a "diffractive reading" (see Barad 2007, 71–94) of Foucault's idea of a government of things as a way of taking up and adding constructively to new materialist concerns. The recent interest in materiality and non-anthropocentric modes of thought invites a fresh look at Foucault's work (as well as other accounts that have come to be labeled "old materialist"). The following chapters provide evidence that Foucault's analytic frame shares many ontological and political commitments articulated up to this point. However, I will also argue that compared to the different strands of new materialism I have discussed so far, Foucault's conceptual tools provide elements for a more convincing way of addressing the issue of ontological politics.

PART II

Elements of a More-Than-Human Analytics of Government

While the first part of this book was devoted to analyzing the theoretical strengths and weaknesses of various strands of new materialist scholarship, the second will discuss some elements of a "more-than-human" analytics of government in Michel Foucault's work. My starting point is Barad's claim that Foucault does not develop a dynamic concept of materiality that takes account of the materialization of human as well as nonhuman bodies.[1] This critical observation is shared by many other theorists. Already in the 1990s, Paul Rutherford had stressed that Foucault failed to see that the operations of biopower consist in the "'making-up' of *both* people and *things*" (1999, 44; emphases in original). Rutherford notes that the regulation of the population requires the management of the environment that provides the living conditions for the human species.[2] In a similar vein, Nigel Thrift has argued that Foucault's account is "curiously devoid of thingness" (2007, 56). He thinks this "tragic omission" might be "part of a more general emphasis on language and texts" (ibid. 56) in Foucault's work as a whole or due to a more specific focus on technologies of the self in his later writings. In Thrift's view, Foucault focuses too much on discourse and language instead of exploring the materiality and the liveliness of things.[3]

Indeed, Foucault rarely pursued this line of research. The notion of biopower remains intimately linked to the disciplining and regulation of human bodies, defining a "set of mechanisms through which the basic biological features of the human species became the object of a political strategy" (Foucault 2007a, 1; see also 1978, 141–42). There are at least two mutually reinforcing factors contributing to this analytic focus on the figure of the human. First, Foucault's critical investigations of the nexus between power and knowledge, prominent in his work in the 1970s, concentrate on the operations of the human sciences. While Foucault sometimes addresses aspects of the life sciences (anatomy, physiology, clinical medicine) that might challenge any neat separation between nature and culture, he was principally concerned with the power effects of "disciplines" like psychology, criminology, or pedagogy. The second reason for Foucault's preoccupation with human life is a certain inconsistency and asymmetry in his account, as he takes for granted that there are crucial epistemological differences between the natural and the human sciences. By following his teachers Gaston Bachelard and Georges Canguilhem in this respect, he tended to underestimate the relevance of the natural sciences for a genealogy of power.[4] In contrast to the "dubious sciences" (see Foucault 1980a, 109), by which he meant the human sciences, Foucault credited the natural sciences with a high-level "epistemological profile" (ibid.).[5] As Rutherford rightly notes, Foucault's "attitude towards the natural sciences was not developed in a manner fully consistent with his own analysis of the relation between power and knowledge" (1999, 61, note 7; 2000, 119). However, it is still possible to productively use and extend Foucault's analytic framework to explain power effects in the natural sciences. Joseph Rouse has pointed to "extensive parallels" (1987, 212) between the disciplinary power at work in prisons, schools, hospitals, and factories and the construction and manipulation of laboratory objects (ibid., 209–47; 1993, 137–62).

In the following, I would like to make a similar argument by using and extending elements in Foucault's work to provide theoretical resources for an "irreductionist approach to the conduct of government" (Asdal 2008, 124). The next three chapters offer a reconsideration of the main points of criticism Barad and many other new materialist scholars raise against Foucault: (1) the undertheorized relation between discursive practices and material phenomena, (2) the privileging of the social, and (3) the persistent anthropocentrism. I argue that it is possible to discern conceptual tools in Foucault's work to address these concerns. The notion of the dispositive successfully grasps the complexities of material-discursive entanglements; Foucault's understanding of technology exceeds the domain of the social; and the concept of the milieu systematically takes into account more-than-human practices. Thus, Foucault's work contains elements of new materialist thought, captured in the idea of a government of things which he briefly outlines in his lectures at the Collège de France.[6]

4

Material-Discursive Entanglements

Grasping the Concept of the Dispositive

With the notion of the dispositive, I find myself in a difficulty which I haven't yet been properly able to get out of.
(Foucault 1980b, 196, translation modified)

As is well known, Foucault's lecture series of 1978 and 1979 at the Collège de France demonstrate that up until well into the eighteenth century the problem of government occupied a central position in European societies (Foucault 2007a; 2008a). The term circulated not only in political tracts but also in philosophical, religious, medical, and pedagogic texts. In addition to management by the state, government also addressed problems of self-control, guidance for the family and for children, management of the household, directing the soul, and other issues (Foucault 2007a, 88; 2000b, 341; see also Sellin 1984; Senellart 1995). Taking up this historical meaning of the term, Foucault distinguishes "the political form of government" from the "problematic of government in general" (2007a, 89), understanding the former as a subgroup of the latter.

While in the course of the lectures Foucault focuses on the "genesis of a political knowledge" (ibid., 363) of governing human beings, he also discerns a more comprehensive understanding of governmental practices—encapsulated in the idea of a government of things. Contrary to many interpretations, Foucault's work on government exceeds a concern for an anthropocentric ethics and forms of (human) subjectivation to analyze the relationalities that connect and separate humans and nonhumans. As I will show, the conceptual proposal of a government of things makes it possible to arrive at a relational account of agency and ontology that is better equipped than many variants of new materialism to tackle the theoretical questions and political issues at stake in contemporary societies.

This chapter argues that in his work from the mid-1970s on, Foucault provides the conceptual tools for a material-discursive understanding of government that goes beyond practices of guiding human subjects.[1] I first elucidate the emergence of the notion of a government of things in Foucault's lectures at the Collège de France of 1977–78. The next section discusses the concept of "economic government" (Foucault 2007a, 33–34) in Foucault's work, linking it to the idea of an "administration of things." In the third part I highlight Foucault's material-discursive understanding of government as "arranging things," captured in the notion of the *dispositif*. This section presents Foucault's distinctive use of this term and spells out its ontological, technological and strategic dimensions. The last part contrasts Foucault's notion of the dispositive with his understanding of the archive and the episteme on the one hand and with current neomaterialist usages of "apparatus" and "assemblage" on the other.

"The Intrication of Men and Things"

In the 1978 lectures at the Collège de France, Foucault traces the genealogy of governmentality from classical Greek and Roman days, via early Christian pastoral guidance, and up to state reason and police science, while the 1979 lectures focus on the study of liberal and neoliberal forms of government. Distinguishing the governmental rationality that emerges in the sixteenth century from the idea of politics suggested in Machiavelli's *The Prince*, Foucault discusses an early modern tract on the art of government by Guillaume de la Perrière. It contains a "curious definition" (2007a, 97) according to which government is conceived of as "the right dispositions of things arranged so as to lead to a suitable end" (ibid., 96).[2] Foucault stresses that the reference to "things" is decisive in this definition, as it distinguishes government from sovereignty. While the latter is exercised "on the territory, and consequently on the subjects that inhabit it" (ibid.), the former operates with and on "things." According to Foucault, de la Perrière's notion of a "government of things" (ibid., 97) does not relate to an additional domain of government apart and separate from the government of humans. Rather than restaging "an opposition of things and men," it relies on "a sort of complex of men and things" (ibid., 96). It is worth quoting the whole passage:

> The things government must be concerned about, La Perrière says, are men in their relationships, bonds, and complex involvements with things like wealth, resources, means of subsistence, and, of course, the territory with its borders, qualities, climate, dryness, fertility, and so on. "Things" are men in their relationships with things like customs, habits, ways of acting and thinking. Finally, they are men in their relationships with things like accidents, misfortunes, famine, epidemics, and death. (Ibid.)

There are several important points to be noted here. Firstly, Foucault's interpretation of the art of government suggests a very particular understanding of "things." The term covers both material entities (like "wealth" or "territory") and discursive elements (like "customs" or "ways of thinking"), and it includes "matters of fact" as well as "matters of concern" (see Latour 2004a).³ To signal this semantic opening toward a more comprehensive and historically informed account, "things" appear in inverted commas. Proposing a relational understanding of "things," Foucault does not conceive of two stable and separate ontological spheres—"humans" and "things"—that interact with one another. Rather, he puts the emphasis on the constitutive bonds that separate and connect them. The qualification "human" or "thing"—and the political and moral distinctions between them—is itself an instrument and effect of the art of government, and does not mark its origin or point of departure. Thus, the government of things does not rely on a foundational sorting of subjects and objects. On the contrary, Foucault questions the idea that contrasts active subjects with passive objects. He employs the term "subject-object" (2007a, 44, 77) to address the phenomenon of the population as on the one hand a material body, "on which and towards which mechanisms are directed," and on the other "a subject, since it is called upon to conduct itself in such and such a fashion" (ibid., 42–43). In this perspective, the art of government determines what is conceived of as active and passive, as mobile and inert. Also, it establishes and enacts the boundaries between socially relevant beings and forms of existence that are deprived of legal and political protection and are reduced to "thingness."⁴

Secondly, since there is no pre-given and fixed ontological borderline between humans and things it is possible to state that "humans" are governed as "things." While medieval forms of government sought to direct human souls toward salvation, modern government treats human beings as "things" as a way of achieving more mundane ends. By this Foucault does not mean a global and all-pervasive process of "reification" (see Panagia 2019, 716–7); on the contrary, human interests, sensations, and affects are essential facts that political reason—a rational knowledge that no longer relies on a divine order of things or the principles of prudence and wisdom—has to take into account. In his comprehensive history of the arts of government, Michel Senellart underscores this historical transformation that distinguishes the modern concept of government from the principle of sovereignty: "The government of things replaces the older government of souls and bodies. The question is no longer, as it was with the Christian authors, about the legitimate use of power; nor is it the one raised by Machiavelli of the exclusive appropriation of power. The question is now about the intensive use of the totality of forces available. So, we note a passage from the right of *power* to a physics of *powers* [passage du droit de *la* force à la physique *des* forces]" (Senellart 1995, 42–43; emphases in original).[5]

While sovereignty focuses on the individual will and legal subjects, government works on empirical data: on geo-physical phenomena (climatic variables, water supply, geological structures, architectural design, etc.) as well as on bio-demographic facts (birth and death rates, health status, accidents, employment, etc.) (Foucault 2007a, 104). By statistically aggregating men on the level of populations, they finally became calculable and measurable and could be conceived of as physical phenomena themselves: a "social physics," in the words of the nineteenth-century sociologist Adolphe Quételet (see Ewald 1986, 108–31). The governor has to take into account the passions and interests of the "multitude" in the same way as he takes into account the climate and the territory, and he has to govern them according to their own nature.[6] Given this "physical" perspective, it would be a mistake to endorse a fundamental political distinction between humans and "things." As Foucault puts it, "to govern means to govern things" (2007a, 97).

Thirdly, Foucault sees this "intrication of men and things" (ibid., 97) made explicit in the metaphor of the ship that often appears in early treatises on government. From Cicero to Thomas Aquinas, the government of a state is compared to steering a ship (Sellin 1984, 363; see also

Senellart 1995). To direct a ship means to be responsible for the sailors, but it also involves "taking care of the vessel and the cargo" and reckoning with "winds, reefs, storms, and bad weather" (Foucault 2007a, 97). The ship is, according to Foucault, a political symbol that stresses the specificity of the art of government. It creates and mobilizes the space in which humans and things are assembled, without possessing or mastering it. It is a "floating space, a placeless space, that lives by its own devices, that is self-enclosed and, at the same time, delivered over to the boundless expanse of the ocean" (Foucault 1998b, 184–85).

Without explicitly mentioning it, Foucault here draws on the etymology of government. The verbs "regere" and "gubernare" originally denoted the direction of a ship, "guvernaculum" meaning the helm. This political imagination is still present in the eighteenth century, when in 1777 Adelung defines "government" ("Regierung") in the following terms: "to determine the direction of a movement according to one's will and to preserve it in this movement" [*"die Richtung einer Bewegung nach seinem Willen bestimmen und in dieser Bewegung erhalten"*] (quoted in Sellin 1984, 372; emphasis in original). To illustrate this definition he refers to the following metaphors that set in motion nonhuman matter: "To govern a ship, to govern the chariot, the shaft, the horses in front of the chariot" [*"Ein Schiff regieren. Den Wagen, die Deichsel, die Pferde vor dem Wagen regieren."*] (Ibid.; emphasis in original).[7]

Quesnay's Principle

The idea of a government of things took shape in a historical constellation that sought to "rationalize" political decision-making by the increasing uptake of scientific knowledge and technological expertise in governmental practices. In his lectures at the Collège de France of 1980–81, Foucault distinguishes between several "modern ways of reflecting upon government-truth-relations" (2014a, 16), encompassing a time period that extends from the state reason of early modernity to socialist realism in the twentieth century. While government is always intimately tied to "the manifestation of truth," as it necessitates "knowledge of the order of things and the conduct of individuals"(ibid., 4–5), he notes a decisive historical transformation in the relation between truth and government from the eighteenth century onwards.

Political economy, which emerged as a distinctive field of knowledge around that time, introduces the question of truth and the principle of self-limitation into the art of government. As a consequence, it is no longer important to know whether the prince governs according to divine, natural, or moral laws; rather, it is necessary to determine the "nature of things" (Foucault 2007a, 49) that defines both the foundations and the limits of governmental action. The physiocrats were the first to put forward the idea of an "economic government" (ibid., 33) that respects and follows the "natural course of things," affirming their autonomy and self-regulatory competences. The government of things, as they advocated it, seeks to reduce or even eliminate authoritarian and arbitrary forms of rule. It connects to the idea of democratic self-organization, where the distance between those governing and the governed approaches zero:

> Governors and governed will be as it were actors, co-actors, simultaneous actors of a drama that they perform in common and which is that of nature in its truth. Summarizing things considerably, this is Quesnay's idea, the physiocratic idea: the idea that if men were to govern according to the rules of evidence, it would be things themselves, rather than men, that govern. Let us call this, if you like, Quesnay's principle, which, despite again its abstract and quasi-utopian character, was of great importance in the history of European political thought. (Ibid., 14)

"Quesnay's principle" marks the starting point of Ben Kafka's (2012) instructive genealogy of the idea of an "administration of things" (Foucault 2007a, 49). In the following, I use Kafka's argument to complement and expand on Foucault's brief remarks on the shifting relations between the "exercise of power and manifestation of the truth" (Foucault 2014a, 13). Kafka argues that in the eighteenth century we still find a clear political distinction between the government of men and the administration of things. One example is the essay *On Public Happiness* (1772) by the Marquis de Chastellux: "[I]n our time, the term police can be understood as the government of men as distinct from administration, which rather designates the government of property" (cited by Kafka 2012, no page number). In this essay, as in the eighteenth century in general, the two political tasks are conceived of as complementary and combinatory. No more than a few years later, they appear to be in possible conflict with or contradiction to one other. Louis de Bonard states in his book on *Primitive Legislation* (1802): "In the modern state, we have perfected the administration of things at the expense of the administration of men, and we are far more preoccupied with the material than the moral" (cited by Kafka 2012, no page number). Given the limited resources of the state, what is needed according to this reasoning is a political choice to be made prioritizing the government of men at the expense of the administration of things.

The most famous proposal to resolve this conflict in the nineteenth century is often attributed to Saint-Simon, but it was actually Auguste Comte who suggested that we should replace the government of men with the administration of things.[8] His objective was to base politics on a sound foundation that systematically excluded any form of despotism. While earlier political thinkers had mostly tended to associate arbitrariness with absolutist governments, for Comte any form of government was susceptible as long as it was based on prejudice, superstition, or religion rather than "positive" principles (Comte 1998, 106–8; Kafka 2012, no page number). In this perspective, political decision-making has to be guided by scientific expertise to allow the development of a democratic society that puts an end to political struggles and social conflicts. It is in this post-revolutionary context in the first decades of the nineteenth century that Comte proposes this famous formula: "The government of things replaces that of men. It is then that there is really *law* in politics, in the true and philosophic sense attached to this expression by the illustrious Montesquieu" (Comte 1998, 108; emphasis in original). Following Montesquieu, Comte suggests a wide understanding of "things." As Kafka reminds us, the "things" invoked here are "res" "in the most general sense of res that is: objects, but also beings, matters, affairs, events, facts, circumstances, occurrences, deeds, conditions, cases, and so forth" (Kafka 2012, no page number). In this inclusive understanding it is "things" that govern human affairs. The formula of a government of things then refers to a *genitivus subjectivus*. As Montesquieu writes: "many things govern men: climate, religion, laws, the maxims of the government, examples of past things, mores, and manners" (Montesquieu 1989, 310 cited by Kafka 2012, no page number).[9]

Comte's proposal to replace the government of men by the government of things sought to substitute the arbitrary power of individuals and collectives with the rule of law and scientific reason. It envisions the end of politics, as a technological and scientifically informed mode of administration will finally displace political controversies. This idea relies on a broad notion of things and their relations with men, including nature, culture, customs, and religion. These relations are based on a rational and intelligible order that can be grasped by empirical science and objective knowledge. Thus, the art of government "implies the constitution of a specialized form of knowledge [. . .] of this truth, and this specialization constitutes a domain that is not exactly specific to politics, but defines rather a set of things and relations that must, in any case, be imposed on politics" (Foucault 2014a, 14–15).

This comprehensive concept of a government of things gets lost in the course of the nineteenth century with the rise of Marxism, which is hostile to the idea of a pacification of social cleavages and political controversies by scientific knowledge and technological expertise. While Friedrich Engels took up the formula that the government of men needs to be replaced by the administration of things, he gave it a completely different meaning, subsuming Comte and Saint-Simon under the rubric of "utopian socialism." In Engels' eyes, the socialist revolution would render the state unnecessary as its only function is to maintain class rule and to secure the dominant relations of production (Kafka 2012, no page number). Marxism inaugurates a different understanding of the relations between government and truth based on the vision of "universal awareness" (Foucault 2014a, 15). While Comte's idea of progress depended on experts and their knowledge, Engels had no need for them as the proletariat would consciously manage "things" once class domination was over and the state had become obsolete.[10] In this perspective, political government relied on ideologies and false knowledge of the real state of affairs, a problem to be overcome by the proletarian revolution: "Strip off the masks, discover things as they happen, become conscious of the nature of the society in which we live, of the economic processes of which we are the unconscious agents and victims, become aware of the mechanisms of exploitation and domination, and the government falls at once" (ibid., 15).

However, Engels also altered an important element in this appropriation of Comte's formula. While Comte conceived of "things" as the subject of government (since "many things govern men," according to Montesquieu's formula), Engels referred to them as the objects of governmental action. The idea of a government of things now operates as a *genitivus objectivus*. It is this narrow understanding of things that shaped twentieth-century Marxism and real socialism. As Kafka stresses, Lenin's *The State and Revolution* relied on Engels' vision, and it also found its way into Bukharin and Preobrazhensky's *ABC of Communism* (1920), which states: "The government of men will be replaced by the administration of things—the administration of machinery, buildings, locomotives, and other apparatus" (cited by Kafka 2012, no page number).

As Foucault shows, Quesnay's principle underwent several mutations in the past centuries. While the exercise of political power has always required knowledge of "the means of governing both these things and these people" (Foucault 2014a, 5), the idea of a government of things has oscillated between a *genitivus objectivus* and a *genitivus subjectivus*—resulting in a rather restricted and a more comprehensive understanding of "things."

The Dimensions of the Dispositive

According to Foucault, the formula of a government of things defines a mode of power very different from sovereignty: "it is not a matter of imposing a law on men, but of the disposition of things, that is to say, of employing tactics rather than laws, or, of as far as possible employing laws as tactics; arranging things so that this or that end may be achieved through a certain number of means" (Foucault 2007a, 99). This dispositional mode of power does not operate by prohibiting, suppressing, or giving orders but rather by attending to an order of things that it contributes to bringing into existence; instead of constructing and steering mechanically, it coordinates and orchestrates dynamic material arrangements. In an interview, Foucault further clarified this concept of government as assembling and composing materialities. He states that government seeks to structure "the possible field of action of others" (Foucault 2000b, 341). It is characterized by "a mode of action that does not act directly and immediately on others. Instead, it acts on their actions [...]. It operates on the field of possibilities in which the behavior of active subjects is able to inscribe itself" (ibid., 340–41).

This relational and performative understanding of assembling and arranging complexes of humans and things is well captured in a notion Foucault frequently used in his work from the mid-1970s on: the *dispositif*. It occupies a crucial role in *Discipline and Punish* (1979), in *The History of Sexuality, Volume I* (1978), and in Foucault's lectures at the Collège de France (see, e.g., 2003; 2005; 2006a; 2007a; 2008; 2014a). In English translations of Foucault's work, *dispositif* is rendered variously and inconsistently as "deployment," "apparatus," "device," "system," "organization," "mechanism," and "construct" (see, e.g., Foucault 1978; see also Burchell 2006, xxiii). While there is certainly a considerable overlap between the meanings of each of these terms and Foucault's use of *dispositif*, they tend to highlight only a selective part of the semantic field or even occlude important etymological ties and conceptual dimensions of the term. Following Jeffrey Bussolini's proposal (2010), I therefore suggest the English term "dispositive" as a better way of grasping the semantic richness and conceptual specificity of *dispositif*.[11]

Foucault seems to have used the term "dispositive" for the first time in his lectures at the Collège de France in 1973–1974, entitled *Psychiatric Power* (see, e.g., Foucault 2006a, 13, 63, 81), in order to describe the workings of disciplinary power and the role of the asylum as a "curing dispositive" (ibid., 164, translation modified; see Elden 2017, 112).[12] *Discipline and Punish*, originally published in 1975, already makes extensive use of the notion to analyze the Panopticon and the "multiple dispositives of 'incarceration'" (Foucault 1979, 308, translation modified). In an interview following the publication of the book, Foucault invokes the notion of the dispositive to address the question of whether a particular method informed his historical investigations. He explained that he shifted his analytic attention from the search for the unsaid, the hidden, or the repressed to explicit strategies and conscious organization, and advocated replacing "the logic of the unconscious" by "a logic of strategy," focusing on "tactics with their dispositives" (1996a, 149, translation modified; Rabinow 2003, 49–50).[13]

In another interview, conducted two years later, Foucault clarifies again the meaning and the methodological function of the term dispositive. It was certainly no coincidence that this interview was initiated by a circle of Lacanians, whom Foucault challenged with his call to go beyond the "logic of the unconscious." He proposed the following definition, which spells out three distinctive components:

> What I'm trying to pick out with this term is, firstly, a thoroughly heterogeneous network consisting of discourses, institutions, architectural forms, regulatory decisions, laws, administrative measures, scientific statements, philosophical, moral and philanthropic propositions—in short, the said as much as the unsaid. Such are the elements of the dispositive. The dispositive itself is the system of relations that can be established between these elements. Secondly, what I am trying to identify in this dispositive is precisely the nature of the connection that can exist between these heterogeneous elements. Thus, a particular discourse can figure at one time as the programme of an institution, and at another it can function as a means of justifying or masking a practice which itself remains silent, or as a secondary re-interpretation of this practice, opening out for it a new field of rationality. In short, between these elements, whether discursive or non-discursive, there is a sort of interplay of shifts of position and modifications of function which can also vary very widely. Thirdly, I understand by the term "dispositive" a sort of [. . .] formation which has as its major function at a given historical moment that of responding to an *urgent need*. The dispositive thus has a dominant strategic function. (Foucault 1980b, 194–95; translation modified, emphasis in original)

In distinguishing between the three dimensions of the dispositive, Foucault draws on the complex etymological trajectory of the French word *dispositif*. It was first used to refer to the enacting terms of a legal decision, later to the deployment of troops in war, and finally it signified a technical device or an apparatus. According to the *Dictionnaire historique de la langue française* (2006, 1101), the term was originally part of the legal vocabulary designating the final words of a judgment in which a court's decision was announced; they brought the legal decision into being. In the eighteenth century, the word entered military language, referring to strategies putting to work "the totality of means arranged [*disposés*] consistent with a plan" (ibid., 1101). In the nineteenth century, the term acquired its contemporary sense: the "way in which the organs of an apparatus are arranged [*disposés*]" (ibid., 1101; Behrent 2013,

87–88). Thus, the word's etymology contains three dimensions that are regularly evoked in English translations, and it is crucial to grasp their interplay if we want to understand Foucault's interest in the notion: an "ontological" meaning, a technical reading and a strategic sense.[14]

Ontologically, the dispositive is a "network" (*réseau*) (Foucault 1980b, 194; translation modified) consisting of a heterogeneous ensemble of discursive and non-discursive elements, material and semiotic entities, without any neat separation between them—in fact, the distinction "doesn't much matter" (ibid., 198; see also Deleuze 1992a, 160).[15] It is a composite of things that seems to include virtually anything from discourses and institutions to bodies and buildings. The dispositive assembles the elements it consists of and is itself the result of this process of "formation" (Foucault 1980b, 195). It is the relational web that binds together these elements, defining their positions and giving them a particular form and shape. Thus, the dispositive is not "an already given object" (Foucault 2007a, 118) but rather the outcome of a particular historical set of regulated practices, seeking to calculate and manage future events and aleatory developments.[16] In medicine, for example, the diagnosis of a (pre)disposition points to (often inherited) risk factors that increase the chances of developing certain diseases in the future—calling for supervision and control of bodily processes in the present.[17]

The dispositive enacts a double movement.[18] On the one hand it mobilizes things, placing them "at one's disposal," defining them as instruments, resources, or means to achieve specific objectives (see Link 2008). One example of this dynamic is provided by the medical anthropologist Lawrence Cohen (2005), who has appropriated the term "bioavailability" from pharmacology. This term denotes "the selective disaggregation of one's cells or tissues and their reincorporation into another body (or machine)" (Cohen 2005, 83). It addresses the rise of transplantation medicine and the technical and normative challenges that go along with the fact that more and more human tissues "became available for extraction from one body followed by infusion or implantation into others" (ibid., 83). The term "bioavailability" seeks to investigate how medical technologies and forms of care are intimately tied to a neoliberal regime of entrepreneurship and economic government, opening up spaces for commercialization and exploitation (see ibid., 85).

On the other hand, the dispositive positions things as "disposable." It enacts lines of differentiation and establishes practices of disregard that make it permissible to discriminate, exclude, or even kill humans and nonhumans considered as useless, unproductive, or dangerous: as "surplus life" (Murphy 2017, 135–45), or "life devoid of value" (Binding and Hoche 1975). This aspect has been highlighted by Tara Mehrabi (2016) in her study of Alzheimer's disease. She proposes the concept of "killability," in order to study how masses of transgenic fruit flies must die in order to promote experimental research on the disease. Mehrabi addresses the question of what might constitute a killable body and how the borders between life and death are permanently drawn and reenacted in the research process: "human and animal becoming in the lab is a relational process that does violence as a constitutive part of knowledge production, as it enacts particular forms of life as killable" (Mehrabi 2016, 54).

The second dimension of the dispositive is *technological*, putting the accent on the "onto-creative aspect" (Bussolini 2010, 100): "Each dispositive has its way of structuring light, the way in which it falls, blurs and disperses, distributing the visible and the invisible, giving birth to objects which are dependent on it for their existence" (Deleuze 1992a, 160; translation modified).[19] Dispositives are defined by how they produce and maintain the differential positions of their elements. They establish a distinctive network that allows certain materializations to emerge rather than others. However, a dispositive is not a stable and closed technological setup but rather a dynamic "ensemble" characterized by "shifts of position and modifications of function" (Foucault 1980b, 195). It is a mobile and morphing arrangement characterized by the structural relations between the heterogeneous elements that make up the dispositive. To be sure, these "functions" are not determined or defined by the "needs" or "demands" of an already existing system (as in classical functionalist theory); quite on the contrary, they are permanently being reworked and modified in the course of the dispositive's operations—a process Foucault calls *functional overdetermination* as the (unintended) effects of its operations enter "into resonance or contradiction" (ibid., 195; emphasis in original) with other effects, so that the "elements" of the dispositive are permanently being redefined, redeployed, and readjusted.

These processes of adaption and modification go beyond a classical imaginary of an always already given socio-material topography characterized by distinctive micro- and macrolevels and their interplay; rather, the political terrain and its conditions of contestability are charted by forces and flows.[20] This idea of a permanent recombination and re-articulation of heterogeneous elements within a relational network comes close to what Gilbert Simondon's philosophy of technics conceives of in terms of a "montage" (2017, 251). In Simondon's work technology is not a material object or the product of thought, but rather an incessant process of adjustment and repair—a practical activity that "most naturally continues the function of invention and construction" (Simondon 2017, 255). His thinking evades ontological dualisms between spirit and substance, human and machine, form and matter to attend to the mobile dynamics that make up and modify specific kinds of individual entities. Simondon's approach to these processes of "individuation" stresses their indeterminacy and unfinished nature, and also analyzes them in terms of power and potentiality: "[T]he technical world offers an indefinite disposability [*disponibilité*] of groupings and connections. For what takes place is a liberation of the human reality that is crystallized in the technical object; to construct a technical object is to prepare a disposability" (ibid., 251, translation modified; LaMarre 2013; Delitz 2014; Lipp 2017, 113–15).[21]

The third aspect of the dispositive is its "*strategic* objective" (Foucault 1980b, 195; emphasis added).[22] Dispositives exist insofar as they address a specific demand or "urgency." They are driven by a "perpetual process of *strategic elaboration*" (ibid.; emphasis in original) that makes it possible to enroll and mobilize the unintended or negative effects within a new strategy. Foucault illustrates this process with the example of the "dispositive of imprisonment." While imprisonment appeared to be the most humane and rational way of handling the problem of illegalities at the beginning of the nineteenth century, it produced "an entirely unforeseen effect":

> the constitution of a delinquent milieu [. . .]. What happened? The prison operated as a process of filtering, concentrating, professionalising and circumscribing a criminal milieu. From about the 1830s onwards, one finds an immediate re-utilisation of this unintended, negative effect

within a new strategy which came in some sense to occupy this empty space, or transform the negative into a positive. The delinquent milieu came to be re-utilised for diverse political and economic ends, such as the extraction of profit from pleasure through the organisation of prostitution. This is what I call the strategic completion (*remplissement*) of the dispositive (Ibid., 195–6; translation modified)

Thus, the strategic objective and the existing form of the dispositive are always marked by a distance—a difference that is not simply the result of underachievement or a sign of imperfection; rather, it becomes a vector in the transformation of the dispositive (Brauns 2003, 44). It is exactly this "tactical polyvalence" (Foucault 1978, 100) or "variable creativity" (Deleuze 1992a, 163) that allows for the flexibility and dynamics of the dispositive and makes it possible to circumvent a functionalist bias (see Foucault 2007a, 118). As Foucault argues, the dispositive is "a matter of a certain manipulation of relations of forces, either developing them in a particular direction, blocking them, stabilising them, utilising them, etc." (Foucault 1980b, 196) Importantly, this concept of strategy does not originate in the "decisions" or "interests" of an individual or collective subject but informs power relations that are "both intentional and non-subjective" (Foucault 1978, 94; see also 1980b, 206).

Foucault illustrates this idea of a "strategy without a subject"[23] by using another example: the attempts made in early nineteenth-century France to attach workers in the first heavy industries to their workplaces. He refers to a number of diverse and heterogeneous tactics that mobilized material and semiotic, human and nonhuman entities. They ranged from pressuring workers into marriage and providing new housing options, via the emergence of philanthropic discourses, to constructing schooling facilities for children. These highly diverse tactical measures resulted in "a coherent, rational strategy, but one for which it is no longer possible to identify a person who conceived it" (1980b, 203). Importantly, these measures and instruments were not "imposed" (ibid., 204) by particular individuals or social classes; rather, they "met the urgent need to master a vagabond, floating labour force. So the objective existed and the strategy was developed, with ever growing coherence, but without it being necessary to attribute it to a subject" (ibid., 204; Foucault 1978, 94–5; 1994b, 16–19; see also Hubig 2000).[24]

However, this does not mean that dispositives simply respond to crises, trying to solve pre-existing problems. There is a "rule of double conditioning" (Foucault 1978, 99) at work: the dispositive impacts on the strategy as much as the strategy informs the dispositive. As Foucault states, it is possible to call strategies of power "the totality of the means put into operation to make a dispositive work or to maintain it [*pour faire fonctionner ou maintenir un dispositif de pouvoir*]" (Foucault 2000b, 346; translation modified). To analyze this "double process" (Foucault 1980b, 195) or the "reciprocal relation of production" (1980b, 203) at stake here, Foucault proposes the concept of "instrument-effect" (Foucault 1978, 48). The dispositive is not exterior to the problem (or independent of a diagnosed "urgency" or "crisis"); rather, it is simultaneously the effect of a particular problematization and an instrument designed to respond to it.[25]

According to Foucault, dispositives are characterized by modes of contestation and forms of "counter-conduct" (Foucault 2007a, 201). He seeks to capture this "agonistic" (see Foucault 2000b, 342) character of dispositives by referring to "a certain plebeian quality or aspect" (Foucault 1980c, 138), arguing that "the existence of a 'plebs' [is] the permanent, ever silent target for dispositives of power" (ibid., 137; translation modified). Foucault rejects an understanding of the plebs as a "real sociological entity" (ibid., 137) or a transhistorical figure and foundation for political revolts; rather, it is conceived of "as a centrifugal movement, an inverse energy, a discharge" (ibid., 138) that materializes in certain arrangements of bodies. This understanding of the plebs does not exclusively refer to human collectives or social categories but seeks to grasp human-nonhuman alliances and material forces that are addressed and targeted in the operations of the dispositive. Foucault stresses the theoretical but also political importance of this account: "This point of view of the plebs, the point of view of the underside and limit of power, is thus indispensable for an analysis of its dispositives; this is the starting point for understanding its functioning and developments" (ibid., 138; translation modified; Foucault 2000b, 346–47).

Beyond Archive, Apparatus, and Assemblage: Conceptualizing Ontological Politics

The rise of the notion of the dispositive in Foucault's conceptual vocabulary marks a complex play of theoretical continuity and rupture. The episteme and the archive played an important role in Foucault's earlier work (see Foucault 1970; 1972), as they both constituted the historical apriori of particular discursive events of an epoch while at the same time operating as a general structure that allowed these discourses to emerge in the first place. The same is true for the dispositive. However, there are two important differences. The archive focuses on a "system of discursivity" (Foucault 1972, 129) that determines what could be said at a particular epoch, while the episteme "defines the conditions of possibility of all knowledge" (Foucault 1970, 168). Thus, both concepts remain within the horizon of discourse. By contrast, Foucault conceives of the dispositive as "both discursive and non-discursive" (Foucault 1980b, 197; see also Hubig 2000).[26] A second difference concerns the strategic character of the dispositive, stressing the co-constitution of power relations and fields of knowledge. The dispositive consists in "strategies of relations of forces supporting, and supported by, types of knowledge" (Foucault 1980b, 196). The interest in the strategic dimension leads to a different account of history. Foucault no longer accentuates historical disruptions and discontinuities, as exemplified by the sequence of different epistemes and archives, but rather conceives of historical processes as driven by agonistic forces and strategic re-workings of dispositives.

Giorgio Agamben (2009) has suggested that the term dispositive, and its Latin precursors *dispositio* and *disponere*, are renderings of the Greek term *oikonomia*, meaning the administration of the *oikos*, of the family and its goods and well-being, or more generally management. It relates to "a set of practices, bodies of knowledge, measures, and institutions that aim to manage, govern, control, and orient [. . .] the behaviors, gestures, and thoughts of human beings" (Agamben 2009, 12). However, Foucault's reading of the dispositive exceeds Agamben's focus on the human and his theological framing of the concept, as it is anchored in an analytics of government that seeks to steer and direct processes of life beyond human existence.[27] While Agamben sets up an opposition between "living beings" and "dispositives" (ibid., 13, translation modified) and suggests an external and negative relationship in which the life of individuals is "contaminated" (ibid., 15) by the workings of dispositives, Foucault's use of the term stresses its ontological and technological dimensions.[28]

Foucault's genealogy of the dispositive of sexuality is a good example of the interplay between the technological, strategic, and ontological dimensions of the concept. In *The History of Sexuality, Volume 1* Foucault contests in two ways what he calls the "repression hypothesis," the Freudian-Marxist idea that Western societies denied or suppressed sexuality from the seventeenth century on due to the rise of capitalism and the hegemony of the bourgeoisie. First, Foucault rejects the idea of originary sexuality as something that came to be constrained and must now be emancipated. He also criticizes the interpretation that the dispositive of sexuality primarily serves class oppression, claiming that sexuality is not something universally given, differently regulated, and known in concrete societies. On the contrary, he argues, "sexuality" is a historical figure that emerged in the nineteenth century and then became a privileged object of knowledge in various disciplines. The dispositive of sexuality arranges and aligns a set of social behaviors, bodily functions, and institutional practices, thereby governing and controlling individuals and their bodies (see Foucault 1978, 107; Behrent 2013, 88; Elden 2016, 53–59). Secondly, Foucault holds that "a technology of sex" (1978, 123) was invented by the bourgeoisie to produce their own distinct kind of discourses, sensations, and truths, thus affirming the body rather than negating it: "The primary concern was not repression of the sex of the classes to be exploited, but rather the body, vigor, longevity, progeniture, and descent of the classes that 'ruled'" (ibid.). Thus, Foucault argues that "sexuality" is a bourgeois innovation, a means of self-affirmation to constitute its "class body." Only later on, in the course of the nineteenth century, did the dispositive come to operate on the social body as a whole, where, as a hegemonic instance, "in its successive shifts and transpositions, it induces specific class effects" (ibid., 127; Foucault 2003, 31–34).[29]

The strategic importance of the notion of dispositive for Foucault's work becomes even clearer as he neatly dissociates the term from apparatus in his writings. The conceptual distinction is already present in the lectures at the Collège de France of 1973–1974 (Foucault 2006a) and in *Discipline and Punish* (1979).[30] While in these earlier texts Foucault uses dispositive in a sense that is sometimes close to the technical meaning of mechanism or apparatus, he already hints at a "more philosophically complicated sense" (Elden 2017, 142). This particular conceptual profile takes shape in the first volume of *The History of Sexuality* and informs Foucault's subsequent understanding of the term. Foucault consciously and consistently distinguishes the notion of the dispositive from the more limited and circumscribed concept of the apparatus, which remains within the realm of sovereignty and state power and focuses on instrumental use (see, e.g., 1978, 86; 89; 95). This understanding of apparatus instructs Foucault's lectures on governmentality at the Collège de France when he discusses the "dispositives of security," distinguishing them from "governmental apparatuses" (*appareils*) in the narrow sense (see Foucault 2007a, 108).[31]

Thus, in Foucault's conceptual vocabulary apparatus (*appareil*) is not synonymous with dispositive (*dispositif*) or interchangeable with the latter term; they are "related concepts, such that apparatus is a distinct subset of dispositive" (Bussolini 2010, 94).[32] This conceptual priority of the dispositive is also theoretically important. Foucault engages critically with traditional political science as far as it focuses on sovereignty and the state as a military-administrative apparatus, but he also distances his concept from Althusser's work on "ideological state apparatuses" (Althusser 1971; 2014). While Althusser sought to expand the scope of state theory by taking into account knowledge production and subjectivation processes, the analysis still remained centered on the state.[33] Foucault's usage of the term dispositive, then, represents an explicit conceptual choice that is obscured when both *appareil* and *dispositif* are translated into English without any differentiation as "apparatus."

The notion of the dispositive opens up the analysis to the strategic relations of forces instead of focusing on the structural organization of state power. It seeks to investigate "the support which these force relations find in one another, thus forming a chain or a system, or on the contrary, the disjunctions and contradictions which isolate them from one another; and lastly, as the strategies in which they take effect, whose general design or institutional crystallization is embodied in the state apparatus" (Foucault 1978, 92–93; see also Bussolini 2010, 93–94).[34] In contrast to the notion of the dispositive, apparatus often refers to the *static collection* of instruments, machines, tools, parts, or other equipment of a given order of things rather than to their *strategic composition*: "Apparatus might be said to be the instruments or discrete sets of instruments themselves—the implements or equipment. Dispositive, on the other hand, may denote more the arrangement—the strategic arrangement—of the implements in a dynamic function" (Bussolini 2010, 96).

There are similarities as well as differences between Foucault's notion of the dispositive and the concept of apparatus in Barad's agential realism. As we saw in the last chapter, Barad's account proposes a move from a static and stable to a performative and dynamic understanding that makes it possible to account for the boundary-making practices of the apparatus (see, e.g., Barad 2007, 170). According to Barad, apparatuses do not just "change in time; they materialize (through) time" (ibid., 203), they are "not located in the world but are material configurations or reconfigurings of the world that re(con)figure spatiality and temporality as well as (the traditional notion of) dynamics" (ibid., 146). Thus, agential realism brings to the fore the innovative and productive dimension of the apparatus, emphasizing its role in "agential cuts" and "intra-actions." However, the stress on the radical contingency of the apparatus and the "ever-changing relations of power" (ibid., 237) does not address adequately the question of how apparatuses are stabilized and consolidated in practice. While Foucault seeks to circumvent any "internal and circular ontology" (Foucault 2007a, 247–48; see also 354) to account for a situated and strategically informed analysis, Barad's analysis tends to disentangle the governmental dimension from the operations of the apparatus.

The concept of the dispositive can also be usefully contrasted with the notion of assemblage (*agencement*) originally proposed by Deleuze and Guattari. This term puts the accent on ontological composition and creativity, and plays a central role in new materialist scholarship that rejects anthropocentric notions of agency. In vital materialist accounts, as we saw, assemblage denotes "ad hoc groupings of diverse elements, of vibrant materials of all sorts" (Bennett 2010a, 23). Bruce Braun has noted that the use of the English word assemblage to translate Deleuze and Guattari's French notion of *agencement* only partly captures the significance of the term. While the former is restricted to a collection of things, *agencement* "relates the *capacity to act* with the *coming together of things* that is a necessary and prior condition for any action to occur, including the actions of humans" (Braun 2008, 671; emphases in original). While this is certainly an important clarification of the fluid and mobile compositions that the term evokes, the strategic dimension the dispositive articulates is not adequately addressed by the conceptual alternative assemblage/*agencement* for two reasons.

First, the accent is put on ontological heterogeneity. Assemblages are often defined as sets of practices that connect a diversity of entities, giving rise to new collectives and unknown configurations of space and time (see, e.g., Ong and Collier 2014, 4). In this sense, dispositives could be "considered a type of assemblage, but one more prone to (in the sense of anticipating, provoking, achieving and consolidating) re-territorialisation, striation, scaling and governing" (Legg 2011, 131). While an assemblage indiscriminately includes nonhumans as well as humans, the notion of the dispositive takes into account the differential boundaries between these heterogeneous elements. Thus, in contrast to the former, the latter term "gives more of a sense of the ongoing *integration* of a differential field of multiple elements" (Anderson 2014, 35; emphasis in original).[35]

Secondly, the notion of assemblage is mostly associated with emergence, innovation, and creation. By contrast, the dispositive "places the emphasis on the movements of stabilization that tend to put heterogeneous elements into order" (Silva-Castañeda and Trussart 2016, 495). While the Foucauldian term is also attentive to the processual dimension of ontologies, stressing how dispositives are permanently rearticulating and transforming their conditions of existence, it is still animated by an interest in how order is re-stabilized and reenacted.[36] This attention to the strategic dimension entails an important analytic advantage, as it circumvents a dualistic approach by examining processes of stabilization and lines of contestation within a single analytic frame. This methodological suggestion is in line with Foucault's claim that power and resistance cannot be separated and his idea of an "immediate and founding correlation between conduct and counter-conduct" (Foucault 2007a, 196). In this light, critique and contestations are not (only) negative and reactive counterparts; rather, forms of dissent and deviance might inform, reform, and/or transform an existing dispositive: "[L]ooking through the lens of Foucault's *dispositif* highlights that there is not necessarily antinomy between disruptive lines and stabilizing ones;

or, put differently, between contestation and institutionalization" (Silva-Castañeda and Trussart 2016, 504; Raffnsøe et al. 2016, 287–291).[37]

To sum up, comparing the concept of the dispositive with the notion of the apparatus and the assemblage brings out a quite illuminating contrast. While the latter terms tend to focus on ontological and technological questions, only the former explicitly articulates these dimensions together with strategic concerns. The concept of the dispositive captures the interplay of ontological, technological, and strategic issues in order to address the problem of "ontological politics," paving the way for a more materialist approach to government. The next chapter explores in more detail Foucault's understanding of technology, while Chapter 6 discusses the strategic role of the milieu within a government of things.

5

More-Than-Social Configurations

Expanding the Understanding of Technology

We frequently speak of the technical inventions of the seventeenth century—chemical, metallurgical technology—yet we do not mention the technical invention of this new form of governing man, controlling his multiplicity, utilizing him to the maximum, and improving the products of his labour, of his activities thanks to a system of power which permits controlling them. (Foucault 2007b, 146)

The old societies of sovereignty made use of simple machines—levers, pulleys, clocks; but the recent disciplinary societies equipped themselves with machines involving energy, with the passive danger of entropy and the active danger of sabotage; the societies of control operate with a third type, computers, whose passive danger is jamming and whose active one is piracy and the introduction of viruses. (Deleuze 1992b, 6)

As we saw in the last chapter, it is possible to discern a more-than-human political analysis in Foucault's work, encapsulated in the idea of a government of things. In a complementary move, he extends the traditional understanding of technology. Instead of reserving the term exclusively for manipulating and mobilizing things in a literal sense, Foucault's vocabulary also applies it to human affairs—or rather it operates across the dividing line between the human and the nonhuman. In this understanding, the concept of technology is a central interpretative resource to analyze governmental practices and their complex dynamics. Against Barad and many others who claim that Foucault remains within the humanist grid, he actually cautions that "we need to avoid 'man' or 'human nature' if we want to analyze social systems and human systems" (Foucault 1994c, 103). While the concrete role of objects, devices, and infrastructures in governmental practices often remains obscure in Foucault's historical work, I will argue that his concept of technology makes it possible to grasp the political matter of a government of things.[1]

The first part of this chapter presents Foucault's understanding of government as a technological invention. The analysis focuses on the material and innovative dimension of governmental practices, and undermines any systematic distinction between their emergence and technical breakthroughs in a narrower sense. The second section distinguishes Foucault's notion of technology from social constructivism and technological determinism on the one hand and from Marxist and humanist accounts of technology on the other. In the third part I discuss two technological metaphors and models for imagining the political structures and processes to which Foucault implicitly refers in his history of governmentality. The clock and the steam-engine governor display distinctive rationalities of ruling, affecting the matter of politics. The fourth section explores the concept of "technologies of security" (Foucault 2007a, 59). Foucault introduces these technologies as a specific feature of liberalism and spells out their role as feedback devices in the eighteenth century, when the idea of "invisible hands" and "checks and balances" informed innovations in physics and engineering but also in economic and political life. In the fifth section I analyze how, in the nineteenth century, this technological account of government was extended to biological understandings of evolutionary theory and physiological processes of regulation, finally giving rise to the program of cybernetics in the twentieth century.

Government as a Technical Invention

The concepts of technology (*technologie*) and technique (*technique*) appear in Foucault's work from the 1950s on. He frequently employed the two terms interchangeably, but there is still a certain coherence and systematicity in his usage. As Michael C. Behrent (2013, 58–60) notes, two points are particularly striking. First, while "technique" figures in Foucault's earliest writings, "technology" only emerges in his work from 1974 onwards, when his research focus shifted to what he then called "technologies of power." Second, the overall frequency with which he engaged with both terms increased significantly after the mid-1970s. Especially in his lectures at the Collège de France of 1978 and 1979 on the history of governmentality and in his subsequent work, "technology" and "technique" acquired a central significance. The two terms not only provided Foucault with a way of connecting material artifacts and infrastructures with governmental rationalities, but also linked practices of political government to forms of self-government—or what he finally came to call "technologies of the self" (Foucault 1997a).[2]

It is quite surprising that the central role of the notion of technology in Foucault's writings is mostly overlooked or ignored by commentators (Matthewman 2011, 66).[3] While there is a substantial amount of literature on "technologies of the self" (see e.g., McKinlay and Starkey 1998; Kelly 2013; Demenchonok 2018), the broader and more general understanding of technology remains largely unaddressed. In his later work Foucault explicitly argues for expanding the meaning of the term to rearticulate it within an analytics of government. He proposes to distinguish between several distinct technologies:

> (1) technologies of production, which permit us to produce, transform, or manipulate things; (2) technologies of sign systems, which permit us to use signs, meanings, symbols, or signification; (3) technologies of power, which determine the conduct of individuals and submit them to certain ends or domination [. . .]; (4) technologies of the self, which permit individuals to effect [. . .] a certain number of operations on their own bodies and souls, thoughts, conduct, and way of being, so as to transform themselves in order to attain a certain state of happiness, purity, wisdom, perfection, or immortality. (Foucault 1997a, 225)

While Foucault distinguishes analytically between these technologies, he still insists that empirically they "always overlap one another, support one another reciprocally, and use each other mutually as means to an end" (Foucault 2000b, 338). Thus, the interplay of the different technologies constitutes "'blocks'" or "regulated and concerted systems" that adjust and align the capacity to use or modify things with processes of communication, power and self-formation (ibid., 337–39). We might define this careful and comprehensive "coordination" (ibid., 338) of distinct technological modes and relationships as a government of things, as it articulates more or less coherent regimes and "ways of doing things" (Foucault 2008a, 42).[4]

Foucault's "wide sense" (Foucault 2000a, 364) of technologies follows from his diagnosis of an important reduction of the semantic field. He observes that the understanding of technology is often restricted to "hard technology, the technology of wood, of fire, of electricity" (ibid., 364). In contrast to this "very narrow meaning," he stresses that "government is also a function of technology: the government of individuals, the government of souls, the government of the self by the self, the government of families, the government of children and so on" (ibid., 364). While this statement seems to reproduce the traditional dichotomy placing material devices, artifacts, machines and infrastructures on one side and social institutions, political regimes and cultural systems on the other, Foucault actually proposes an integral concept of *tekhnē* as "a practical rationality governed by a conscious goal" (ibid., 364). He takes up the Greek root of the term "technology" that links it to the arts, crafts or skills combining material instruments and social practices. This "guiding concept" (ibid., 364) then makes it possible for Foucault to "bypass the boundary between the social and the material, the human and the nonhuman" (Altamirano 2014, 12; see also Rooney 1997).[5]

Foucault's concept of technology takes up Heidegger's understanding of *tekhnē* as "standing reserve" (*Bestand*) (Heidegger 1993 [1954]). In Heidegger's work technologies do not only define a particular set of devices or procedures; they are not just a means for specific aims but rather what makes specific means possible in the first place. In this reading, technologies configure material and non-material entities as resources that can be stored, mobilized, and distributed (Dean 1996, 57–61; Seibel 2016, 29–31). However, while Foucault shares Heidegger's assertion of an ontological connection between humans and technology, he rejects the underlying idea of a genuine subject not (yet) affected by technology: "For Heidegger, it was on the basis of Western *tekhnē* that knowledge of the object sealed the forgetting of Being. Let's turn the question around and ask ourselves on the basis of what tekhnai was the Western subject formed and were the games of truth and error, freedom and constraint, which characterize this subject, opened up" (Foucault cited in Gros 2005, 523; Dorrestijn 2011, 225–26).[6]

Grasping the Materiality of Technologies: Beyond Social Constructivism and Technological Determinism

Foucault's understanding of technologies conceives of them in terms of knowledges and practices, without restricting their analysis to isolated artifacts or systems of objects. The term not only refers to devices, machines, or applications of scientific knowledge but also defines a mode of calculating, regulating, and intervening, a practical concern of controlling future events. This broad concept of technology seeks to circumvent two possible pitfalls.

First, Foucault counters the claim that technical artifacts and devices are socially "shaped" or "constructed" (MacKenzie and Wajcman 1985; Bijker et al. 1987).[7] He avoids pre-analytical distinctions between the social and the technological or the micro- and the macro-level as the social cannot be isolated from the technological. Rather than focusing on influences and causal links between two separate entities, the social is conceived of as a technological invention that emerged in the nineteenth century (see, e.g., Donzelot 1984; Ewald 1986).

Second, Foucault was careful to also distinguish his work from technologically determinist accounts. According to this perspective, changing social and political relations are the simple effect or a straightforward outcome of technological developments and innovations. To expose the limits of this form of analysis, he discusses the example of a historian of the Middle Ages[8] who showed that

> at a certain moment it was possible to build a chimney inside the house—a chimney with a hearth, not simply an open room or a chimney outside the house; that at that moment all sorts of things changed and relations between individuals became possible. All of this seems very interesting to me, but the conclusion that he presented in an article was that the history of ideas and thoughts is useless. What is, in fact, interesting is that the two are rigorously indivisible [. . .]. It is certain, and of capital importance, that this technique was a formative influence on new human relations, but it is impossible to think that it would have been developed and adapted had there not been in the play and strategy of human relations something which tended in that direction. What is interesting is

always interconnection, not the primacy of this over that which never has any meaning. (Foucault 2000a, 362)

In addition to providing a critical engagement with social constructivism and technological determinism, Foucault's concept of technology also "depends on the violation of a multiple system of taboos" (Gordon 1980, 238), as it challenges two (sometimes linked) lines of analysis and critique: the Marxist understanding of power and the humanist concern with technology.

Foucault's technological reading of modern government draws substantially on Marx's insights in *Capital* while at the same time dismissing functionalist, economistic, or state-centered concepts in Marxist theory. Already in *Discipline and Punish*, Foucault notes that disciplinary technology allows for a "recoding of existence" (1979, 236) that fundamentally differs both from physical repression and from ideological manipulation. This recognition of a historical transformation of power relations becoming more technological is indebted to Marx's analysis of the organization of industrial labor and the regulation of space and time within the capitalist factory regime (Lustig 2014, 76; Kammler 1986, 149–50). While Foucault critically exposed the dogmatic and determinist tendencies in "scholastic Marxism" (Foucault 1985, 3), he repeatedly stressed that he "follow[s] these essential indications" (see Foucault 2007c, 158) laid out by Marx.[9] As we saw in the last chapter, Foucault rejects the Freudian-Marxist idea that Western societies suppressed sexuality from the eighteenth century on due to the rise of capitalism. Instead, he emphasizes the creative and inventive dimensions of power—an analytical perspective that can be found "between the lines of the Volume II of *Capital* [. . .], or at least the sketch of an analysis" (ibid., 158 emphasis in original). However, Foucault's account of technology also exceeds Marx's understanding of power as originating in the sphere of production (by ownership of the means of production) and his humanistic concept of man as an essentially productive being (Foucault 1994d, 470; see Behrent 2013, 83–84).

This technological understanding of power also differs from the humanist critique of instrumental or technocratic reason. There has been a permanent concern in political and social theory that technology is intrinsically alien to the political sphere. This critical tradition extends from Carl Schmitt's thesis that technical reasoning distorts political decision-making (Schmitt 2007), via Max Horkheimer's and Herbert Marcuse's critiques of the nexus of technological organization, social domination, and "instrumental reason" (Horkheimer 2012; Marcuse 1991), to Jürgen Habermas' understanding of science and technology as the "ideology" of industrial societies (Habermas 1970). It draws on a strict line of division that separates the human sciences from the natural sciences, reserving technological questions for the latter. Foucault broke methodologically with this line of reasoning by refraining from asking how politics is suppressed, inhibited, or concealed by technological matters, posing instead the question of how it is permanently produced, and transformed by technologies (Dorrestijn 2011, 224–25; Seibel 2016, 23–26; see also Gordon 1980; Dean 1996, 52–53).

From very early on in his work, Foucault sought to question and subvert the boundary between the political and the technical, the mundane and the scientific. Already in the 1960s he stated, against the humanist critique of technology: "the effort undertaken by people of our generation is not to make claims for man *against* knowledge and *against* technique, but precisely to show that our thought, our life, our way of being, and even our most everyday ways of being belong to the same systematic organization and are thus subject to the *same* categories as the scientific and technical world" (Foucault 1994e, 518; emphases in original; translated by Behrent 2013, 67). In this reading, technology is not conceived of as an illegitimate expansion into the domain of the social or a problematic transgression of disciplinary boundaries; rather, it is an essential part of social life, also exceeding the category of the social. In *Discipline and Punish*, Foucault illustrates this more-than-social understanding of technologies with the example of military regulations of the eighteenth century. In the manuals that prescribe various modes of aligning bodies and weapons, specifying how exactly to use a rifle by distinguishing different stages and corporal engagements, he discerns a "meticulous meshing" (Foucault 1979, 153) of human bodies and technological objects. In this process soldier and rifle are fused into a single body constituting "a body-weapon, body-tool, body-machine complex" (ibid., 153).[10]

Foucault's account opens up a whole range of new questions to be asked about how technologies enact "a matrix of practical reason" (1997a,

225).[11] They do not provide a silent background or a secret basis, nor do they serve as a simple resource for governmental action; on the contrary, they actively (re)configure political and moral practices. According to Foucault, "things" can operate as political and moral actors since they allow for certain practices rather than others. In this light, morality is conceived of as "a set of physico-political techniques" (1979, 223; see Matthewman 2013, 286). Apart from "[t]he *body-object articulation[s]*" (ibid., 152; emphasis in original), Foucault's analysis focuses particularly on spatial arrangements and architectural designs. One example is the transformation of the hospital at the end of the eighteenth century, in which spatial distributions (isolated beds and the circulation of air) operated as "the agent and the instrument of cure" (2007b, 149). Similarly, prison walls and cells act as moral agents by working on the body and the soul of the deviant subject, at least partly replacing or complementing human action: "[S]upervisors do not have to exert force—this is assured by the materiality of things" (Foucault 1979, 239; see Matthewman 2013, 286).[12]

From the Clock to the Governor: Metaphors, Models, and Matters of Politics

The government of the state always relied on the state of technological devices and developments. Technological metaphors and models have been employed to imagine and arrange political structures and processes, displaying distinctive rationalities of governing and ruling.[13] In his lectures at the Collège de France in 1978 and 1979, Foucault insists that the concept of governmentality differs decisively from medieval understandings of power, as it draws neither on the sacred will of God nor on a cosmic order of things: It "is something without a model, which must find its model" (2007a, 237). Foucault might have overstated his case here by ignoring the manifold metaphorical and practical transfers between the emerging natural sciences and the modern understanding of politics.[14] However, in other passages in these lectures Foucault also accepts some kind of parallelism or at least "contemporaneousness" (ibid., 296) between the development of *raison d'état* and police science on the one hand and the natural sciences on the other. On this account, "politics would be to the art of government something like what *mathesis* was to the science of nature in the same period" (ibid., 286; see Seibel 2016, 54).[15]

Georges Canguilhem (2008a) has shown that Descartes' understanding of the human body as a mechanical system had important implications for imagining how politics operates. It replaced the individual will of the sovereign with a new mechanics of power. In Descartes' work "the technological image of 'command' (a type of positive causality by a dispositive or by the play of mechanical connections) substitutes for the political image of commandment (a kind of magical causality; causality by word or by sign)" (Canguilhem 2008a, 86; translation modified; see also Pasquinelli 2015, 84). The idea and ideal of a highly complex, technological system that works without frictions and failures became the paradigm for the sovereign disposition of humans and things—an interconnected ensemble characterized by accurate simultaneousness, precision, and efficiency. It was the mechanical clock that from the fourteenth to the late eighteenth century not only served as a model for scientific investigations of nature but also became an emblem of authority and political order.

The idea of clockwork already guided Hobbes' philosophy and his understanding of the political body of the *Leviathan* as a machine (Hobbes 1962 [1651]).[16] As Otto Mayr (1986) has shown in his remarkable study *Authority, Liberty, and Automatic Machinery in Early Modern Europe*, this metaphor also informed the cameralist and mercantilist designs of monarchical power and remained a recurrent topos of government well into the age of enlightened absolutism. The frequent uptake of the clockwork metaphor in political treatises symbolizes the disciplinary regime and its "'mechanics of power'" (Foucault 1979, 138). The governmental apparatus of this time is conceived of as a complex of chains that link causes and effects, with the sovereign on the top as the "mainspring" that allows everything else to move.[17] The idea of a clockwork state originated in mechanist philosophy and defined a static model, in which the governed only figured as passive subordinates and "cogwheels" (Seibel 2016, 55–57).[18]

Liberal government, which emerged in the second half of the eighteenth century, stressed the limits of disciplinary steering. The population became the target of a different governmental regime, a "technical-political object of management and government" (Foucault 2007a, 70), necessitating new and more complex forms of "regulation" to maximize its biological and economic productivity, its health, and its wealth. Liberal government operates not by suppressing or limiting the inherent fluctuations and dynamic variations of the governed entity, but seeks to hedge them, to moderate and control them. Instead of connecting passive cogwheels, governmental technologies attend to an active, highly mobile and difficult-to-steer collective subject whose productivity will be hampered rather than enhanced by too much intervention. Foucault's analysis engages extensively with the central argument of (classic) liberalism, which postulated that governmental interference with the operations of the economic and social systems was harmful (Foucault 2007a; 2008a). According to this view, these systems exhibit mechanisms of self-regulation that maintain themselves in equilibrium at an optimal state due to certain interior properties, while interventions from outside are detrimental: "Whenever a given social variable (e.g. the balance of trade, the price of hogs, or the number of available laborers) would deviate from the equilibrium value, the general public (as statistical entity), motivated by ordinary self-interest, would respond to this automatically in such a manner as ultimately to counteract the deviation" (Mayr 1971a, 2).

The model for this flexible and situational control mechanism could be found by the end of the eighteenth century in a new technological device: the steam-engine "governor," an invention inspired by instruments employed to regulate windmills. This centrifugal device was designed by James Watt and his associate Matthew Boulton at the end of the eighteenth century and installed on the Watt steam engine—a machine considered as inaugurating a new age and a cornerstone of the industrial revolution. The "governor" made it possible to control the steam inflow (to a valve) by means of a flying ball, and its design was further refined in the following decades. Soon the "governor" not only became an integral part of every stationary steam engine, it also illustrated the undeniable practical benefits of the principle of self-regulation for a larger public. It worked as a feedback device that modifies and adapts its operations in response to certain pieces of information and stimuli. The "governor" exemplified the function that Adam Smith ascribed to the "invisible hand" of the market as it takes care of a dynamic adjustment of variable relations of forces, resulting in the self-regulation of the system as a whole. It contributed to the ideal of a self-organized society that does not follow the orders of a central authority but operates via the checks and balances of its various constitutive parts (Mayr 1986, 164–80; Seibel 2016, 59).

The technological ideal of self-regulation was at the heart of the liberal concept of order. From the end of the eighteenth century on, the well-organized state is no longer imagined as a timeless clockwork but rather in the form of a dynamic feedback mechanism exemplified by the steam-engine governor. Instead of seeking the perfect arrangement of the state, which is conceived of as an invariant system whose order is eternal, the problem of governing contingencies emerges: a dynamic and infinite process that necessitates continuous attention, permanent registering, and flexible reactions. Given the fundamental impossibility of supervising the totality of economic processes and social practices, the liberal state then no longer draws on transcendent principles but reinvents itself as "the regulator of interests" (Foucault 2007a, 346; Seibel 2016, 57–61).[19]

It is by no means accidental that Foucault's lectures at the Collège de France in 1978 and 1979 focused on liberal authors of British provenance. The metaphor of "balance" was employed to a degree in the philosophical, economic, and political literature in Britain that was never reached in Continental Europe. However, liberalism did not invent machines with a feedback mechanism. Their history goes back to classical antiquity, which used "liquid-level regulators like the float valves in automobile carburators and in bathroom water tanks" (Mayr 1986, 190). The first modern self-regulating system invented originally in modern Europe was probably the thermostatic regulator designed by Cornelis Drebbel at the beginning of the seventeenth century to maintain constant temperatures in chicken incubators. Methods for regulating windmills by avoiding excessive speed followed in the eighteenth century (ibid., 190–93).[20] As Mayr demonstrates, feedback devices were cultivated and further developed in Britain while they were largely ignored in Continental Europe well into the eighteenth century. The principle of the feedback loop has not only informed material practices, giving rise to ever more refined technical devices, but also philosophical debates and economic thought. It promoted ideas of self-regulation, and dynamic equilibrium that inspired some of the basic conceptual innovations of political and economic liberalism: "the 'checks and balances' of constitutional government and the 'supply-and-demand' mechanism of the free market" (ibid., xviii).

While the invention and improvement of feedback systems in technological practices helped to overcome traditional inflexible modes of control in eighteenth-century Britain, the principle of *laissez-faire* operated as a critical yardstick for evaluating and finally superseding mercantilist policies. Such mechanisms operated in different domains and often developed independently from each other. However, there still seems to be substantial common ground between technological concepts of feedback and social and economic ideas of self-organization: "In all these cases, systems controlled by a rigid program began to be replaced by systems with the property of self-regulation, capable of maintaining their own equilibrium without external direction through suitable arrangements of their internal processes" (Mayr 1971a, 22).[21]

Technologies of Security

To address this "simultaneous emergence" (Mayr 1971a, 22) of feedback mechanisms in philosophical writings and practical applications, in economic thought, and in technological devices, Foucault introduces the concept of "technologies of security" (2007a, 59) as a distinctive feature of liberalism.[22] Technologies of security operate "within reality, by getting the components of reality to work in relation to each other" (ibid., 47). They are not limited to regulating "men's behavior" and do not discriminate between acting on humans and on nonhumans, but rather address the "interplay of reality with itself," going beyond a narrow understanding of "matter" (ibid., 47).

Foucault distinguishes three important dimensions of technologies of security within liberal government. First, they consist in "the management and organization of the conditions in which one can be free" (Foucault 2008a, 63–64). Technologies of security are designed to "let things happen" (2007a, 45) by determining the requirements and circumstances under which circulations and exchanges can take place. They seek to protect the permanently endangered naturalness of the population, as well as its intrinsic forms of free and spontaneous self-regulation.[23] Thus, technologies of security represent the very opposite of the disciplinary system. While the latter assumes a prescriptive norm, the former take the empirically normal as a starting point. Technologies of security replace disciplinary constructivism, which assumes "an empty, artificial space that is to be completely constructed" (ibid., 19), with a pragmatic realism, working on a "number of material givens": "flows of water, islands, air, and so forth" (ibid., 19). Also, they operate by a "progressive self-cancellation of phenomena by the phenomena themselves" (ibid., 66). Rather than adjusting reality to a predefined should-be value, they seek to adapt the regulatory efforts to the differential normalities that characterize the governed reality. They do not draw an absolute borderline between the permitted and the prohibited, but rather specify an optimal middle within a given spectrum of variations (ibid., 55–63; see also Terranova 2009).

Second, technologies of security rely on different domains of scientific knowledge and technical expertise. They operate within the space of the uncertain and contingent, aiming at anticipating and controlling future events by a "rationalization of chance and probabilities" (Foucault 2007a, 59). Technologies of security address "series of events or possible elements, of series that will have to be regulated within a multivalent and transformable framework" (ibid., 20). Also, they exhibit a "centrifugal" tendency that integrates an increasing number of elements, "allowing the development of ever-wider circuits" (ibid., 45). Within this new governmental rationality that emerged in the eighteenth century, statistics is transformed from a science of the state to a "main technical factor" (ibid., 104) that focuses on the mathematical distribution of events, e.g., as the average rate of diseases, accidents, births and deaths.[24] The dynamics of the population as well as the determinants of the economy became visible as empirical regularities (ibid., 104). These technical instruments of statistical calculation and mathematical quantification also made it possible to supervise and evaluate governmental practices in terms of their efficiency and effectiveness, submitting them to a cost-benefit calculus that critically examines investments and interventions to determine their success and failure (Seibel 2016, 61–64).

Third, technologies of security conceive of "natural processes in the broad sense" (Foucault 2007a, 45) as an ensemble composed of interacting elements that are not valued as either good or bad in themselves, but understood as governed by an internal logic and open-ended dynamics. They focus on relations between elements instead of working on isolated entities (see ibid., 47), seeking to adjust and accommodate the assembled reality to achieve certain ends. Technologies of security pay attention to the "poly-functionality" (ibid., 19) of elements that might be used for different and contradictory purposes and strategies. They aim at coordinating processes of self-organization—articulating a second-order regulation or a regulation of self-regulations (Klauser et al. 2014, 873–74).[25]

Let us briefly note that Foucault's concept of technologies of security endorses an understanding of the governor as both a directing political position and a material entity for regulating technical systems (see Seibel 2016, 22). In the following section I identify an important modification and extension of the idea of feedback mechanisms in the nineteenth century, which will finally allow for the emergence of cybernetic forms of government that rely on communication, command, and control.

From Physics to Biology and Beyond: Towards a Cybernetic Government

According to Foucault, the government of things is defined by a relational and reflexive mode of power that takes into account mechanisms of self-regulation and self-control. It operates as "a set of actions upon other actions" (2000b, 341)—without requiring that the governor or the governed are human beings. In a paper entitled "On Governors," published in 1868, the British physicist James Clerk Maxwell introduced an important innovation into the debate on social and technical feedback mechanisms.[26] Maxwell's interest in governors was motivated by the issue of dynamic stability, and he proposed to rigorously distinguish them from "moderators." He suggests that most machines that were formerly known as governors—like Watt's centrifugal device—should rather be conceived of as moderators. Maxwell observes a common limitation of all these mechanisms: they are unable to maintain constant speed, as their corrective action (e.g., increase in resistance or reduction of steam supply) is directly proportional to excess speed. Genuine governors must, according to the criteria Maxwell advocates, possess an additional mechanism that enables them to translate any output error into a corrective action that increases steadily until the output error has entirely disappeared.

Maxwell specifies that while a moderator acts "directly on the machine," a governor "sets in motion a contrivance which continually increases the resistance as long as the velocity is above its normal value, and reverses its action when the velocity is below that value," and it "will bring the velocity to the same normal value whatever variation [. . .] be made in the driving power or the resistance" (Maxwell 1868, 271). While both regulate, moderators only make it possible to slow down a machine without actually governing it. The governor, by contrast, "instead of being applied directly to the machine, is applied to an independent moving piece" (ibid., 274). This produces the astonishing result that "the position of the machine is the same as if no disturbance of the driving power or resistance had taken place" (ibid., 275). Governors in this reading are machines within machines, and they operate by and as an autonomous intermediary that directs the machine instead of slowing it down (Dotzler 2004, 181–83).[27]

Beyond physics, the principle of feedback control also inspired biological theory and evolutionary thinking in the nineteenth century.[28] Around the same time as Maxwell wrote his classic text on governors, the French biologist Claude Bernard introduced the concept of a regulatory mechanism that controls the vital functions of an organism and maintains stability in health (1957 [1865]). As Canguilhem shows, Bernard takes up the old Hippocratic idea of some kind of spontaneous mechanism or natural medication that corrects or compensates for the illnesses that might affect an organism. According to this reasoning, "there is in every organism an inborn moderation, an inborn control, an inborn equilibrium" (Canguilhem 2012, 72)—a regulatory apparatus that physiologist Walter B. Cannon described in a book published in the first third of the twentieth century with the programmatic title *The Wisdom of the Body* (Cannon 1963 [1932]). Cannon also introduced the scientific term "homeostasis" to account for the capacity of the organism to react and adapt to changing physiological conditions (Canguilhem 2012; see also Tanner 1998; Rieger 2003).

Bernard's and Cannon's works exposed the idea of a "machine within a machine," as they build on physiological knowledge that there "exist organic functions whose purpose is to control other functions and thus, by regulating certain invariants, to enable the organism to comport itself as a whole" (Canguilhem 1988, 82). Thus, technological machines with feedback control found their counterpart in the concept of a self-regulating body that ensures homeostasis. This conceptual transfer across disciplines inspired the development of a general theory of regulation, which sought to describe dynamic processes in very diverse fields of knowledge from biology and engineering via politics to economics. It ultimately gave rise to the cybernetic revolution after WWII.

It was Maxwell's crucial conceptual distinction between moderators and governors that paved the way for imaging cybernetic machines by aligning communication and control. It made it possible to design flexible technological systems that self-adjust without any need for a steering intervention from outside. Maxwell's concept of a governor that relies on channels of communication within the technological system acquired significant relevance with the emergence of the computer and electronic devices. Norbert Wiener, who coined the term cybernetics to designate a new domain of knowledge after WWII, explicitly credited Maxwell's text as a pioneering work (Wiener 1948, 19).

While the proponents of cybernetics discovered Maxwell's paper nearly a century after its first publication, the term already figures in a text the French physicist André-Marie Ampère published in the first half of the nineteenth century, entitled *Essai sur la philosophie des sciences, ou exposition analytique d'une classification naturelle de toutes les connaissances humaines* (Ampère 1834). In this essay Ampère calls for a new science that he terms "Cybernétique," "deriving from the word kybernētike that was initially used narrowly to denote the navigation of a vessel, but already with the Greeks assumed a much wider meaning of an *art of government in general*" ("l'art de *gouverner* en general") (Ampère 1834, 140–41; emphasis in original). The ambition of the essay is to classify the totality of human knowledge of its time. Within this broader context, Ampère links cybernetics to a "theory of power" conceiving both as dimensions of "proper politics" ("politique proprement dite"). Politics thus understood consists in permanent regulation, specifying and redefining the objectives and taking up the knowledge of humans and things to arrive at a form of government that operates self-referentially and immanently. Cybernetics here defines a general form of government that seeks "to examine the different systems in relation to the principles on which they rely" ("à examiner les différents systèmes relatifs au principe même sur lequel il repose") (ibid., 141). Thus, for Ampère politics operates by regulating complex systems of humans and things, of interacting forces and reciprocal causalities that need to be governed to minimize crises and to lead to "an improvement of the social" ("amélioration de l'état social") (ibid., 141; see Vogl 2004, 67–68; Wolf 2008, 462–65).

Foucault does not engage in depth with the history of cybernetics or cybernetic theories and concepts. He only briefly discusses the meaning of *kubernētēs* in some of his lectures at the Collège de France. The lectures of 1980–81 focus on the Christian techniques of the spiritual direction of the soul. Foucault mentions that the *kubernētēs* is conceived as a "governor," but this governor is "(not so much]) someone who guides the conduct of the one being directed according to a precise and considered technique, [as] his alter ego, his representative, witness, guarantor, and surety before God and with regard to God" (2014a, 256). Still, Foucault was well aware that this form of spiritual guidance was linked to a more comprehensive meaning of *kubernētēs* designating the person responsible for the direction of a ship.[29] As we saw in Chapter 4, the navigation of a ship (e.g., the polis or the church) is a metaphor often used in classical texts to address the challenges of governing complexes of humans and things. Foucault takes up this nautical image in his lectures at the Collège de France of 1981–82, discussing the idea and the implication of "piloting" (*pilotage*). He points out that "piloting" informs "three types of techniques [. . .]: first, medicine; second, political government; third, the direction and government of oneself. In Greek, Hellenistic, and Roman literature, these three activities (curing, leading others, and governing oneself) were regularly analyzed by reference to the image of piloting" (Foucault 2005, 249). Foucault is probably referring here to the multiple connections between the arts of medicine, navigation, and politics described in the writings of classic antiquity, for example in the texts of Plato, Hippocrates, and Quintillian (see ibid., 267, note 7).[30]

While these references appear marginal and the discussion unsystematic, it is sensible to expand on Foucault's engagement with cybernetics to further flesh out the analytical frame of a government of things.[31] As we will see in Chapter 8, the concept of "environmentality," proposed in his analysis of neoliberal governmentality, provides a useful tool for examining contemporary modes of neocybernetic control. Before we turn to this, the next chapter discusses Foucault's understanding of the milieu, which cuts across ontological distinctions between human and nonhuman or organic and inorganic.

6

Beyond Anthropocentric Framings

Circulating the Idea of the Milieu

I think that we arrive at this idea that is essential for the thought and organization of modern political societies: that the task of politics is not to see to the establishment within men's behavior of the set of laws imposed by God or necessitated by men's evil nature. Politics has to work in the element of a reality that the physiocrats called, precisely, physics, when they said that economics is a physics. When they say this, they are not aiming so much at materiality in the, if you like, post-Hegelian sense of the word "matter," but are actually aiming at the reality that is the only datum on which politics must act and with which it must act. (Foucault 2007a, 47)

The notion of the milieu is becoming a universal and obligatory mode of apprehending the experience and existence of living beings; one could almost say it is now being constituted as a category of contemporary thought. (Canguilhem 2008b, 98)

Donna Haraway has argued that Foucault's critical project did not sufficiently destabilize anthropocentrism and remained confined to analyzing relations among humans. According to this extremely influential account, Foucault's work is seriously hampered by a "species chauvinism" (Haraway 2008, 60; see also 2012, 107) that curtails its analytic and critical value. In a similar vein, Nicole Shukin has claimed that Foucault's insight into the working of biopolitics "bumps up against its own internal limit at the species line" (Shukin 2009, 11). In this perspective, concepts like governmentality and biopolitics are centered on human populations, unable to address the complexities of human-nonhuman relationships or the impact of governmental rationalities and technologies beyond human collectives.

This chapter will advance an alternative reading of Foucault's work. It builds on elements of his writings to spell out the contours of a more-than-human concept of biopolitics. I will start by analyzing Foucault's work on genetics and heredity. While some scholars have claimed that his account is based on a developmental understanding of temporality and an organic concept of the body, I will suggest that Foucault actively engaged with contemporary genetics and molecular biology to put forward a material-semiotic concept of life. The second part explores the significance and the dimensions of the notion of "milieu" in Foucault's lectures on governmentality at the Collège de France. After reconstructing his brief genealogy of the term, the third section demonstrates that "milieu" constitutes an integral element in the emergence of a liberal governmentality in the eighteenth century that seeks to govern the aleatory. The next part further investigates the figure of the population, showing how liberal government not only targets collective life but also enacts a "vital power" that draws on and imitates the forces of the living. The fifth section proposes a concept of biopolitics informed by an understanding of the milieu as a medium of government. This non-anthropocentric framing of biopolitics no longer exclusively addresses human individuals and populations, but attends to the co-constitution of humans and nonhumans.

"The Algorithms of the Living World"

It is very instructive to read Foucault's account of liberal technologies of security discussed in the last chapter alongside his writings on the molecular biology of his time, especially his reception of the work of Georges Canguilhem and François Jacob. Foucault sees modern genetics, like liberal governmentality, as characterized by the calculation and control of the aleatory. It is animated by the precarious dynamics Jacob calls "the logic of life." This logic, however, goes far beyond life "as we know it," suggesting new modes of theoretical engagement and empirical analysis: "Listen to the luminous lesson of F. Jacob: 'Life is no longer investigated in laboratories today. One no longer seeks to grasp its contours. [Biology] endeavors only to analyze living systems, their structure, their function, their history [. . .]. To describe a living system is to refer as much to the logic of its organization as that of its evolution. Today it is the algorithms of the living world that interest biology'" (Foucault 1994c, 103; Jacob 1973, 299–300).

While some scholars criticize Foucault's work for advocating the idea of an integral body and a linear reading of historical processes (Haraway 1997, 11–12; see also Barad 2007, 200), Foucault was actually well aware of the limitations of an organic and developmental understanding of life. He argues that "living systems" (Foucault 1994c, 103) no longer subscribe to a "metaphysics of life" (ibid, 103) but have to be conceived of in terms of program and code—categories that transcend neat divisions into organic or inorganic, semiotic or material, artificial or natural. This informational understanding of the body and life also shaped Foucault's account of heredity and genetics, which were recurrent topics from his first writings to his final works. Stuart Elden has pointed out that Foucault took up the question of heredity when he discussed mental illness as far back as 1953, and part of the material found its way into his first book *Mental Illness and Psychology* (1987 [1954]). The theme of heredity continues to be present in his work of the 1960s and 1970s, focusing on issues of racial purification, degeneration, crime, and sexuality, but it was also important in clarifying his genealogical account stressing emergence, provenance, lineage, and birth (Foucault 1998c; Elden 2017, 10). The engagement with genetics also informed the last text Foucault authorized for publication before his death in 1984, a revised foreword to Canguilhem's *The Normal and the Pathological* (1991).[1]

In fact, when presenting his candidacy for the chair at the Collège de France in 1969, Foucault envisioned a comprehensive study of the knowledge (*savoir*) of heredity that would go far beyond the focus on human genetics:

> It developed throughout the nineteenth century, starting from breeding techniques, on through attempts to improve species, experiments with intensive cultivation, efforts to combat animal and plant epidemics, and culminating in the establishment of a genetics whose birth date can be placed at the beginning of the twentieth century. On the one hand, this knowledge responded to quite particular economic needs and historical

conditions. Changes in the dimensions and forms of cultivation of rural properties, in the equilibrium of markets, in the required standards of profitability, and in the system of colonial agriculture deeply transformed this knowledge; they altered not only the nature of its information but also its quantity and scale. On the other hand, this knowledge was receptive to new developments in sciences such as chemistry or plant and animal physiology. (Foucault 1997b, 7)

It is well known that Foucault soon redirected his research focus to the prison system and crime. However, this does not mean that he completely abandoned this project. On the contrary, Foucault's interest in questions of heredity and genetics informed his work on biopolitics and the notion of the milieu. His research proposal very much anticipates Haraway's emphasis on the "breeding system" (Haraway 2008, 53) and its importance for the human-nonhuman interface. It also challenges the interpretation of the concept of biopolitics as "seriously outdated and incapable of taking account of the new technoscientific practices that continually rework the boundaries between the 'human' and the 'nonhuman'" (Barad 2007, 65). In fact, when he directly engages with genetics Foucault comes up with an understanding of the body and life that is very close to Haraway's reading of the cyborg (Haraway 1991, 149–82).

Foucault's first text directly discussing the impact of modern genetics is a review devoted to Jacob's book *The Logic of Life: A History of Heredity*, originally published in 1970.[2] Together with Jacques Monod, Jacob was among the first to call the genome a "program," a code for directing the cell, thereby taking up insights from cybernetics developed by Norbert Wiener and Claude Shannon in the 1950s.[3] From the 1960s onwards, this metaphorical transfer of the informational paradigm to genetics made it possible to comprehend genes as "letters" or "words" of a molecular alphabet. Jacob's work confirmed the idea that hereditary transmission occurred via the communication and interpretation of commands contained in the DNA code (Sarasin 2009, 379–84; Talcott 2014, 263–64).

Foucault's review praises what he describes as "the most remarkable history of biology that has ever been written" (1994c, 104), inviting us to rethink "life, time, the individual, chance" (ibid., 99). For Foucault, genetics gave rise to a revolution in the order of knowledge that proceeded by deception and disappointment: "[M]olecular biology has discovered that connections between nucleic acids and proteins in the nucleus are as arbitrary as a code" (ibid., 100). The review stresses the importance of populations and the series in biological research, going beyond the interest in individuals and singularities: "One must no longer dream of life as the grand, continuous and attentive creation of individuals. One must think the living being as the calculable play of chance and reproduction" (ibid., 103).

This account of genetics also affirms the significance of informational concepts in understanding genetic processes: "There was consultation of the program, sending of instructions via messengers, translation of the instructions, execution of the orders" (ibid., 102). In this reading DNA is a code without semantics, a language without an interpreter, as "the interpreters [. . .] are the reactions themselves: there is no reader, there is no sense, only a program and a production" (ibid., 103). In contrast to genetic determinist accounts which regard DNA as "the book of life" and human (and nonhuman) life as an expression of genomic structures, Foucault puts the accent on the performativity of life, undermining any stable conceptual distinction between nature and nurture or genotype and phenotype.[4] Instead of contrasting language and matter, form and substance, he conceives of genes and cells as material-semiotic entities: they are organic systems as well as "small machines" (ibid., 99) and "calculators" (ibid., 103). This "cybernetic" account of genetics is encapsulated in the surprising idea of "a biology without life" (ibid., 103). Thus, Foucault not only repeats and restates the critique of anthropocentrism already put forward in *The Order of Things* (1970; Sarasin 2009, 384–93); he now also proposes a concept of life that seeks to go beyond organic understandings of bodies and the opposition of vitality and matter.

Some years later Foucault gave this informational understanding of genetics a new twist when he discussed *De la biologie à la culture* by Jacques Ruffié (1976). Foucault's review focuses on what biology has to say about human races—a question that preoccupied him at the time, as he had just finished his lecture series of 1975–76 at the Collège de France. In *Society Must Be Defended* (2003) he investigates the genealogy of the category of race, and argues that in Western societies a particular kind of racism based on biological knowledge emerged from the eighteenth century onwards. In line with Ruffié's argument, Foucault emphasizes that "race" does not correspond to a stable prototype but denotes instead "an ensemble of variations" (Foucault 2014b, 129) within a population. Accordingly, populations are defined by molecular characteristics rather than by morphological features—a fundamental displacement that finally makes it possible to dissolve the idea of "human races" (ibid., 129). For Foucault, the idea of races as "raw and definitive biological facts" (ibid., 129) is misguided. Instead of assuming separate and isolated races with distinctive and invariant features that supposedly constitute humanity in its diversity, he stresses historical variations and mutual dependencies within and between populations—understood as biological aggregates that are made and unmade by evolutionary processes. In this relational reading, modern genetics takes into account historical interconnections to conceive of humanity as "a *pool* of intercommunicating genes,'" in the words of the geneticist Ernst Mayr, whom Foucault quotes approvingly (ibid., 129; emphasis in original). This informational understanding of evolutionary processes enables Foucault to envision an affirmative biopolitics that ultimately breaks with the racist past: "a 'bio-politics' which would not be one of divisions, self-preservation, and hierarchies but of communication and polymorphism" (ibid., 129; Sarasin 2009, 393–96).[5]

The third text that directly addresses genetic knowledge is a modified version of Foucault's Introduction to the English translation of Canguilhem's *The Normal and the Pathological* (1991), entitled *Life: Experience and Science* (1998a [1984]). Here again, as in the review of Jacob's book published nearly fifteen years earlier, Foucault refers to the notion of information to account for then contemporary molecular biology and genetics. He sees the importance of Canguilhem's book (written in 1943 and republished in 1966 in a substantially expanded version) in the way it reformulates in informational terms "the old question of the normal and the pathological": "codes, messages, messengers, and so on" (Foucault 1998a, 475–76). Following Canguilhem, Foucault focuses especially on the notion of error in this understanding of life: "[A]t the most basic level of life, the processes of coding and decoding give way to a chance occurrence that, before becoming a disease, a deficiency, or a monstrosity, is something like a disturbance in the informative system, something like a 'mistake'. In this sense, life—and this is its radical feature—is that which is capable of error" (ibid., 476; Sarasin 2009, 396–402; Talcott 2014, 256–63).

For Foucault concepts and language define relational modes of interaction where living beings and "their" milieu co-emerge: "Forming concepts is a way of living and not a way of killing life; [. . .] it is to show, among those billions of living beings that inform their milieu and inform themselves on the basis of it, an innovation that can be judged as one likes, tiny or substantial: a very special type of information" (ibid., 475; translation modified). As we will see, it is exactly this "cybernetic" understanding of the milieu that goes beyond Canguilhem's account and helps to further elaborate Foucault's idea of a government of things as it cuts across distinctions between human and nonhuman, organic and non-organic.

The Genealogy of the Milieu

Foucault undertakes a brief genealogy of the milieu in his lectures of 1978 and 1979 at the Collège de France. Like technologies of security, the milieu is an integral part of liberal governmentality as it addresses "the problem of circulation and causality" (Foucault 2007a, 21). Foucault argues that the concept and practice of the milieu first appeared in the eighteenth century, and he distinguishes three different elements in its emergence (see de Vries 2013).[6]

First, the milieu makes a technical appearance in Western European towns in the later years of the eighteenth century (see Foucault 2007a, 21). The interplay of different factors—the suppression of the city walls to facilitate trade and economic exchange, the increase in the urban population raising serious health issues, and the challenges of crime prevention—made it necessary to supervise and administer the movements of human and nonhuman entities: "[I]t was a matter of organizing circulation, eliminating its dangerous elements, making a division between good and bad circulation, and maximizing the good circulation by diminishing the bad" (ibid., 18). Foucault notes that the term milieu did not (yet) appear in the programmatic texts and systematic reflections of the architects and town planners of that time; however, the "technical schema" (ibid., 21) of the notion quite obviously informed their practices and the actual modifications of urban spaces "even before the notion was formed and isolated. The milieu, then, will be that in which circulation is carried out" (ibid.; see also Foucault 2000c, 150).

Secondly, the idea of the milieu emerges in physics, operating as an explanatory resource to account for "action at a distance" in the work of Isaac Newton and his followers: "It is therefore the medium of an action and the element in which it circulates" (Foucault 2007a, 20–21; Altamirano 2014). Foucault draws on the work of Canguilhem and his conceptual history of the milieu (see Canguilhem 2008b, 98–120; Foucault 2007a, 27, note 37). Canguilhem shows that while Newton did not use the notion (he employed the term "fluid" instead), it had a strategic importance in the search for a solution to a problem in then contemporary mechanics, the question of how one body sets another in motion without having any direct physical contact (Canguilhem 2008b, 99). According to Canguilhem, Newton came to understand "fluid" as an intermediary element located between two bodies (e.g., light source and eye) that exist in the medium in which they move: "they are situated in the middle of it (*au milieu de lui*)" (ibid., 99).

Newton's mechanistic account dismisses any substantive idea of the milieu as a surrounding or background, in order to understand it in terms of dynamic forces. Thus "the notion of the milieu is an essentially relative one. When we consider separately the body that receives an action transmitted by the milieu, we forget that a *milieu* is a medium, *in between two centers*" (ibid., 100; emphases in original; see also Gabbey 2002). However, Newton's concept of the milieu also exceeds contemporary mechanics insofar as it goes beyond corpuscular categories of knowledge. It incorporates a heterogeneous set of elements into his account of natural phenomena, material as well as immaterial, corporeal as well as incorporeal: "It is not simply the category of matter that loses its integrity here, the categories of mind, and even of life, lose their strict boundaries: after Newton, we no longer have a concept of matter, but only a concept of milieu" (Altamirano 2014, 21; cf. 19–21).

Thirdly, very much like the trajectories of the idea of self-regulation discussed in the last chapter, the mechanistic notion of the milieu was subsequently taken up and transformed in the discipline of biology constituted in the eighteenth century (Foucault 2007a, 20). It became a central concept used to account for the relationship between living entities and their natural habitat While the French biologist Jean-Baptiste Lamarck adopts the mechanical sense of milieu as the totality of forces that act on an organism from the outside,[7] this model of explaining life in terms of a complex of causal determinations originating in its environment is increasingly displaced by an altered meaning of the milieu. This conceives of organisms as actively creating and transforming the milieus they inhabit. Thus, research in biology tends to focus on the interactions of living beings and "their" milieu, e.g., as a precarious balance between the "*milieu extérieur*" of the outer world and the "*milieu intérieur*" of the organism, as Claude Bernard (1957 [1865]) put it in the second half of the nineteenth century. While according to Lamarck the organism is directly shaped by the milieu, Foucault credits Darwin with establishing the population as a kind of mediating agency or a "meta-milieu": a "medium between the milieu and the organism" (2007a, 78). In this reading the effects of the milieu pass through the population (ibid., 77–78), indicating "the difference between a relation of the physical type and a relation of the biological type" (Canguilhem 2008b, 111). While in the former the organism functions mechanically within delimited spaces, the latter takes into account that "it is characteristic of the living that it makes its milieu for itself, that it composes its milieu" (ibid., 111; Altamarino 2014; Muhle 2008, 140–53).[8]

Governing the Aleatory

While Foucault draws on Canguilhem's reconstruction of the notion of the milieu, he also departs from his teacher's historical interpretation as he particularly attends to how the milieu is set in motion within new forms of calculation and control—a perspective that is absent in Canguilhem's account (Sprenger 2019, 78–82). Foucault only alludes briefly to the theoretical debate about the notion of the milieu in physics and biology, as his discussion focuses on the "pragmatic structure" (2007a, 21) it gave rise to. He links the interest in controlling circulations that emerged in the eighteenth century to a new configuration of power that he distinguishes from sovereignty on the one hand and discipline on the other.

As we saw in the last chapter, Foucault conceives of technologies of security as the precondition of the liberal concept of freedom, which ensures "the possibility of movement, change of place, and processes of circulation of both people and things" (ibid., 48–49). While "sovereignty capitalizes a territory, raising the major problem of the seat of government," and while "discipline structures a space and addresses the essential problem of a hierarchical and functional distribution of elements, [...] security will try to plan a milieu in terms of events or series of events or possible elements, of series that will have to be regulated within a multivalent and transformable framework" (ibid., 20). Technologies of security are not guided by a static ideal or closed plan but oriented to an open future that is not exactly predictable or entirely controllable (ibid., 20).

Foucault presents the Western European city as a paradigmatic example of how circulations are fostered and managed from the eighteenth century on within this governmental regime. From the very beginning, he stresses that the issue of circulation poses the problem of governing complexes of humans and things; it comprises a heterogeneous ensemble of material artifacts and discursive arrangements, of artificial infrastructures and natural environments, addressing "the fine materiality of human existence and coexistence, of exchange and circulation" (ibid., 339). By circulation, Foucault understands the "material instruments" (ibid., 325) that enable certain mobilities: roads, rivers, canals, bridges, public squares, etc. However, the term is not restricted to material entities but also includes what Foucault calls "circulation itself": "the set of regulations, constraints, and limits, or the facilities and encouragements that will allow the circulation of men and things" (ibid., 325). Briefly, circulation is conceived of as material as well as semiotic; it is a technical as well as a social issue. Furthermore, circulation encompasses both the mobility of humans and the movements of goods and nonhuman organisms. The milieu systemically integrates the geographical, climatic, or hydrographic conditions of human existence and the coordinates of social life (see Foucault 2003, 245).[9] Thus, the milieu articulates the link between a naturally given space and an artificially constructed space, without systematically distinguishing between them. It is "a set of natural givens—rivers, marshes, hills—and a set of artificial givens—an agglomeration of individuals, of houses etcetera." (Foucault 2007a, 21)[10]

In this understanding, the milieu is more than an "environment," a "background" or a "surrounding" in which individuals and populations live and evolve; it is an interactive space, a relational network that constitutes the elements of which it consists as much as it is itself their endpoint or outcome. While the milieu is the object of regulations and adjustments, it also exhibits self-regulatory capacities that have to be respected and fostered. It defines an "intersection between a multiplicity of living individuals working and coexisting with each other in a set of material elements that act on them and on which they act in turn" (Foucault 2007a, 22). Here, Foucault quite clearly recognizes the idea that agency is not exclusively a property of humans; rather, agential forces originate in relations between human and nonhuman entities.

However, the milieu not only defines a spatial constellation but also re-configures existing temporalities. It relates "to a series of possible events; it refers to the temporal and the uncertain which have to be inserted into a given space. The space in which a series of uncertain elements unfold is [. . .] roughly what one can call the milieu" (ibid., 20). In this sense, the milieu is also "in the middle" between past events and an open future. It is not so much the target or the object of government but rather its "medium" (Balke and Muhle 2016, 18; Sprenger 2019). The milieu defines a spatial-temporal arrangement of a hybrid ensemble of material entities mobilized to control future trajectories and to achieve specific goals (in the case of urban circulation with regard to hygiene, trade, traffic, surveillance, crime prevention, etc.). It articulates "the problem of the treatment of the uncertain" (Foucault 2007a, 11), bringing to light "natural" phenomena that cannot be completely controlled but still possess self-regulatory tendencies that governmental practices could draw on (de Vries 2013).

This spatial-temporal account of the milieu not only makes it possible to address the phenomena of circulation, but also eschews any simple and uni-directional concept of causality or exclusive focus on human agency. According to Foucault, the milieu "is an element in which a circular link is produced between effects and causes, since an effect from one point of view will be a cause from another" (ibid., 21). In other words: what is a cause and what is an effect depends on the mobile and relational network of circulations that makes up a milieu. This concept of the milieu renders problematic any neat or a priori separation of causes from effects, and disturbs linear and straightforward ideas of causality—an account very much in line with Barad's reminder that causal relations do not preexist but rather are produced in agential materializations (see Barad 2007, 236).

Thus, the concept of the milieu attends to the multiple, complex, recursive, and always mobile relations between interdependent elements and entities of all kinds mobilized in "circulations" (see O'Grady 2013,

253–56). Given this relational nature and the contingent arrangements of material networks, milieus are not characterized by fixed ties between singular causes and isolated effects, but rather by a "correlation" (Foucault 2007a, 11) or an "unstable co-causality" (Terranova 2009, 234; O'Grady 2014, 516) as elements and the milieus they inhabit emerge together. Processes of circulation are defined by serialities that lend themselves to the generation of statistical knowledge and the calculation of probabilities, and which make it possible to establish recurring patterns and structural regularities (see Foucault 2007a, 19). The central point here is the intimate relationship between the government of circulations and the production of knowledge that seeks to transform uncertainties into risks and probabilities. But this will to knowledge goes beyond acquiring information on the specific materialities of the milieu and "all who live in it" (ibid., 21); it also actively shapes and transforms the milieu to produce some outcomes rather than others. Thus, the milieu is not a given and pre-existing time-space but the material condition and the technical medium of government: "What one tries to reach through this milieu, is precisely the conjunction of a series of events produced by these individuals, populations, and groups, and quasi natural events which occur around them" (ibid., 21).

Before analyzing how the milieu is designed as a "field of intervention" (ibid., 21) within liberal government, let us first note that Foucault's interest in material circulations and infrastructures is not limited to the lectures on governmentality at the Collège de France.[11] The arrangement of (urban) spaces and architectural designs was a longstanding and lasting interest in his work, the most famous instance of this being the panopticon (see, e.g., Foucault 1979; 2000c). In the early 1970s, Foucault had been working collaboratively with Deleuze, Guattari, and many others on town planning, housing, and related questions (Elden 2017, 168–77; Usher 2014).[12] The problem of circulation already figured in a talk delivered in 1974 (see Foucault 2000c), where Foucault analyzed the field of "urban medicine" that emerged in the eighteenth century. On this occasion, he argued that urban medicine had established a "politico-medical system" (ibid., 146) that aimed at "controlling circulation. Not the circulation of individuals but of things and elements, mainly water and air" (ibid., 147–48): a "medicine of things" (ibid., 150). In his later interviews, too, Foucault stressed the centrality of the problem of circulation. In an interview on architecture, he pointed out how "bridges, roads, viaducts, railways" (Foucault 2000a, 354) have allowed for "a certain allocation of people in space, a *canalization* of their circulation, as well as the coding of their reciprocal relations" (ibid., 361; emphasis in original).[13] It is exactly this interest in "coding" and "canalizing" circulations that characterizes liberal forms of government.

Liberal Governmentality and Vital Politics

As we saw in the last chapter, liberal governmentality operates as "the regulator of a milieu, which involved not so much establishing limits and frontiers, or fixed locations, as, above all and essentially, making possible, guaranteeing, and ensuring circulations: the circulation of people, merchandise, and air, etcetera." (Foucault 2007a, 29). Liberalism endorses the general idea that it is indispensable to let "things follow their course" (ibid., 48). Interventions of any kind, then, have to respect the fact that "reality develops [. . .] according to the laws, principles, and mechanisms of reality itself" (ibid.). Given this interest in controlling circulations, Foucault conceives of liberalism not as an economic theory or a political ideology but as a specific art of governing. It has its target in the epistemic figure of population, and it relies on political economy as the principal form of knowledge.

Thus, the concept of the population occupies a central role in liberal government: "the population and milieu are in a perpetual living interrelation, and the state has to manage those living interrelations between those two types of living beings" (Foucault 2000d, 415–6; translation modified). The population constitutes an "absolutely new political personage" (Foucault 2007a, 67) that was only discovered in the eighteenth century.[14] According to Foucault, it is characterized by a double ambiguity. Apart from operating as a "subject-object" (ibid., 44; 77) in governmental practices (see Chapter 4), it also articulates a specific bio-social status: "while being woven from social and political relations [it] also functions as a species" (ibid., 22). In the texts of the political economists, the population is no longer defined in legal terms "as a collection of subjects of right" confronted with the sovereign will but rather as "a set of processes to be managed at the level and on the basis of what is natural in these processes" (ibid., 70). The population, then, represents a new biopolitical figure, incorporating the "emergence of the problem of 'naturalness' of the human species within an artificial milieu" (ibid., 21–22).[15] This understanding of the population "as a kind of thick natural phenomenon" (ibid., 71) is an integral element of the government of things that operates not above or against nature but rather by and through nature: "We have a population whose nature is such that the sovereign must deploy reflected processes of government within this nature, with the help of it, and with regard to it" (ibid., 75; see also Terranova 2009).

In fact, liberal government not only targets the life of populations but also defines a "vital power" that "exerts a positive influence on life" (Foucault 1978, 137), drawing on and imitating the power of the living.[16] As Maria Muhle argues, Foucault follows Canguilhem's account of the history of biology, which sees the originality of biology as a discipline in the way it establishes living matter as something characterized by the production of specific vital norms—norms that have to be strictly distinguished from the laws of the natural sciences. In this reading, biology breaks with two forms of reductionism that understand the organism either as fully determined by an external, physical environment or as the result of (the Aristotelian idea of) an inner *telos* that is inscribed in every single organism. Canguilhem conceives of the vital norms of an organism as variable and historical. Organisms do not simply inhabit or adapt to their milieu but rather constitute and modify it—thereby transforming themselves, too. Given this relational account of the milieu, Muhle stresses that life is not only the object of governmental power but also its mode of operation; liberalism seeks to govern populations by attending to their "inner" norms (Muhle 2008, 252–60; Quélennec 2011).[17]

The liberal principle of the "natural" indeterminacy of circulations sets clear limits to any attempt to repress or restrict economic processes. However, political economy introduces a very specific concept of nature that no longer conceives of it as "an original and reserved region" (Foucault 2008a, 15). Rather, it discloses "the existence of phenomena, processes, and regularities that necessarily occur as a result of intelligible mechanisms" (ibid., 15). For liberals, nature is not a material substratum to which governmental practices are applied, but rather their "permanent correlative" (ibid., 16). Political economy allows for a "governmental naturalism" (ibid., 61; see also 15) that works on a terrain which co-emerges with the practices of government: "Nature is something that runs under, through and in the exercise of governmentality. It is, if you like, its indispensable hypodermis. It is the other face of something whose visible face, visible for the governors, is their own action. Their action has an underside, or rather, it has another face, and this other face of governmentality, its specific necessity, is precisely what political economy studies" (ibid., 16). Thus, liberalism does not seek to reduce direct state interventions but rather removes certain areas and issues from the political realm—treating them as natural, technical, or economic questions.

The liberal principle of "frugal government" (ibid., 28) is itself an important operator extending and intensifying governmental interventions. The invention of the "free" market or "free" circulations of goods, labor, and capital "required a large political apparatus to render certain circuits possible and other circuits impossible" (Salter 2013, 9). It has nothing to do with "setting free" natural processes or removing artificial constraints; rather, it requires the active construction and willful creation of a system of "governing less in which the negative consequences of the free market, such as food shortages, became nonpolitical problems that the market itself should solve" (ibid., 9).

Foucault's account of liberal governmentality displaces the idea of freedom as a natural fact or an anthropological constant; rather, in liberalism the "freedom of behavior [. . .] serves as a regulator" (2008a, 65). Liberalism sets in motion a very specific concept of freedom that privileges mobility and movement and is utterly dependent on mechanisms of security: "[F]reedom is nothing else but the correlative of the deployment of dispositives of security. [. . .] it is in terms of this option of circulation, that we should understand the word freedom, and understand it as one of the facets, aspects, or dimensions of the deployment of dispositives of security" (Foucault 2007a, 48–49; translation modified). This focus on the government of circulations has two particular analytic strengths. First, it stresses how circulations are systematically linked with rationalities and technologies of controlling and channeling them. Secondly, it goes beyond approaches that explore how some movements are possible or encouraged while others are effectively impossible or prohibited by circumventing the mobility/immobility dichotomy. Foucault's understanding of circulations redirects analytic attention away from (im)mobilities to address the milieu beyond and before concrete movements and circuits take place—a generative matrix that allows them to emerge in the first place, and that operates according to the principle "to make move and let stop" (Salter 2013).[18]

Elements of a More-Than-Human Biopolitics

Foucault's analysis of the strategic role of the milieu within liberal governmentality coincides with an important theoretical shift in his work. In the lectures of 1978 and 1979 at the Collège de France, he defines "liberalism as the general framework of biopolitics" (2008a, 22). This results from the self-critical insight that his analysis in *Discipline and Punish* (Foucault 1979) and *The History of Sexuality, Volume 1* (Foucault 1978) had exclusively addressed processes involving population regulation and the corporeal disciplining of individual bodies.[19] The analytics of government, focusing on circulations of humans and "things," makes it possible to revise the original research agenda. Foucault now not only draws attention to the production of individual and collective bodies but also proposes a more comprehensive concept of biopolitics that takes into account the entanglements of human and nonhuman entities.

This broader understanding of biopolitics materializes in the lectures at the Collège de France in 1978 when Foucault discusses Jean-Baptiste Moheau's *Recherches et considerations sur la population de la France* (1994 [1778]), describing the author as "the first great theorist of what we could call biopolitics" (Foucault 2007a, 22).[20] Moheau's book was one of the first to assert that knowledge about the state of the population is essential for governmental work. It deals with a variety of factors that determine the characteristics of a population, encompassing "physical" elements as well as "political" or "moral" ones without neatly distinguishing between them (Cole 2000, 38–39).

Foucault builds his argument on the central importance of the milieu by quoting approvingly Moheau's insight that government means to "govern the physical *and* moral existence of their subjects" (Foucault 2007a, 23; quotation taken from Moheau, emphasis mine). Although Moheau's book does not mention the term, it still proposes "a political technique that will be addressed to the milieu" (ibid., 23). According to Moheau even climatic or geographical conditions are subject to governmental calculations and interventions, as it is their objective to alter "the course of things" (ibid., 22).[21] It is quite clear that the governmental rationality Moheau proposes is limited neither to the domain of the social nor to directing human beings; rather, it articulates an extensive understanding of "things" and their "circulations" that targets the relations between the natural and the artificial, the human and the nonhuman. In Moheau's political imaginary the sovereign deals less with (human) nature than "with the perpetual conjunction, the perpetual intrication of a geographical, climatic, and physical milieu with the human species insofar as it has a body and a soul, a physical and moral existence" (ibid., 23).

The reformulation of the concept of biopolitics within a more-than-human analytics of government entails two important dimensions. First, we see a move beyond an understanding of biopolitics as limited to the physical and biological existence to the more complex idea of a government of things. The latter acts on the milieu as it provides the "point of articulation" (ibid., 23) between the "natural" and the "artificial," the physical and the moral. It exceeds disciplining and regulating individual and collective bodies to include processes of subjectivation and moral problematizations. In this light, human nature is not an invariant anthropological universal but is differently articulated within practices—practices that are conceived of as more-than-human processes.

There is a second difference between the concept of biopolitics Foucault had favored in earlier publications and the one he sketches in the governmentality lectures. Not only does the government of things relate to the interplay of physical and moral, biological and non-biological issues, but the biological can only play out in a certain milieu. In this perspective, neither nature nor life is a self-evident or stable entity or property. Rather, Foucault refers to "a multiplicity of individuals who are and fundamentally and essentially only exist biologically bound to the materiality within which they live" (ibid., 21). In this perspective, life is not a given but depends on material conditions of existence within and beyond biological processes.[22] This understanding of the milieu not only disrupts conventional ideas of exteriority and externality but also urges humans to recognize alterity in their own humanness. Thus, nonhuman doings are the precondition for humans to emerge and to exist.[23]

The idea of a government of things suggests an extensive understanding of biopolitics that attends to the multifold entanglements of humans and nonhumans. However, the notion of biopolitics has mostly been used in a much more limited sense since Foucault's death, exclusively addressing "phenomena peculiar to the life of the human species" (Foucault 1978, 141). Paul Rabinow and Nikolas Rose have famously argued that the notion of biopower is not applicable in nonhuman contexts, as it pertains to "more or less rationalized attempts to intervene upon the vital characteristics of human existence" (Rabinow and Rose 2006, 196–97; Rose 2001; 2007). According to this interpretation, the operations of biopower rely on modes of subjectivation and forms of biosociality that pertain to humans alone—which means that animals and other nonhumans cannot be subjected to biopower.

In the past twenty years, however, a growing number of scholars have sought to extend the interpretative frame of the concept. In *When Species Meet* (2008) Haraway emphasizes the multiple interrelations of human and animal reproductive practices, showing how the international trade in livestock and the management of breeding practices also shaped the histories of eugenics and genetics (2008, 53). Proposing the concept of "companion species," Haraway points to the unstable boundaries between species, arguing that human and nonhuman life are mutually constitutive and intimately intertwined. Thus, to restrict the operations of biopower to human individuals and populations only revives outdated forms of anthropocentrism and obscures the way in which biopower permanently transgresses the human-nonhuman species divide. Haraway not only rejects the assumption that nonhuman animals cannot be conceived of as subjects (see also Despret 2006; 2008), but also asserts the historical and ongoing interaction between and systematic coproduction of humans, animals, and technologies (Haraway 2008; see also Sarah Franklin 2007).

Pointing to the limitations and weaknesses of animal rights theories, Cary Wolfe's *Before the Law: Humans and Other Animals in a Biopolitical Framework* (2013) offers another way of interrogating human exceptionalism by referring to Foucault's notion of biopolitics. In Wolfe's reading, Foucault's concern with the government of life rather than with human individuals and populations and his focus on "forces" and "bodies" instead of political subjects enables him to go beyond the humanist frame. In a critical dialogue with the works of Agamben and Esposito, Wolfe points to the repressive and deadly effects of this power over life that takes "the animal" as its primary object. He concludes that current practices of factory farming "must be seen not just as political but as in fact *constitutively* political for biopolitics in its modern form. Indeed, the practices of maximizing control over life and death, of 'making live,' in Foucault's words, through eugenics, artificial insemination and selective breeding, pharmaceutical enhancement, inoculation, and the like are on display in the modern factory farm as perhaps nowhere else in biopolitical history" (Wolfe 2013, 46; emphasis in original; see also Shukin 2009; Collard 2012; Asdal et al. 2017).[24]

Similarly, in *Cloning Wild Life: Zoos, Captivity, and the Future of Endangered Animals* (2013) Carrie Friese shows how cloning endangered animals "is embroiled in a biopolitics that links humans and animals through both the traffic in techniques and in corresponding human-animal relations through which both species become together" (2013, 14). Friese points out the technical and semiotic transfers between the practices of cloning endangered animals and human embryonic stem cell research. These practices of co-constitution gave rise to two "different but interrelated biopolitical regimes" (ibid., 14), undermining any attempt to restrict biopolitics to the sphere of human subjects. In addition to Friese's work on cloning animals, scholars have employed the Foucauldian notion of biopolitics to investigate breeding practices (Holloway and Morris 2012), biodiversity and species preservation policies (Youatt 2008; Chrulew 2011; Srinivasan 2014; Braverman 2017), environmental and agricultural management (Lorimer and Driessen 2013; 2016), and the role of databases in governing nonhuman life (Braverman 2014).[25]

While this line of research has provided ample empirical evidence of how productively the Foucauldian conceptual apparatus could be used to study various forms of governing animal and plant life, the perspective advanced in this book differs from this scholarship in one important respect. Even if the conceptual proposal of a government of things also suggests a more comprehensive and complex understanding of biopolitical processes, I am not primarily interested in expanding Foucault's analysis into research areas he did not deal with in his own work. Rather, my reading provides a take on the question of the nonhuman in Foucault that resembles Jeffrey T. Nealon's recent engagement with his work. Instead of "extending his analyses into animal formations and institutions that he *did not* study (contemporary corporate farming practices, genetic manipulation, the companion animal phenomenon, etc.)," Nealon attends to "neglected formulations concerning animality and the emergence of biopower in Foucault's own work" (Nealon 2016, 141; emphasis in original). While other scholars have focused on what they found missing or inadequately dealt with in Foucault's account of biopolitics, Nealon seeks to disclose what is already present but often overlooked in the reception of Foucault's work.[26]

In a similar vein, in this book I draw on conceptual tools and methodological suggestions Foucault put forward to further develop and sometimes distort them, in order to outline an analytic approach that is attuned to new questions and contemporary problems. While it is necessary to extend the "bios" in question from humans via animals to plants, it is even more important to rethink the object and the medium of biopolitics. I seek to spell out the trajectories of contemporary biopolitical practices that draw on a modified understanding of the milieu in the next part of this book.

PART III

Toward a Relational Materialism

As we have seen in the preceding chapters, Foucault's concept of the dispositive and his understanding of technology and the milieu give substance to the analytic frame of a government of things. It successfully disturbs the preoccupation with the guidance of human individuals and collectives and shifts attention to the material infrastructures and political matters informing governmental practices.[1] To be sure, a more-than-human analytics of government is not a project actively pursued by Foucault himself. Neither does it assume a definite form in the studies of governmentality that took off in the early 1990s (see, e.g., Burchell et al. 1991; Barry et al. 1996; Walters 2012). This "direction for research" (Foucault 1981b, 253) developed a broad understanding of technologies and rationalities to analyze a wide range of social systems and political institutions. However, the studies of governmentality so far did not sufficiently attend to the empirical status, ambivalent role, or contested identity of nonhuman entities and material infrastructures or to the "societal dimensions of a 'hard' technical system" (Woolgar and Neyland 2013, 29; see also Agar 2003, 4001; Michael 2017, 76).

To confront and correct this "relative blindness of Foucauldian approaches to the non-human elements in political assemblages" (Nimmo 2008, 91), the next part of this book proposes to align an analytics of government in the wake of Foucault with insights from science and technology studies (STS), especially actor-network theory (ANT) and feminist and postcolonial technoscience.[2] Chapter 7 takes up and extends the call for a "relational materialism" articulated by John Law and Annemarie Mol (Law 1994; Law and Mol 1995; Law 1999; Mol 2013; see also Bodén et al. 2019). Law and Mol draw on STS work to argue that "materials" of all sorts (machines or humans, social institutions or the natural world) do not pre-exist their relations but are interactively (or intra-actively) constituted. This "ruthless application of *semiotics*" (Law 1999, 3; emphasis in original) goes beyond an interest in signs and language to propose that "materials" acquire their properties and attributes by engaging with other materials: a "material semiotics" (Law 2004; see also Haraway 1991) or "a semiotics of things" (Latour 1996, 375). Borrowing from and "bending" (see Law 1994, 96) Foucault's insights, the project of a relational materialism also attends to the strategic dimension of ontologies, suggesting that "strategy is recursively and reflexively implicated in the performance of materiality" (Law and Mol 1995, 274; see also Law 1994, 96).[3]

The alignment of a Foucauldian analytics of government and STS work makes it possible to better grasp the current regime of power. While government has always operated by mobilizing nonhumans and steering socio-material environments, there is nevertheless an important transformation to be observed in the present. As I argue in chapter 8, contemporary forms of government tend to focus on the social, ecological, and technological conditions of life as a privileged field of intervention instead of targeting directly the individual or collective body. They seek to design and modulate distinctive milieus by cutting through ontological divides between the technological and the natural or the human and the nonhuman. The environment is conceived of less as an originary domain or the natural surroundings, and itself becomes a technological project. Stephanie Wakefield and Bruce Braun note that this form of government entails a specific operational mode of "arranging things." It involves "not just the government *of* integrated and highly technologized socio-ecological systems, but government *through* such systems, such that it is no longer clear that government in any way seeks to produce subjects as it did before. Government, from this view, is as much about managing circulation and modulating flows as it is about molding individuals" (Wakefield and Braun 2014, 5; emphases in original).[4]

Following Brian Massumi (2009), Ben Anderson (2010; 2011b), Jennifer Gabrys (2014), and Erich Hörl (2017; 2018), I propose to call this governmental regime "environmentality." In exploring this new configuration of power and expanding on Foucault's diagnosis, I will also extend the critique of a peculiar lack of historical and political sensitivity in new materialist scholarship. I argue that the environmental mode of government deploys essential analytical and conceptual components of new materialist discourse like the focus on the agency of nonhumans and the emphasis on contingency, complexity, and non-determinant nature. Contemporary neoliberal regimes draw on biological, geological, or technological variants of the nonhuman, seeking to secure or promote distinctive forms of human life. As we will see, nonhuman organisms, geological forces, and technological artifacts are neither marginalized and ignored nor conceived of as passive and inert; rather, their "doings" are actively co-opted and captured for a diverse set of political, ecological, and economic strategies.

7

Aligning Science and Technology Studies and an Analytics of Government

This chapter further elaborates the conceptual proposal of a government of things by drawing on and engaging with recent developments in STS to directly address political questions and normative issues.[1] Combining the conceptual and empirical strengths of STS work and an analytics of government makes it possible to arrive at a "more fully materialist theory of politics" (Braun and Whatmore 2010, x).[2] This move contributes to a problematization[3] of politics as an exclusively human domain characterized by conflicts of interest or common decision-making while simultaneously avoiding certain limitations and blind spots of new materialist work, especially the turn to ethical and aesthetic questions (at the expense of political concerns) and the uptake of forms of scientism as a way of bolstering epistemological truth claims.

In the following, I will explore four important dimensions of this theoretical hybrid: the empirical investigation of ontologies, the analytic focus on practices, the normative proposition of a more-than-posthuman account, and the critical preference for an experimental approach to science and politics.

A Turn to Political Nontologies

New materialisms share with STS work and Foucault's analytics of government an interest in ontological questions. However, this common ground gives rise to very different understandings of the "real." New materialist literature is often characterized by a "'reification of inquiry'" where ontology is transformed from "a kind of study to a kind of thing to be studied" (Sismondo 2015, 442). It tends to endorse an "invigorated realism: reality is more real than you thought!" (Woolgar and Lezaun 2015, 465), presenting it in transhistorical, transcendental, and universal terms (Bruining 2013).[4]

In contrast to new materialist scholarship, STS research does not "ground new explanatory agencies in 'matter'" (Woolgar and Lezaun 2015, 466; note 2; see also Mol 2013; Abrahamsson et al. 2015). It also differs from a traditional philosophical account that posits a unitary substance and is concerned with the metaphysical question of the fundamental nature of being. In contrast to this reified and reduced understanding of ontology, work in STS proposes a shift from epistemological perspectivism to ontological multiplicity. Instead of assuming a singular reality on which there are multiple perspectives, this line of research approaches ontological issues by "deflating the original concept, both in making it mundane and by making it multiple" (Sismondo 2015, 442; see also Lynch 2013). STS scholars are not interested in the general question of "being" but are rather concerned with specific processes of "becoming" or situated modes of "doing." They study how realities come into existence and how they are done in practices—mostly without even employing the term "ontology" in their work (see Woolgar and Lezaun 2013, 324; Jensen 2015, 18–20). If they do, they use "the term 'ontology' in this wilfully counterintuitive, playfully anti-philosophical way" (Mol 2013, 380; Woolgar and Lezaun 2015, 465) to empirically investigate its trajectories and transformations.

Ontology in this sense is not limited to discovering or disclosing the real; quite on the contrary, it seeks to "to interrogate the whatness of things" (Woolgar and Lezaun 2015, 465) by exploring how specific ontological entities and realms emerge and how they are stabilized by agentic forces. STS work aims to overcome the modernist split between epistemology and ontology in order to examine how a distinctive set of practices brings about singular ways of being by investigating "mundane, contingent enactments of particular orderings of similarity and difference, without supposing from the outset that the world (or the specific matter of concern) is either unitary or fragmentary" (Lynch 2013, 458; see also Law 1994). This theoretical interest in practices necessitates a metaphorical inventory, conceptual tools, and a methodological vocabulary different from realist or (social-) constructivist accounts. Instead of "representation," "perspective," or "construction," the focus is on "intervention," "performance," and "enactment" (see, e.g., Mol 1999, 77; Mol 2002, 32–33, 41; Law 2009, 151).

Foucault's genealogical work shares the STS focus on practices and the problematization of traditional philosophical understandings of ontology. Both lines of research conclude that the question of the make-up of the world is not a theoretical but a practical issue, open to empirical and historical investigation "from the angle of what 'was done'" (Foucault 1998d, 462).[5] Foucault puts forward a nominalist frame of analysis that expresses a systematic skepticism toward theoretical propositions which posit a universal and coherent unity beneath historical variation.[6] In this light, universals no longer provide the starting point for analysis but appear as the effects of practices—an "ensemble of more or less regulated, more or less deliberate, more or less finalized ways of doing things" (ibid., 463; see also Foucault 1997c, 317). They are not monolithic entities that undergo historical modulation so much as a network of heterogeneous elements that cannot be boiled down to an underlying essence or an individualized species.[7]

Foucault's work starts with the methodological assumption that ontological entities as such, organizing and founding certain practices, do not exist. However, the rejection of classical ontological claims is not coupled with social constructivism or philosophical relativism; rather, it gives rise to a historical and praxeological account that serves to analyze how processes of objectivation and subjectivation are orchestrated in "mutual development and [. . .] interconnection" (Foucault 1998d, 460), that is to say how they come into being via coordinated material-discursive practices. Thus, Foucault proposes a distinctive "choice of method" (Foucault 2008a, 2; see also 1998d, 462) to analytically grasp how dynamic ensembles of matter and meaning emerge—what he came to call "transactional realities" (see 2008a, 297):

> [I]nstead of starting with universals as an obligatory grid of intelligibility for certain concrete practices, I would like to start with these concrete practices and [. . .] pass these universals through the grid of these practices. [. . .] I start from the theoretical and methodological decision that consists in saying: Let's suppose that universals do not exist. And then I put the question to history and historians: How can you write history if you do not accept a priori the existences of things like the state, society, the sovereign, and subjects? (Foucault 2008a, 3; see also ibid., 20)

STS research and Foucault's analytics of government both shift the emphasis from epistemological or theoretical questions to practical issues, multiplying the term "ontology." Or rather: the two lines of research start from "nontology," in order to investigate ontological "constitution," "saturation," or "determination." In Foucault's genealogical work it is exactly this analytic focus on "non-existence" that animates the analytics of government (see e.g. 1998d, 463). The supposition that "universals do not exist" provokes the question of how a distinctive "domain of things" has come into being: "What the conditions of this emergence were, the price that was paid for it [...], its effects on reality" (ibid., 460). In contrast, ANT has for a long time focused on mapping and charting different networks without attending to the question of how they are sustained and animated (Sunder Rajan 2006, 290, note 26; 20). It has analyzed power "chiefly as an effect or emergent property of networks, rather than the means of their production" (Nimmo 2008, 91). ANT has usually investigated the emergence of distinctive realities or forms of world-making, their struggles and successes, without adequately addressing the marginal, the excluded and the failed. It was only recently that ANT took up the "nontological" question more explicitly. As John Law and Marianne Lien note, ANT in its early days did not put forward a concept of ontological multiplicity that also encompassed "not quite realised realities" (2012, 363). The longstanding research focus on emergence has resulted in reproducing and reviving instead of destabilizing and undermining dominant practices of Othering. As Law and Lien stress, it is important to "attend not just to ontologies enacted, but also to their shadowland of alterities" (ibid., 373).

Thus, Foucault's analytics of government and STS both "turn to nontology" (Farías and Waller 2016). There is, however, a crucial difference in how the common theoretical preference for a processual understanding of ontology materializes in concrete methodological and conceptual choices. While STS work examines "empirical ontologies" (Marres 2013, 435; see also Law and Lien 2012), focusing on singular and local practices and how they are stabilized and reproduced in the present (see Sismondo 2015, 445), Foucault explores "historical ontologies" (see Foucault 1997c, 318). The latter's genealogical project investigates comprehensive regimes of practices or "practical systems" (ibid., 317) and large-scale historical transformations.[8]

The analytic frame of a government of things proposed in this book seeks to combine these complementary concerns, to grasp the global within the local, and the macro within the micro.[9] This double-edged "ontology of the present" (Foucault 1994g, 148) enables us to better consider the political dimensions of ontological questions. While it puts the emphasis on indeterminacy and contingency, it also attends to strategic arrangements and recursive patterns. Within this relational ontology it is the regime of practices that establishes modes of existence and not the other way around. In this "derivative" account (see Sismondo 2015, 446), ontologies are nothing more (but also nothing less) than relational effects enacted in practices.[10]

However, as there are multiple enactments of reality, it is necessary to analyze the norms articulated within practices. Mol's recent work (2013) on Dutch diet and eating practices offers one innovative way to address normative questions. She introduces the notion of the "ontonorm" to capture how normativities are embedded in particular dietary techniques and how these practices enact food and the human body very differently. By distinguishing the advice to *mind your plate* from the "marginalized alternative, *enjoy your food*," she inquires "how to value contrasting versions of reality. Which version might be better to live with? Which worse? How, and for whom?" (Mol 2013, 381; emphases in original) Instead of judging reality on the basis of a predefined normative catalogue, this empirical investigation makes it possible to trace how norms come into being as part of practices and to evaluate how they differ (e.g., food as "fuel" vs. "pleasure").

Given the multiplicity of practical enactments, it is possible to endorse some ontological configurations rather than others ("more just," "greener," etc.) as they are always already part of the "unseen and anomalous generativity of practices" (Law and Lien 2012, 373). The empirical sensibility for contingencies and heterogeneities is connected to the political interest in articulating alternative "worldings" (Haraway 2016, 76) or different versions of a "cosmopolitics" that seeks to determine what entities may participate (and how) in the composition of a shared world (Latour 2004b; Stengers 2005; de la Cadena 2010; Blaser 2016; Dányi and Spencer 2020; see also Michael 2017, 125–28). As Mol makes clear, it is precisely the political question of how things come into being (and stay in being) that distinguishes the new materialisms from work in STS: "As long as ontology is taken to be stable and singular, it may either be within reach or out of reach, but *good* and *bad* have nothing to do with it. If, by contrast, realities are adaptive and multiple, if they take different shapes as they engage, and are engaged, in different relations, then questions of ontological politics become important" (Mol 2013, 381; emphases in original; Sismondo 2015, 446; Law 2017).[11]

Rejecting Scientism, Displacing Agency: A Material Account of Relationality

In putting the accent on the praxeological and performative dimensions of ontologies, the conceptual proposal of a government of things advances an alternative understanding of agency. It diverts attention from individual actors and their capacities to conditions of emergence and modes of doing. In an instructive article published more than twenty years ago, Emilie Gomart and Antoine Hennion (1999) explicitly link ANT's interest in networks with a Foucauldian understanding of the dispositive. Gomart and Hennion start by stressing important common points between ANT and some contemporary theories of action (especially ethnomethodology and interactionism). Both lines of research seek to describe human and nonhuman action symmetrically and share an ambition to go beyond worn-out dualist oppositions that still dominate social analysis, characterized by a conceptual framework that juxtaposes agency vs. structure, activity vs. passivity, freedom vs. determination (1999, 222–24). While pointing out common ground and parallel endeavors, Gomart and Hennion still insist that ANT proposes a style of analysis that significantly transgresses the more conventional critique of the subject of action or the focus on distributed agencies that these theories of action endorse:

> ANT cuts a thread that is crucial to the theory of action: the link between action and an (albeit) distributed actor. [. . .] To put it bluntly, the approach does not undo the model of human action, but allows the cognitive capacities of humans to migrate to objects. These in turn become efficient, intelligent, co-ordinated, or "purposive." [. . .] By contrast, ANT seeks to describe the composition of heterogeneous elements in networks which produce emerging action from an indeterminate source. (Ibid., 224–5)[12]

According to Gomart and Hennion, ANT enacts a more radical move compared to even the most sophisticated and original theories of action as it displaces the question "who acts" in favor of asking "what occurs" (see ibid., 225). Instead of just expanding the traditional concept of action (to include nonhuman entities like technologies, animals, etc.), ANT invites us to rethink the conditions and constellations of agency—also raising the question of whether "agency" is the right notion to address the issues at stake. In advocating a theoretical move to overcome the problematics of action, the authors explicitly take up the Foucauldian concept of the dispositive, stressing its productive and constitutive dimensions that go beyond forms of prohibition and restriction: "[T]he power of the 'dispositif' rests on the proliferation of new competencies that it lets emerge" (ibid., 221).

While many new materialists also emphasize the importance of "emergence" and "event," they often remain bound to the problematics of action instead of displacing it. As we have seen, both OOO and vital materialism consider agency as a quality of material existence. While OOO situates agency directly in the intrinsic capacities of objects that allow them to "act," vital materialism considers agency to be distributed across a range of heterogeneous entities and processes, originating in the "intelligent vitality or self-organizing capacity" (Braidotti 2013, 60) of matter. Agential realism is—in spite of its self-designation—more attuned to subverting the agential imagination as it attends to the performative enactments of matter. In line with STS work, Barad conceives of agency in terms of "doing" rather than as "attributes": as "*a matter of intra-acting; it is an enactment, not something that someone or something has*" (2007, 178; emphasis in original). Hence, with the notable exception of agential realism, new materialists seek to revise and expand the traditional concept of agency while leaving intact its crucial premises and preconditions.[13]

Sebastian Abrahamsson and his colleagues (2015) convincingly develop this line of critique in their engagement with Bennett's idea of "thing power." They show that theories which conceive of action as a capability of individual entities are informed by a political narrative that foregrounds liberal freedom and choice.[14] This voluntaristic understanding of action is the counterpart and complement of scientific research methods that stress determinate causal effects. These two accounts mutually constitute and stabilize one another, enacting a hierarchical topography that privileges some materialities over others by isolating "primary entities (effectors, actors) from their secondary contexts (confounding variables, mitigating circumstances)" (ibid., 14). In contrast, Abrahamsson et al. propose to circumvent the link to action in order to focus instead on multiple practices and modes of "doing": "Things, so we seek to stress, neither 'cause' effects nor 'act' all by themselves. Materialities work in concert; they are relational" (ibid., 14).[15]

The link between new materialist concepts of agency and a positivist understanding of science is by no means accidental. While STS work investigates "situated knowledges" (Haraway 1991, 183–201), seeking to destabilize and undermine the sovereignty of epistemological truth claims, new materialist accounts sometimes endorse an authoritative and universal concept of science. The focus is on scientific "breakthroughs" and "discoveries" that provide the background or basis for new materialist thought. This holds true for the role of Bohr's quantum mechanics in agential realism or early twentieth-century biology in Bennett's vital materialism (Lettow 2017, 109–11). For Catherine Malabou it is "the revolutionary discoveries of molecular and cellular biology" (2016, 431) that have opened up new political horizons. According to this reading the recent advances in epigenetics and regenerative medicine articulate the potential of a "resistance of biology to biopolitics" (ibid., 438), as they reveal hitherto unknown forms of transforming and reprogramming bodies. This "surprisingly scientistic" (Willey 2016, 998; 2017) frame of analysis risks fueling familiar narratives of scientific progress in new materialist literature, thus re-erecting rather than challenging epistemological truth claims. Scientific knowledge often serves as a self-evident, true, and irrefutable foundation for formulating ontological and political claims, while its contested and provisional status is rarely acknowledged or analyzed.[16] New materialist engagements with science are characterized by an "'ebullient' mode" (Fitzgerald and Callard 2015, 11) that tends to take theoretical statements and empirical results from science "as more-or-less true—with little contest or context, and in the absence of a sense of the wider, often fierce, epistemological and ontological debates *within* those sciences" (ibid., 11; emphasis in original; Braun 2015, 3–4; see also Bruining 2013, 162–3).[17]

The theoretical hybrid of a government of things not only goes beyond the scientism of many new materialist accounts, it also provides a more convincing understanding of relationality. It diverts analytic attention from simple capacities and inherent qualities to complex networks and arrangements, taking relations to be primary and originary forces instead of secondary or derivate processes. However, this theoretical move does not imply a normative judgment or preference for an "affirmative relationality" (Dolphjin and van der Tuin 2012, 127) as it is important "not to lapse into a political romanticization of relation" (Hörl 2017, 7). The shift from individual actors and their capabilities to the dispositives that allow for certain emergences rather than others does not mean that relations are per se "good" (or "bad"). As evaluative processes can only take place once the results of empirical investigations are available, the question cannot be answered beforehand but depends on the materialities of the concrete relations. The analytic frame of a government of things advances a *material relationalism* that differs from OOO's rejection of relationality as well as from vital materialism's and agential realism's "relational enthusiasm" (ibid., 7). Rather than embracing relationality as such and conceiving of relations as something with a fixed and stable normative value, the analytics of government advocated in this book suggests a mobile and material understanding of relationality. This makes it possible to address the following questions: how do particular entities emerge, in what contexts do they operate, what effects do they (co-)produce? In this relational-materialist account relations are evaluated in terms of their material texture as they might strategically mobilize and exploit distinctive alliances.[18] As Erich Hörl notes, contemporary governmental practices and technologies "reduce, regulate, control, even capitalize relations to an enormous extent" (ibid., 8). Referring to Nigel Thrift's concept of "augmented relationality" (Thrift 2008, 165), he diagnoses a "neoliberal-capitalist destruction of the relation" (Hörl 2017, 8) that reterritorializes relations to "calculable, rationalizable, exploitable ratios" (ibid., 8).[19]

So, analytically and normatively it is important not to generally oppose a sensibility for relationality to essentialist accounts but rather to empirically investigate the specific forms these relationalities assume.[20] As we saw in Chapter 4, Foucault's notion of the dispositive explicitly addresses the strategic dimension of these compositions. It also cautions against the theoretical claim to be "(re)turning to the material" or "bringing the material back in." This new materialist call is, seemingly paradoxically, articulated in a historical situation where the material in contemporary regimes of government is less and less conceived of as "inanimate stuff" (Bennett and Loenhart 2011, 2) or as "passive and immutable" (Barad 2003, 801), but rather as "vibrant matter." Current governmental practices draw on and enroll the "agency" of non-human nature in processes of "digitalization," "molecularization," and "informatization" (Anderson 2011a). As Andrew Barry has stressed, one central dimension of contemporary modes of government is to assess the performativity and "eventfulness" of materials, seeking to monitor and manage them. In this light, the focus on material agency does not go beyond human control and governmental technologies; rather, it is a central element in them. Barry gives two "broad reasons" (2013, 14) for this new interest in the "doings" of matter, which are bound up with the operational mode of current dispositives:

> First, in the context both of the escalating costs of raw materials and energy, and of the demands of consumers and industry, there is an abiding emphasis on assessing the *performance* of materials [. . .]. Second, material assemblages are the object of a growing range of *regulatory* requirements governing such issues as environmental waste, biosecurity, safety and energy use [. . .]. The production of information about materials is therefore intimately associated with the growth of national and transnational regulatory zones, regimes that govern, measure and monitor the impact of materials on both persons and the physical environment. (2013, 14; emphases in original)[21]

One central contribution in STS that brings together empirical investigations in material agencies and an interest in governmental practices is Steve Woolgar and Daniel Neyland's work on *Mundane Governance* (2013). In several case studies, these authors analyze how practices of governance and accountability connect to ordinary activities such as waste recycling, car driving, and passing through airport security, especially addressing the objects and technologies implicated in these processes. The book states that "STS inclined re-conceptualizations of objects and technology can offer new understandings of the nature and practice of governance" (Woolgar and Neyland 2013, 3). Woolgar and Neyland propose an innovative concept of governance attentive to ontological concerns that takes up an STS-based conceptualization of objects and technologies. In order to understand governmental processes, they argue, it is necessary to "focus on political constitution at the level of ontology" (ibid., 21). Their objective is to develop "an ontologically sensitive analytic framework" (ibid., 21) that investigates how "various people, things, processes [are] drawn together or even constituted through governance relations" (ibid., 30).

Woolgar and Neyland convincingly show that the process of ontological constitution is about how an entity comes to have and maintain certain properties or characteristics. They propose a complex conceptual framework affirming that the distinction between the human and the nonhuman is part and parcel of processes of ontological determination. However, despite these programmatic considerations, the narratives Woolgar and Neyland present are largely animated by human action and focus on the social domain. While they are right to argue, against new materialist scholarship, that "current emphases on materiality tend to bestow entities with a form of agency, which distracts from an investigation of how entities get to be material in the first place" (ibid., 37, note 11), the contribution of "things" to ontological constitution remains an open question in their account. The insight that the ontological status of entities is not a given but a practical accomplishment, shifting the accent from "being" to "doing," is not accompanied by any analysis of how the (nonhuman) entities contribute to this process of achieving and maintaining ontological stability and durability (Marlin 2014).

This problem points to the role of nonhumans within an analytics of government—a question I will address in the next section.

A More-Than-Posthuman Account

The analytics of government proposed in this book differs substantially from how many strands of new materialism (and beyond) put forward posthumanist commitments. While the rejection of anthropocentric modes of thought is certainly a theoretical accomplishment, there are at least three caveats that need to be addressed in advancing posthumanist claims.[22] They relate to historical, analytic, and normative questions closely associated with a more-than-human agenda, but are rarely addressed in new materialist scholarship.

The first challenge concerns the understanding of posthumanism as a decisive historical break. In this reading the contemporary world is composed of material objects, hybrid networks and fluid identities that imperatively demand a different mode of theorizing. Accordingly, many commentators refer to the posthuman as a specific historical condition that is bound to a theoretical reorientation and readjustment (Braidotti 2013; Åsberg 2018). However, conceptualizing posthumanism as a historical moment after the dissolution of "Man" inadvertently reintroduces the human as a once stable category. Ironically, this understanding of the posthuman condition thus tends to reaffirm the concept it seeks to undermine. The categories of the human and the nonhuman, nature and culture, continue to operate unacknowledged even in accounts which proclaim their implosion or erosion. They are evoked as "true descriptors of real distinctions" (Castree and Nash 2006, 502) that actually worked in the past, producing the nostalgic image of a holistic and stable world before the advent of technoscience and the practices of fragmentation, disruption and recombination that arrived with it. As Sarah Whatmore reminds us, we have to be cautious concerning the seductive powers of the "post" tag. Her conceptual preference for the "more-than-human" over the "posthuman" is based on the observation that the analytic interest is in "what *exceeds* rather than what comes *after* the human" (Whatmore 2004, 1361; emphases in original).

Similarly, Bruce Braun (2004) has alerted us to the complicities the excitement about having entered a new age or the shift from a natural to a *"postnatural* world" (Åsberg 2018, 186; emphasis in original) entail. He argues that humanisms and posthumanisms often remain within the same analytical and theoretical horizon, even if they differ in their normative evaluations: "What often goes unnoticed is that by historicizing the posthuman we end up recentring the human: the human is that being that 'once was,' but which has been 'eclipsed' or 'transcended.' Here's the crux: such posthumanisms *require* the figure of 'Man'[. . .] In this sense, posthumanism's fevered celebration of the posthuman is *of a kind* with humanism's mourning of the passing of 'Man'" (Braun 2004, 1354; emphases in original). Thus, the posthuman condition is a theoretical and political gesture instead of a historical event. It does not mark an ending or closure but rather an unfinished project and an incessant opening.

Secondly, there is also an analytic problem to be addressed. The historical framework of rupture and succession is often coupled with a rigid and hierarchical opposition between humanism and posthumanism. Since new materialist accounts are gaining more and more currency within and beyond academia, humanism now tends to be conceived of as a theoretically flawed and finally untenable position. As the proposal to place "Man" at the center of all deliberations is intimately tied to the legacies of colonialism, patriarchy, and capitalism, humanism is regarded "as at best naïve and as at worst dominating and tyrannical" (Murdoch 2004, 1356). In new materialist discourses (and well beyond them), humanism is generally pitched against posthumanism. Apparently, we are confronted with two antagonistic and mutually exclusive frames of analysis.

This binary constellation is unfortunate and unproductive—especially given the new materialist claim to question and transgress dualisms. There is a risk of losing sight of many critical and analytical resources available within humanism once we move to the posthumanist condition. Instead of enforcing borders and sticking to simple oppositions, it is more promising to explore common ground. In fact, it might be the case that "posthumanist critics and commentators have been rather too hasty in jettisoning humanistic perspectives. In consequence they have paid too little attention to critical and emancipatory practices that occur *inside* humanism. In this view, the posthumanist condition can best be understood by *working through* humanist discourses" (Murdoch 2004, 1357, emphases in original; Badmington 2004).

This endeavor of working with and through humanist concerns instead of against them also gives more substance to the critique of anthropocentrism that is often expressed as a very abstract and general charge, without linking it to the problem of Eurocentrism or to postcolonial debates on alterity. As a result the debate tends to homogenize "Man," thereby ignoring the internal fractures and fissures of the concept. Thus, power asymmetries, forms of domination and social inequalities within "the human" are rarely addressed as the focus of interest shifts to human-nonhuman entanglements or assemblages (Lettow 2017, 111; Braunmühl 2018; Neyrat 2018, 19–20). Again, the "Man" that is to be left behind and superseded by a posthuman imagination is inadvertently reaffirmed as something solid and stable (Meißner 2013, 165–66; Garske 2014, 122–24; Ellenzweig and Zammito 2017, 10–11).

The theoretical shifts suggested so far—the critique of the understanding of the posthuman condition as an historical event or decisive break, and the proposal to conceive of humanism and posthumanism as complementary or mutually corrective instead of alternative or exclusive ontologies—prove helpful in addressing a third problem within conventional understandings of posthumanism. The turn to posthumanism is sometimes accompanied by a normative egalitarianism that tends to obscure the de facto privileged role and the planetary power of humans to affect other bodies. What is needed is a "strategic anthropocentrism" (Donaldson 2014, 6) that acknowledges the asymmetrically destructive and oppressive power of humans.[23] While it is certainly important to destabilize and abandon the "anthropological matrix" (Latour 1993, 107), a more-than-human analytics of government goes one step further as it affirms the ongoing responsibility of "Man" for endangering living conditions on the whole planet (Coole 2013, 461; Cudworth and Hobden 2015, 144). Instead of erasing the figure of the human, it acknowledges and attends to the complexities of material entanglements while still endorsing the concept of a strong responsibility that accounts for the vulnerabilities, injustices, and hazards humans inflict on both humans and nonhumans.

This normative challenge is only rarely met by new materialist scholars, as they often produce conceptual inconsistencies and voids that cannot be bridged by the theoretical positions they put forward. One example is Bennett's claim that concepts of human agency and consciousness are part of a misguided modernist imaginary that needs to be corrected by an adequate understanding of the distributedness of agency and responsibility. Despite this, Bennett resorts to registers of human decision-making and individual choice when dealing with political responsibility—a move that is inconsistent with the posthumanist agenda she adopts. As we saw in Chapter 2, the vitalist insight into the complexities of collective becomings results in a radically individualist concept of responsibility. According to Bennett, it is perfectly possible to opt out of assemblages we identify as harmful (a "we" that is tacitly addressed as an exclusively human collective) while it is impossible to assign full responsibility to human individuals and their actions (Bennett 2010a, 37–8).[24] To overcome these normative problems, it is helpful to move beyond the humanism/posthumanism divide in order to explore how to articulate, revise, and extend concepts of responsibility in more-than-human practices.

There is another aspect to this problem. Posthumanist accounts not only tend to homogenize the human but also its alter image: the nonhuman. The latter is often negatively defined by what it is not, by differing from the category of the human. Moreover, the reference to the nonhuman tends to obscure the heterogeneity and diversity of the entities assembled under this category. The classification "nonhuman" might refer to technological artifacts or material infrastructures, to rocks or to apes. It covers a highly diverse spectrum of living and non-living, man-made and "natural" entities. Thus, "the 'nonhuman' world [. . .] is inadequately addressed through a terminology that does not make any distinction other than that of human/nonhuman" (Lettow 2017, 112; Stengers 2010).[25]

The conceptual proposal of a government of things addresses these historical, analytical, and normative questions by cultivating a "more-than-posthuman" (Häkli 2018, 8) account.[26] It encourages the development of novel approaches and vocabularies to meet the dual challenge of questioning and decentering human privilege and power, while still acknowledging the specific accountability and responsibility of human bodies. Seemingly paradoxically, it affirms the importance and indeed necessity of human action to change the devastating social and material situation of the world, while rejecting the very idea of action on which these destructive and violent practices rely. It ensures that human beings are made accountable for the domination, deterioration, and suffering they inflict on both human and nonhuman bodies, without understanding the nonhuman as pure resource or passive victim of human action (see Neimanis 2014, 10). The idea of a government of things contributes to imagining a "humanized posthumanism" that seeks to develop "forms of critical reflection (that is, reworked notions of justice, nature, and humanity) that are appropriate for the entangled ecologies in which we now find ourselves" (Murdoch 2004, 1359).

This relational-materialist concept of posthumanism does not seek to erase the figure of the human but honors Haraway's reminder to "stay with the trouble" (2016), simultaneously superseding and remaining faithful to the humanist legacy. Haraway has always been engaged in a critique of anthropocentrism, while at the same time distancing herself from strong posthumanist claims, stressing the political significance of this double negation:

> I never wanted to be posthuman, or posthumanist, any more than I wanted to be postfeminist. For one thing, urgent work still remains to be done in reference to those who must inhabit the troubled categories of women and human, properly pluralized, reformulated, and brought into constitutive intersection with other asymmetrical differences. Fundamentally, however, it is the patterns of relationality [. . .] that need rethinking, not getting beyond one trouble category for a worse one even more likely to go postal. (Haraway 2008, 17; see Haraway 2004, 49; Haraway in Franklin 2017, 50–1; Meißner 2013, 166–7)

The political quandary of this "more-than-posthuman" frame of analysis is well exposed by Astrida Neimanis in her critical engagement with the legal claim of a "human right to water" promoted by the United Nations General Assembly. Neimanis shows that in human rights discourse, a commitment to social justice is linked to an understanding of water as an indispensable instrument and an exchangeable and quantifiable resource for human life and well-being. There seems to be an either-or constellation, as care for human nature seems to trump the concern for watery nature. Thus, to affirm the discourse of human rights risks negating the "doings" of water. Neimanis argues that in order to overcome this confrontational approach, we need to develop new political imaginaries that expand and complement the right to water paradigm by understanding watery nature as active and continuous with human nature:

> It is possible to be critical of the ontological premises and presuppositions of "human rights" without negating the hard-fought political traction that this social justice discourse has at last garnered in a global context. [. . .] We can recognize that human rights are what we "cannot not want," as Wendy Brown (paraphrasing Gayatri Spivak) has suggested, even as we also imagine something more robust, something beyond the limits of our own selves, alongside them. (Neimanis 2014, 24)

What is at stake here is how to tie the care for equitable, more just relations between humans to the concern for a more comprehensive community—a "we"—of which humans are an integral part. This political challenge informs Foucault's call for a "critical ontology of ourselves" (1984b, 50), which marks the "we" as a space of investigation and negotiation. It does not refer exclusively to a human collective but is rather to be conceived of as an ongoing and open project, an indeterminate question or a moving target. As Foucault puts it:

> [T]he problem is, precisely, to decide if it is actually suitable to place oneself within a "we" in order to assert the principles one recognizes and the values one accepts; or if it is not, rather, necessary to make the future formation of a "we" possible, by elaborating the question. Because it seems to me that the "we" must not be previous to the question; it can only be the result—and the necessarily temporary result—of the question as it is posed in the new terms in which one formulates it. (1984a, 385)

Thus, the conceptual proposal of a government of things advances a "critical posthumanism" (Castree and Nash 2006, 502) that attends to the ways the categories of the human continue to inform political and scientific practices. It is not limited to questioning modern conceptual dualisms or anthropocentric practices but "also suggests a new task of tracking new evocations of the human in response to the refashioned entities and rhetoric of technoscience" (ibid., 502). This political project builds on and extends posthumanist concerns within social movements to challenge contemporary technoscientific culture and capitalist ecologies. It counters a mainstream critique of anthropocentrism with an "insurgent posthumanism" or a "political posthumanism" (Papadopoulos 2018, 95), envisioning forms of mobilization and contestation that go beyond political institutions and the social domain. This politics of matter extends traditional forms of resistance and protest, as it seeks to practically engender new modes of human and nonhuman coexistence. It gives rise to "more-than-social movements" that "attempt to create the conditions for the articulation of alternative imaginaries and alternative practices that bypass instituted power and generate alternative modes of existence" (ibid., 198).

Cultivating an Experimental Ethos

As we saw, the analytic frame of a government of things invites a different style of critical investigation that emphasizes the prospects of experimental modes of doing science and politics. The engagement with experimentation has been a longstanding topic in STS, and contributed to shaping the field from its beginnings in the 1970s and 1980s. STS researchers undertook historical studies on the emergence of experimental cultures (see, e.g., Hacking 1983; Shapin and Schaffer 1985), they investigated the role of experiments in scientific controversies (see, e.g., Collins and Pinch 1982), and they conducted ethnographic explorations of laboratory practices (e.g., Latour and Woolgar 1979; Knorr-Cetina 1981; Traweek 1988). More recently, STS work has also analyzed experimental formats of public participation and involvement (e.g., Callon et al. 2009; Marres 2012; Lezaun et al. 2017). As Javier Lezaun and his colleagues note, STS work has provided ample evidence that experiments have a central role in bringing together science, technology, and the public. It has offered "an expansive account of experimentation as entailing not just a distinctive method of scientific inquiry but also a genre, an apparatus, and a particular form of publicity or sociality" (2017, 200–1).

The ongoing emphasis on the experimental in STS articulates the desire to move beyond more conventional political imaginaries to call for "alterontologies" (Papadopoulos 2018; 2010). However, the focus on the experimental dimension in doing science and politics does not mean to negate or denounce critical investigations, as suggested by prominent new materialist scholars (see the Introduction to this book); rather, it reinvigorates and reinvents critical endeavors by stressing their creative, innovative, and affirmative dimensions. The experimental sensitivity suggests more tentative modes of thinking and relating to human-nonhuman encounters and invites us to imagine new forms of cooperation and co-existence.

As I have shown elsewhere (Lemke 2011a; see also 2011c), Foucault similarly proposes to redirect critical attention from categorical judgments to "experimental" practices (Foucault 1997c, 316) that seek to expand and transform existing normative horizons. He suggests shifting the register of critique from moral inquiries to pragmatic investigations that analyze how specific dispositives operate. Thus, the ambition is to provide a map of contemporary topographies of government. This experimental setting cultivates an inventive and explorative orientation that does not negate but, instead, carefully investigates the normative yardsticks that are part of the social and historical reality to which they critically relate: "When I say 'critical,' I don't mean a demolition job, one of rejection or refusal, but a work of examination that consists of suspending as far as possible the system of values to which one refers when testing and assessing it" (Foucault 1988b, 107).

This experimental critique comprises two seemingly contradictory dimensions.[27] Foucault conceives of experience as both dominant structure and transformative force, as existing background of practices and transcending event, as the object of theoretical inquiry and the objective of moving beyond historical limits.It articulates a specific critical gesture or an experimental "ethos"[28] that "put[s] itself to the test of reality, of contemporary reality, both to grasp the points where change is possible and desirable, and to determine the precise form this change should take" (Foucault 1997c, 316). However, the preference for the experimental is less a theoretical choice than the result of historical experiences with traditional forms of critique that claimed to be "radical" or "global" but were often bound to technocratic visions and teleological trajectories of historical progress: "[W]e know from experience that the claim to escape from the contemporary reality so as to produce the overall programs of another society, of another way of thinking, another culture, another vision of the world, has led only to the return of the most dangerous traditions" (1997c, 316).[29]

This local and tentative understanding of critical engagements captures well central aspects of modern experimental practices. As the historian of science Hans-Jörg Rheinberger reminds us, each experimental system is governed by a play of differences and displacements. Taking up Jacques Derrida's work on deconstruction and his notion of *différance*, Rheinberger argues that an experimental arrangement necessarily goes beyond reproduction and stability as it has to allow for the unexpected and unknown. The meticulous design and comprehensive control of the material set-up is mobilized to generate moments of surprise and uncertainty. Hence, an experimental system always contains a certain form of "excess." It has *more stories* to tell than the experimenter at a given moment is trying to tell with it. [. . .] Experimental systems contain remnants of older narratives as well as fragments of narratives that have not yet been told. Grasping at the unknown is a process of tinkering; it proceeds not so much by completely doing away with old elements or introducing new ones but rather by *re-moving* them, by an unprecedented concatenation of the possible(s). It differs / defers. (Rheinberger 1994, 77–78; emphases in original; Rheinberger 1997)[30]

This experimental gesture is especially important for doing science and politics in the Anthropocene within the "collective experiment of climate change" (Gabrys and Yusoff 2012, 18; see also Krohn and Weyer 1994; Clarke 2014).[31] Jessi Lehman and Sara Nelson (2015) state that as the conditions of life on Earth are endangered and uncertainties about future trajectories proliferate, an experimental orientation is as much needed as it is inescapable. They suggest that we should understand this experimental endeavor as a "political project" that addresses the challenge of negotiating pathways and options for experimental design, defining "their goals in and through struggle" (Lehman and Nelson 2015, 445). Contradicting and subverting the seemingly natural self-transformative dynamics within capitalist regimes and technocratic visions of world-making (e.g., geo-engineering or terraformation), this "politics of experimentation" makes room for "experimental practices in different, more equitable ways" (ibid., 446).

This "experimental ethos" (ibid., 446) requires determining who is part of the process under investigation, who is affected by it, and who might possibly benefit from or be negatively impacted by it. This challenge is particularly demanding as these experimental practices are not only confronted with the difficulty of aligning and negotiating human interests and strategies but also have to attend to multiple practices that include nonhuman entities and processes. Instead of engaging with alternative employments or different reconfigurations of human practices, the experimental mode also extends to problems of accounting for responsibilities, alliances, and solidarities of both humans and nonhumans (ibid., 447).[32] Thus, the practice of experimentation requires and results in building a collective that transgresses the boundaries of the human and explores experimental trajectories identified in democratic processes. This critical ethos is sensitive to asymmetric power relations and to forms of domination and exclusion, but it also attends to practices of "counter-conduct" (Foucault 2007a, 201) that contest and challenge existing forms of government—possibly leading to new material configurations beyond capitalist regimes and technocratic imaginaries. However, as experimental projects are open-ended and uncertain, a positive and productive outcome is not guaranteed; experimental encounters may also result in frustration and failure.

This experimental gesture engenders multiple forms of collaborative inquiry and research that undermine dominant thought styles within individual disciplines. The turn to "experimental entanglements" (Fitzgerald and Callard 2015, 16) explores interdisciplinary exchanges and problematizes hegemonic narratives within the scientific disciplines and established divisions of expertise. It takes up the epistemological and ontological controversies within them and connects them to conceptual, methodological and empirical debates and inquiries in other disciplines (Neimanis et al. 2015).[33]

The impact of this "experimental imperative" (Whatmore 2004, 1362) is not limited to academic settings and research institutions. It involves and promotes forms of cooperation across and between scholars and other publics, questioning and transgressing the boundaries of scientific disciplines. It provides a meeting ground for different forms of expertise and audiences, combining critical, creative, and "radical" engagements with the present. Here, Foucault's analytical interest in mundane "arts of existence," political activism and artistic practices aligns with the recent interest of STS scholars in connecting their work with other spheres of public engagement (see, e.g., da Costa 2010).[34] These developments supplement conventional forms of scientific communication that "rely on generating talk and text with experimental practices that amplify other sensory, bodily, and affective registers and extend the company and modality of what constitutes a research subject" (Whatmore 2004, 1362). These conceptual, normative, and affective resources not only open up the boundaries between individual disciplines and research cultures but also those between the scientific and the non-scientific world, making it possible to reengage diverse publics as producers and not only as consumers of scientific research (Neimanis 2015 et al., 88–90). They provide modes of critical interrogation and intervention that foster a productive dialogue with contemporary efforts to actively redistribute expertise, rethinking the role of knowledge and technology and redefining democratic culture.[35]

Again, we note that STS-inspired work on the public and political dimension of material objects differs significantly from new materialist accounts. While object-oriented ontology is content to cherish the weirdness of "objects," which escapes any attempt to control and steer them, STS scholars investigate how material devices, artifacts and settings are implicated in public participation processes and political agendas (Lezaun et al. 2017). Noortje Marres' work on object-centered strategies of "material participation" is crucial in this context (Marres 2012; 2013). Focusing on environmental politics, Marres analyzes how mundane practices like cooking, heating, or gardening become significant sites for engaging with problems like climate change and issues of sustainability. Her case studies, ranging from ecological kettles to eco-show homes, disclose how material artifacts not only mediate political action but have been explicitly designed to enable and enact political participation. This "experimental ontology" operates by a "deliberate investment of non-humans with moral and political capacities" (Marres 2013, 423) and invites us to reconsider political engagement as intimately linked to forms of material participation, technological innovation and social change.

As Lezaun et al. (2017) note, this broadening of the STS interest in the experimental is coupled with a deepening of the field's original commitment to the principle of general symmetry in investigating science and technology. The research focus on material participation opens up prospects for new modes of experimentation with democratic issues and alternative formats as ways of promoting public understanding and engagement with science and technology. Thus, current STS work is characterized by "a double move in relation to experiments and publics": "By scrutinizing the role of experimentation in social and public life, it unsettles the question of how science, technology, and public relate or should relate to one another in contemporary societies. At the same time, STS researchers adopt experiments as a *resource or instrument for social and public inquiry*, developing their own experimental techniques to probe and perhaps even alter the very meaning of democracy in technological societies" (Lezaun et al. 2017, 201; emphasis in original).

However, in examining experimental modes of political participation and involvement, it is helpful to connect STS work on this issue closer to a Foucauldian analytics of government. Such an integral approach provides a more comprehensive picture of the material conditions of experimental practices that systematically privilege some settings and sites over others. As Nicholas Beuret notes, we need to investigate thoroughly the regulatory patterns and governmental regimes that unevenly configure experimental spaces and capacities: "[T]he resources required to experiment, including spaces opened up for experimentation through regulatory and legislative programs, are far from evenly distributed and have a significant effect on the capacity to experiment and produce autonomy" (Beuret 2019). Thus, starting from the principle of general symmetry, the concept of a government of things explores how the conditions for democratic experiments are differentially produced and distributed.

A similar point could be made with regard to Marres' analysis of material participation. While the empirical case studies convincingly show how the devices and objects in question perform political participation, there is little room for theoretical debate about what makes them matter in the first place. As Paul-Brian McInerney observes, Marres' examples all relate to environmentalist concerns, an already well-established political issue, while her analysis does not address the question of "how material objects play a role in creating the political relevance necessary for material participation to emerge" (2014, 717). In this regard, a relational-materialist account is necessary to investigate the constellations and contexts that enable certain issues (and not others) to become political matters at all. It also helps to explore the multiple trajectories of this "ontologization of politics" (Marres 2013, 423). Material devices and objects may shape public deliberations and contestations in very different and sometimes even contradictory ways, and they could have politicizing as well as depoliticizing effects, e.g., by framing public matters as private concerns.[36]

To sum up, the conceptual proposal of a government of things adds a crucial analytic layer to current STS work on experimentation as it attends to the material conditions of "experimental ontologies" (ibid.,

423). Moreover, it invites us to rethink the focus on processes of subjectivation and technologies of the self that is so prominent in the social sciences and in the reception of Foucault's work. It is necessary to shift our attention from the "exclusive preoccupation with the fabrication of particular kinds of political *subjects*" (Marres and Lezaun 2011, 491; emphasis in original) to the analysis of governmental practices that "provide the intelligibility key for the correlative constitution of the subject and the object" (Foucault 1998d, 463; 1984d, 334). This perspective is not limited to investigating how things shape and affect political subjects, spaces or strategies, but "extends to the political capacities of things in their own right" (Marres and Lezaun 2011, 495)—an analytical frame that is particularly useful in exploring contemporary modes of government.

8

Environmentality

Mapping Contemporary Political Topographies

Le pouvoir est devenu materialiste. (Foucault 1994h, 194)
("Power has become materialist.")

The notion of environmentality[1] was introduced by Foucault in his lecture series at the Collège de France in 1978–79 to define an operational mode characteristic of neoliberal technologies of government that were beginning to take shape at the time (2008a, 261). According to Foucault, the term denotes a "governmentality which will act on the milieu and systematically modify its variables" (ibid., 271; translation modified). It seeks to govern the "environment" of human and nonhuman entities rather than operating directly on "subjects" and "objects" (see Anderson 2010, 232; 2011b, 39). As we will see, this environmentality marks a significant transformation of the classical modes of "governing things" as it engages with new technological formats and enacts an altered concept of the milieu.

Importantly, the Foucauldian understanding of environmentality needs to be distinguished from more specific and delimited usages of the term that implement a "specific optic for analyzing environmental politics" (Agrawal 2005, 226; see also Fletcher 2010; 2017; Bluwstein 2017). This body of research investigates questions such as the constitution of environmentally-conscious subjects (Agrawal 2005; Cortes-Vazquez and Ruiz-Ballesteros 2018) or the role of environmental organizations in the management of a sustainable economy (Luke 1995; see also Rutherford 2011). Going beyond this distinctive thematic focus, Foucault's notion of environmentality proposes a much broader understanding of the environmental. It seeks to capture an essential feature of contemporary neoliberal modes of government: the management of "fluctuating processes" (Foucault 2008a, 259). Hence, environmentality in this sense might refer to strategies addressing environmental objectives but is not limited to this particular policy field; rather, it denotes governmental practices that seek to steer and manage performances and circulations by acting on and controlling the heterogeneities and differences that make up a milieu.

The term "environmentality" first appears in Foucault's discussion of the neoliberal account of criminality in the lecture of March 21, 1979 at the Collège de France.² This lecture focuses on the analysis of the Chicago School and its proposal to extend the economic form of the market to the social field in general. In this neoliberal program, Foucault discerns elements of a new governmental rationality that is less concerned with targeting individual behavior or the deviant population but rather seek to alter material conditions and contexts in order to implement regulatory strategies. According to Foucault, this "environmental technology" (Foucault 2008a, 259) no longer pursues the project of a pervasive disciplinary society. Instead, it displaces "normalizing" or "standardizing" technologies (ibid., 261) to promote the "optimization of systems of difference" (ibid., 259) by operating on the milieu of individuals and populations: "[A]ction is brought to bear on the rules of the game rather than on the players, [. . .] there is an environmental type of intervention instead of the internal subjugation of individuals" (ibid., 260; Gabrys 2014, 34–35; Hörl 2017; Sprenger 2019, 82–84).

Foucault promises to discuss this environmental regime in more detail in the following lectures. However, he never returned to the notion of environmentality. A shortened version of the six pages of preparatory notes for the lectures, in which he outlined his understanding of environmentality, is included as a footnote in the book publication of the lecture series (ibid., 260–61). While the term certainly lacks conceptual clarity and depth, it nevertheless remains "a provocation for thinking" (Gabrys 2014, 35). It provides an abstract analytic grid to investigate how current neoliberal technologies enact and rely on environmental modes of government that are "open to unknowns and transversal phenomena" (Foucault 2008a, 261). Indeed, the notion of environmentality captures a new dispositive of "governing things," putting forward a distinctive ecological understanding of the environmental as both technological and natural: "the problem of the milieu to the extent that it is not a natural milieu" (Foucault 2003, 245; translation modified).

This chapter discusses constitutive elements of this contemporary mode of government. After presenting the rise of a resilient biopolitics and a neo-cybernetic regime of control in the first section, the second part of this chapter focuses on the emergence of a new set of political technologies. These technologies comprise practices that set in motion the capacities of nonhumans to recalibrate, on the one hand, dysfunctional socio-ecological systems and, on the other, mechanisms of vital systems security that seek to sustain the conditions for collective life by anticipating emergencies. I argue that taking up and elaborating Foucault's concept of environmentality contributes to a more complex analytic frame of contemporary political topographies. As I show in the last part of this chapter, this project opens up new avenues for analyzing and criticizing capitalist practices, and it also identifies important limitations of the Anthropocene narrative while generating a productive dialogue between more-than-human accounts and those focusing on a critique of political economy.

To address two possible misunderstandings: By presenting the short sketch of environmental technologies associated with a resilient biopolitics, neo-cybernetic control, probiotic ecologies, and vital systems security in the following sections, I do not mean to argue that they exclusively organize and structure contemporary forms of government. The dispositive of environmentality is certainly not (yet) hegemonic, but co-exists with alternative modes of government. However, it is theoretically and politically significant not only for understanding human-nonhuman relations and regimes of control today but also for assessing tendencies and trajectories of future governmental projects and pathways (see also Lorimer 2017, 28). Similarly, attending to the environmental in contemporary forms of government does not translate into a particular emphasis on historical breaks and decisive ruptures. I do not want to suggest that "older" or "outdated" forms of government that operated by normalizing and disciplining subjects or conceived of objects as inert and passive are simply being abandoned and replaced by more refined and indirect modes of governing. Rather, it seems appropriate to account for the simultaneous interplay of plural and heterogeneous dispositives that operate alongside one another, contradicting or complementing (or contradictorily complementing) one another—defining the criteria of "adequateness" or "outdatedness" as part of their own operations. Thus, a historically sensitive and empirically informed analysis is needed that examines the complexities and specificities of governmental technologies, attending to the "overlappings, interactions, and echoes" (Foucault 1978, 149).

Resilient Biopolitics and Neocybernetics

As Braun (2015) notes, new materialists rarely inquire how ideas of non-deterministic nature and complexity relate to political transformations and economic constellations. There is a remarkable historical conjuncture between the neoliberal critique of concepts of stability, linear development, and homeostasis on the one hand and the rise of the new materialisms that embrace ideas of fluidity, non-linearity, and contingency on the other. This observation of a historical parallelism does not mean reducing new materialist concepts to cultural epiphenomena or ideological expressions of an underlying and more fundamental "logic" of capital. While the idea of a necessary and uni-directional causal connection between new materialist commitments and contemporary capitalism would certainly be simplistic and reductionist, a more historical sensibility is still needed if we want to inquire how neoliberal government and new materialisms co-emerge within determinate but contingent historical constellations, and how the former might have captured or absorbed the critical impulses of the latter (Braun 2015; see also Nelson 2014).

The shift toward an environmental mode of government is linked to the rise of complexity science and the proliferation of the ecological since the 1970s. In this context, the concept of resilience plays a crucial role. Originally adopted in the 1970s by the ecologist Crawford S. Holling (see, e.g., Holling 1973), this idea has in the last few decades become "the dominant paradigm for the administration of life" (Nelson 2014, 2). It informs scientific disciplines and policy arenas as diverse as international development, public health, financial regulation, corporate risk analysis, the psychology of trauma, urban planning, and environmental politics. The resilience discourse makes it possible to redirect highly diverse operational procedures and strategic arrangements toward a common horizon and a single analytic frame, subjecting them to the logic and logistics of crisis. It provides a central conceptual key to rethink governmental practices, organizational processes, and institutional settings in order to address events of emergency and methods of crisis management.

The concept of resilience emerged at the historical moment when the apparent political-economic failure of the Fordist regime of accumulation was joined by a growing insight into the significance of the ecological crisis, famously expressed by the Club of Rome's report on the *Limits to Growth* (Meadows et al. 1972). The critique formulated by new social movements and environmentalist groups demonstrated an increasing awareness of environmental degradation and resource depletion. It also highlighted the interconnections of ecological and economic questions, especially the destructive environmental effects of Fordism and the evident regulatory problems of Keynesianism (Cooper 2008, 15–50; Nelson 2015, 466–71). Holling's work and his adoption of the concept of resilience advocated a specific solution to this double crisis. While many contemporary ecologists and economists understood the earth as a closed thermodynamic system, emphasizing the need for a "steady-state economy" (Daly 1980) to reestablish stability in the light of finite resources and environmental limits, Holling suggested a quite different trajectory. Instead of viewing the environment as composed of stable and static entities that evolve along linear and foreseeable developmental pathways, he envisioned it as an integrated but open complex system whose dynamics escapes prediction. Holling emphasized the role of self-organization within systems and their ability to remain cohesive even after experiencing extreme perturbations. According to this approach systems produce order by actively engaging with their environment, maintaining their constitutive relationships and borders by selecting the input they need for their "autopoeisis."[3] Holling put forward an understanding of the environment in terms of an unpredictable and often turbulent ecological system. However, instead of conceiving of this dynamics as a threat to be minimized or eliminated, he proposed a new way of governing that respects and adapts to it, harnessing "the uncertainty generated by non-linear dynamics as a catalyst for innovation and growth" (Nelson 2014, 5).[4]

In problematizing the conventional notion of stability and the normative ideal of homeostasis, the concept of resilience seeks to grasp how systems are able to retain structural integrity and cohesion even while undergoing extreme stress or situations of shock. If stability refers to the familiar idea of a return to equilibrium, ecological resilience designates "the ability of a system to maintain its structure and patterns of behavior in the face of disturbance" (Holling 2010, 76; 1973, 17).[5] Holling advocates a particular style of government that follows from this understanding of resilience. It breaks with the idea of knowledge-based planning that characterized postwar political and economic strategies by stressing opportunism and preparedness instead of prevention and prediction:

> A management approach based on resilience [. . .] would emphasize the need to keep options open [. . .] and the need to emphasize heterogeneity. Flowing from this would be not the presumption of sufficient knowledge, but the recognition of our ignorance; not the assumption that future events are expected, but that they will be unexpected. The resilience framework can accommodate this shift in perspective, for it does not require a precise capacity to predict the future, but only a qualitative capacity to devise systems that can absorb and accommodate future events in whatever unexpected form they may take. (Holling 1973, 21)

The concept of resilience has come to reorient policies as it shifts the problem from the (re-)establishment of stability to the question of how to support and foster adaptive capacities in uncertain ecologies. It provides "a pervasive idiom of global governance" (Walker and Cooper 2011, 144), redefining the problem of security as a matter of flexible adaptation under conditions of irreducible uncertainty and inevitable threat. Confronted with an environment of disruptive events and incalculable risks, the challenge is how to design systems able to adjust to and benefit from future shocks that are in principle unavoidable and unforeseeable. Resilience has become the normative yardstick to measure individual and organizational fitness to adapt to traumatic experiences and turbulent ecologies, as it possesses two important strengths. First, resilience points to the problems and failures of conventional management practices and traditional post-war policies to address socio-economic change by sticking to a seemingly outdated model of homeostasis and by separating ecosystems from social systems. Secondly, Holling also proposes new ways of mobilizing non-linear processes as the very sources of innovation and dynamism (Nelson 2015).

Conceiving of unpredictable trajectories and events of crisis as drivers for economic growth, neoliberal strategies foster the emergence of new markets from catastrophe bonds to weather derivates and hitherto unknown academic disciplines specialized in the production of knowledge about uncertain environmental futures (Cooper 2008; 2010; Walker and Cooper 2011). Furthermore, complexity theory and the concept of resilience were instrumental in reconfiguring environmental functions and services as lucrative sites for capital investment. They set the stage for a form of neoliberal environmentalism that integrates the concern about natural limits and finite resources into the circuits of capital, converting (environmental) critique of capitalist ecologies into a new mode of capitalist expansion: "The construction of nature as a 'service provider' via resilience-based management techniques has enabled a major shift in the way that capital circulates through non-human natures, pricing and exchanging ecological capacities rather than stocks of material resources" (Nelson 2014, 10; Braun 2015, 9). While this market-based environmentalism recognizes the infrastructural and life-supporting functions of animals, rainforests, and watersheds (e.g., by pollinating crops, sequestrating carbon dioxide, or filtering water), it conceives of these capacities as "natural capital"—assets that must be privatized, monetized and commodified in order to address biodiversity loss, climate change, and the depletion of resources. The concepts of "natural capital" and "ecosystem services" have now become an integral part of mainstream economics and environmental politics, giving rise to a diversity of markets, policies, and programs to identify and assess the economic value of ecological work (Walker and Cooper 2011; Battistoni 2017).[6]

While complex systems theory and second-order cybernetics started as a critique of Keynesian economics and Cold War politics that centered around concepts of homeostasis and stability, they soon became an integral element of a new capitalist regime. The advent of resilience theory marked a stark contrast to the postwar constellation in which classical thermodynamics and mechanistic concepts of equilibrium played a central role. After WWII, they provided the model for economic and ecological organization guided by the regulatory idea of a foreseeable and measurable trajectory that results in the return to the status pro ante after perturbation. Holling's work consists in destabilizing and finally overthrowing this familiar narrative. It represents the beginning of a major transformation in which "the figure of the environment shifts: from the harmony of a natural balance to a churning seed-bed of crisis in the perpetual making" (Massumi 2009, 154; Walker and Cooper 2011).

As Melinda Cooper has shown, the imaginary of homeostasis that governed economic theory for a very long time has in recent decades been superseded by very different concepts of economic growth and innovation that enroll elements of evolutionary biology and complexity theory. Classical liberalism conceptualized the economy as a steady movement from one equilibrium to another—a vision shared by Darwin's concept of nature, where natural selection fulfills a uniform regulatory function as an invisible hand to provide stability and adaptive capacities. By contrast, contemporary neoliberalism tends to disentangle economic prosperity and growth from the norm of the equilibrium and steady-state concepts of evolution. Instead, neoliberals endorse a complexity approach to economic and biological evolution that relies on several basic presuppositions: "first, complex systems evolve best in far-from-equilibrium conditions [...]; moreover, such systems evolve most productively when they are free from external regulation—complex systems in other words prefer to self-organize; and finally, although an individual complex system eventually exhausts its possibilities of further differentiation, there is no essential limit to the evolution of complexity per se. In nature as in economics the law of complexity is one of increasing returns punctuated by periodic moments of crisis" (Cooper 2008, 44).

Thus, the neoliberal agenda entails both a departure from the regulatory idea of equilibrium and a reevaluation of crisis. Crisis is no longer necessarily to be avoided, something exceptional and temporary; it is seen instead as a creative resource and essential precondition of innovation and change. It follows that the strategic imperative is less to prevent events from happening but rather to prepare for a future where crisis is ubiquitous and catastrophic outcomes are to be confronted. Contemporary forms of government redefine and redesign technologies of security in the light of all-pervasive risk and ever-present danger. As security is conceived of as both problematic in normative terms, as it tends to restrict innovation and change, and is practically unachievable in the light of multifold challenges and experiences of crisis (from economic crisis to climate change), this new dispositive governs humans and nonhumans without the promise of a future beyond all crisis (Wakefield and Braun 2014, 4–5). Thus, Braun states that "in important respects resilience is the name for our contemporary form of biopolitics. Viewed through the broadest of lenses, we might posit that resilience is a mode of government proper to neoliberalism, in which government seeks not to punish, nor to prevent or discipline, but rather to modulate 'natural processes' by creating a 'milieu'" (Braun 2014, 61).

However, as Sara Nelson rightly notes, technologies of resilience no longer correspond to Foucault's earlier descriptions of biopolitics (2014, 8). In Foucault's classical analysis the two central elements of biopolitical interventions—the disciplining of the individual body and the regulation of the collective body of the population—are characterized by a stable bond. It connects them and keeps them apart as two distinct areas, making it possible to generate norms within the human sciences and new statistical and demographic knowledge. Biopolitics in this sense takes the form of a convergence of disciplinary mechanisms focused on the individual with "regulatory mechanisms" that intervene at the level of the population to "establish an equilibrium, maintain an average, establish a sort of homeostasis, and compensate for variations within this general population and its aleatory field" (Foucault 2003, 246).

By contrast, mechanisms of resilience exploit and foster differences and deviances. They "seek to capitalize on alterity rather than mitigate it" (Nelson 2014, 8). Technologies of resilience no longer claim to predict or prevent, but seek to adapt to and accommodate disruptive future events. Instead of relying on quantifiable rates of occurrence, statistical averages or normal curves of distribution, they refer to qualitative properties, structural patterns, and complex relationalities that define a system. This environmental mode of governing seeks to control the comprehensive natural and technical infrastructures of life, addressing not just biological life but the material conditions required to sustain and foster certain "modes of life" (see Foucault 1997d, 137) or "forms of life" (Foucault 1997e, 164).[7]

Thus, environmentality is defined by the simultaneous proliferation and denaturalization of the ecological (Hörl 2017). While there is today hardly any field or object that cannot be framed in ecological terms, the concept is increasingly decoupled from any reference to nature. This revised understanding of ecology unsettles the difference between nature and technology: "[T]he concept of ecology is pluralized and disseminated; it is outlined and consolidated as the concept of non-natural ecologies; it even mutates into technoecology" (Hörl 2017, 2). This denaturalized and technologized concept of ecology allows for a new "environmental culture of control" that integrates elements that were previously considered to be objects of government, rendering "environmental even what used to be called *Umwelt* or 'environment'" (Hörl 2017, 5).[8]

Hörl distinguishes between "three major phases" (ibid., 9) in the development of this environmental mode of government. The first starts around 1900 and extends to first-order cybernetics after WWII. It relied on the concepts of "control," "information processing," and "communication," and was mainly concerned with the problem of adaptation, equating rationalization with increased control. The second phase initiated by second-order cybernetics in the late 1960s was fueled by innovations in computer science, earth systems research and evolutionary theory. It focused on the problem of learning, especially mobilizing forms of self-control and autopoiesis. The third, neo-cybernetic, phase began around the turn of the millennium, significantly extending the concept of the "ecological" that now comprises both the natural and the technological sphere. This phase is characterized by the rise of a new set of technologies that seek to measure and control environmental forces, ranging from bio- to nano- and geotechnologies, including sensorial as well as algorithmic practices.[9] The main problem of this environmental mode of government is "the capture and the control, the management, the modulation of behavior, of affects, of relations, of intensities, and forces by means of environmental (media) technologies whose scope ultimately borders on the cosmic" (ibid., 10).[10]

Let me briefly present two examples to illustrate how this environmental rationality informs governmental practices. Braun (2014) discusses the fact that current car models often contain a new technical device that measures fuel efficiency. While the conventional fuel gauge simply signaled when it was time to refill the tank, the new gas consumption meter targets the act of driving itself, encouraging the driver to "economize" the use of fuel: "Crucially, this does not occur through law, nor through confinement or discipline, but rather through acting on the *environment* of the driver, in this case through the design of the apparatus itself. Indeed, the driver is configured here less as a thinking subject than as part of a relay circuit within the car itself" (ibid., 53; emphasis in original). According to Braun, we are witnessing a shift from the subject to the "car-driverassemblage" (ibid., 52) that necessitates going beyond the focus on individual will or rational choice. It is possible to conclude that the subject here is neither conceived of as an essence that predates power relations nor as their effect, but rather as a milieu of power. The drivers are not simply reacting to certain stimuli and signals from their environment; rather, "these signals are also increasingly registered at a preindividual, affective level, such that the responses of the individual increasingly approximate the 'automatic' responses of a machine" (ibid., 54).[11]

Another example of this environmental mode of governing is Jennifer Gabrys' case study (2014) on the Connected Sustainable Cities (CSC) project developed by MIT and Cisco. Her analysis shows that the project's vision of a smart and sustainable city seeks to reconcile environmental sensibility with economic growth. As the project relies heavily on participatory digital media and computer networks for its design of a "greener" future, Gabrys notes a tendency to displace the emphasis on governing the life of individuals and populations toward a "biopolitics 2.0" (ibid., 37). She takes up and further elaborates Foucault's notion of environmentality, understanding it "not as the production of environmental *subjects* but as a spatial–material distribution and relationality of power through environments, technologies, and ways of life" (ibid., 32; emphasis in original). This environmental mode of government, Gabrys argues, does not eliminate "individuals" and "populations"; rather, it displaces them as distinctive and isolated subject-objects of government. While "the environmentally responsible citizen" remains an important operator within the Connected Sustainable Cities (CSC) project, it is targeted as a responsive node "through processes that might generate *ambividuals*: ambient and malleable urban operators that are expressions of computer environments. [. . .] Ambividuals are not singularly demarcated or erased, but variously contingent and responsive to fluctuating events, which are managed through informational practices" (ibid., 42–43; emphasis in original). It follows that these "urban operators" are not restricted to human subjects only, as "the articulation of actions and responses occurs across human-to-machine and machine-to-machine fields of action" (ibid., 43).[12]

Probiotic Ecologies and Vital Systems Security

A second set of environmental technologies that enroll nonhuman nature and offer new prospects of promoting and harnessing the self-organizing tendencies of ecological and social systems include probiotic strategies and vital systems security. In an article published in 2017, Jamie Lorimer identifies a highly diverse set of recent practices that introduced "formerly taboo entities into our bodies, homes, cities and the wider countryside" (2017, 28). He proposes the label "probiotic" in a broad sense to distinguish these forms of intervention from conventional practices of health maintenance and environmental management characterized by an antibiotic approach. Examples range from "the rise of 'probiotic' diets [. . .], to established forms of 'organic' farming and 'biological' means for pest control [. . .], to the rise of 'rewilding' in nature conservation [. . .] and practices of 'managed realignment' in flood defense" (ibid., 28). Lorimer's analysis focuses on two empirical cases: the reintroduction of wolves into ecosystems, and the use of worms in health programs to tackle a number of human diseases and conditions. According to Lorimer, strategies of rewilding and reworming are informed by a "probiotic 'environmentality'" (ibid., 28), that differs significantly from traditional nature conservation policies as well as from conventional practices in health and hygiene that address microbes primarily as "pathological" agents.

According to Lorimer, the probiotic turn draws on an ontological regime very different from that of the antibiotic world. First, the strategies it puts forward are "concerned more with the processes of movement, circulation and interaction than with the essential character and composition of its constituent forms" (ibid., 32). Thus nonhuman species are not valued for their charisma, rarity, or authenticity but rather for the functions and services they provide within specific ecosystems or milieus. Secondly, the interventions informed by a probiotic environmentality operate in far-from-equilibrium ecologies. They share this aspect with technologies of resilience, but there is an important difference. While many contemporary strategies are closely associated with "regimes of anticipation" (Adams et al. 2009, 247), preparing for the yet-to-come, probiotic rationalities conceive of the present as already affected by disruptive processes: "Ecologies subject to rewilding and reworming are already tipped into disaster and in need of proactive transformation" (Lorimer 2017, 33). Thirdly, it is the striking absence rather than the excessive presence of certain entities and species that is seen as responsible for disturbances and destructions. It is the lack of "doings" essential for the (desired) operations of specific ecologies that causes dysfunctions in the respective system. The probiotic account suggests "that the absence of the calibrating agencies of keystone species makes the system go awry and causes dysbiosis" (ibid., 33). Fourth, probiotic ecologies channel and modulate the interrelations between species, recognizing their significance for mutual well-being. While these interventions enact a more-than-human ontology that stresses multispecies entanglements, they also revive and restore the anthropocentric matrix as they "remain centred on securing valued versions of human life" (ibid., 34).

To be sure, this new environmental style of management does not abandon control; rather, it entails a distinctive shift of control mechanisms. It enacts a "controlled decontrolling of ecological controls" (Klaver et al. 2002, 14 cited by Keulartz 2012, 60) as it fosters the strategic use of particular species considered vital to provide for desired systemic outcomes. Probiotic strategies give rise to an "enlightened anthropocentrism" (Keulartz 2012, 49), enrolling the capacities of nonhumans to regenerate or reverse dysfunctional socio-ecological systems as means to deliver "services" and "functions" for human futures. Rather than promoting an ethical approach that intrinsically values (biological) diversity, the contributions of animals and other nonhumans are reduced to their functional aspects and their infrastructural role as ecological service providers (Lorimer 2017, 39; see also Keulartz 2012, 65). Lorimer contrasts this environmental mode of government associated with the probiotic turn with biopolitical regimes concerned with the disciplining of the individual body or the regulation of the population: "While wolves [. . .] (and to a much lesser extent) worms are disciplined as individuals and governed as species (biopolitics), those concerned with their reintroduction are primarily interested in their ecological or 'environmental' agencies. As 'ecological engineers' and 'gut buddies' they are valued for their abilities to modulate or recalibrate dysfunctional ecologies" (ibid., 36).[13]

Another example of how probiotic strategies operate is Braun's analysis of contemporary "'eco-cybernetic'" (2014, 50) forms of urban flood management.[14] These environmental policies seek to neutralize the risks of "natural disasters" caused by anthropogenic climate change not by working against nature but by governing through nature and its properties. Like Lorimer, Braun stresses that these strategies are not informed by ethical concerns but by an interest in "the *services* that non-human nature could provide in the face of rising sea levels due to global warming, storm surges, or a combination of both" (ibid., 58; emphasis in original). Rather than adapting urban life to the unpredictable dynamics of environmental events, these strategies seek to include them in governmental practices. Most importantly, these probiotic interventions do not work by an external mode of operation that restricts, modifies, and contains the environmental conditions of human life but rather by aligning, channeling, and enrolling them. In this arrangement, nature is allowed to follow "its own course" while it is simultaneously mobilized as an integral part of a more complex system that no longer distinguishes between natural and social, human, and nonhuman realms:

> One of the advantages of such "natural" design elements is their ability to cancel out *other* natural processes. A network of islands and offshore piers, for instance, is seen to diminish the velocity of waves that accelerate through the Verrazano narrows, dispersing this energy before the waves impact the harbor edge. [. . .] The energy of storms is not to be repelled, but absorbed. [. . .] The "naturalness of nature" appears as an *object and means of government* at a particular historical moment, one in which society and nature come to be seen as a single integrated system [. . .]. (Ibid., 58, 60; emphases in original)[15]

Steven Collier and Andrew Lakoff have investigated how this idea of an overarching systems architecture materializes in the protection of "critical infrastructures" (Collier and Lakoff 2008a). These large-scale and interdependent systems are composed of heterogeneous material elements and include electricity grids, transportation and communication networks, water and food systems, and chains of industrial production. In their collaborative work, Collier and Lakoff argue that the recent political interest in these critical infrastructures is part of a much longer history of governmental interventions in designing and managing infrastructures. However, it also indicates a new biopolitical constellation they define as "vital systems security." Vital systems are material circuits of circulation that provide functions and services deemed to be indispensable for life in contemporary societies. Invented to maintain and enhance population security, they are also prone to unpredictable and potentially disastrous events disrupting the systems critical to economic and social life. Contemporary security policies conceive of vital systems as threatened by a range of possible events, from natural disasters and industrial accidents to terrorist attacks and pandemic events (Lakoff and Collier 2010; Collier and Lakoff 2015; see also Baldwin 2013; Opitz 2016).

The genealogy of vital systems security can be traced back to strategic bombing theory in interwar Europe, which focused on identifying and mapping vital nodes within an enemy's industrial system for future attacks (Collier and Lakoff 2008b).[16] These strategies were further developed and refined during the Cold War. Policy experts and military planners in the US sought to respond to a possible nuclear strike by the Soviet Union and its allies by installing a "distributed system of preparedness that would enable civilian production facilities to withstand an attack and support a viable counteroffensive" (Lakoff and Collier 2010, 249; Collier and Lakoff 2008b). In the 1960s and 1970s, methods of civil defense and nuclear attack preparedness migrated into other policy domains and became integrated into "a more general political technology oriented toward multiple types of threat" (Lakoff and Collier 2010, 257). In addition to responding to military attacks, emergency planning increasingly engaged with natural catastrophes and industrial accidents.[17] Political initiatives to install concrete vital security systems programs seeking to manage and mitigate a diverse set of vulnerabilities date back to the 1990s, but only after the attacks of September 11 did they become a central part of governmental strategies. This involved the passage of legislative acts (e.g., the Pandemic and All-Hazards Preparedness Act), the emergence of new institutional structures and governmental agencies (e.g., the Department of Homeland Security), and the development of forms of expertise and policy frameworks (e.g., systemic risk regulation). By the first decade of the new millennium, the sectors considered to be of strategic importance for vital systems security included "agriculture and food, the defense industrial base, energy, public health, banking and finance, drinking water and water treatment, chemical plants, dams, information technology, postal systems and shipping, transportation systems, and governmental facilities" (ibid., 247; Collier and Lakoff 2015).[18]

Collier and Lakoff regard the "emergence of vital systems security as a significant mutation in biopolitical modernity" (2015, 21). It is characterized by a new set of political technologies that differs from sovereign power on the one hand and classical biopolitics on the other. While the former relates to the political body of the state seeking to secure stability in confronting internal and external threats, the latter is concerned with the social body of a population and mechanisms to improve its well-being and prosperity. Vital systems security targets the technologies and instruments of classical biopolitics that increasingly came to be problematized as sources of vulnerability and risk, setting the stage for a "reflexive biopolitics" (ibid., 21). Like classical biopolitics this seeks to foster the welfare and health of populations, but it does so by addressing a new object: material infrastructures, functions, and services considered to be essential for maintaining collective life.[19] Vital systems security goes beyond traditional forms of population security as it strives to prepare for and govern emergencies of various kinds such as natural disasters, terrorist attacks, pandemic disease outbreaks, and disruptions of critical infrastructures:

> [T]hese two forms of biopolitical security differ in their objects of concern, knowledge practices, and norms. Whereas population security addresses regularly occurring events that are distributed over the population in predictable ways, vital systems security deals with events whose probability cannot be precisely calculated, but whose consequences are potentially catastrophic. Vital systems security does not rely on statistical analysis of past events to generate knowledge about security threats, but rather on the simulation or enactment of potential future events. Its interventions seek to increase the resilience of critical systems and to bolster preparedness for future emergencies. (Ibid., 22)

In this reading, vital systems security is not limited to a concern with the safety of a political or a social body. Rather, it defines a socio-techno-ecological hybrid marked by the interrelations of complex and heterogeneous systems that are threatened by disruption, breakdown, infection (by biological and digital viruses), and vulnerabilities through global interdependencies and environmental exposure—threats that are deemed to exceed the existing arsenal of technologies of security focused on the population (Collier and Lakoff 2008b; 2015; Folkers 2018, 214–21; 343–52). Lakoff and Collier discern a novel way of understanding how to map and manage future disasters by technologies associated with vital systems security. They define "preparedness" as a political technology that "responds to the governmental problem of planning for unpredictable but potentially catastrophic events" (2010, 244), distinguishing it from the principle of precaution. Both share the idea that catastrophic incidents cannot be predicted or calculated; but while the latter seeks to keep the event from occurring, the former "rather assumes that the event will happen" (ibid., 263). "Preparedness" develops a set of operational responses reducing vulnerabilities to ensure that vital systems continue to operate through and after the disastrous event.

Mechanisms of vital systems security are an integral element of the dispositive of environmentality. As we have seen, environmental modes of government shift the biopolitical frame of reference from population security to vital systems security and complement antibiotic with probiotic ecologies. An analytics of government makes it possible to map the selective formats and uneven vulnerabilities environmental interventions enact, attending to the "caesuras within the biological continuum" they inflict (Foucault 2003, 255; see Cavanagh 2014). Like classical biopolitics, its reflexive version sustains and supports certain forms of human and nonhuman co-existence while excluding or marginalizing others or simply suggesting that these may be killed or "let die" (Foucault 2003).[20] In line with the racist legacy Foucault described in his work, its more recent biopolitical variant distinguishes "inferior" from "superior" forms of existence (within the human species and beyond). Contemporary biopolitical practices are still animated by an immunological drive to defend the integrity of the hegemonic socio-techno-ecological systems, securing the life of some while sacrificing the existence of others (Esposito 2008; 2011; Swyngedouw and Ernstson 2018, 14–21).[21] We therefore need to explore how this environmentality (de)values particular forms of existence and how it is articulated with processes of naturalization, racialization, and gendering. The relational-materialist frame of a government of things calls for a critical examination of these ontological and normative "cuts" (to take up Barad's notion) as they differentially valorize certain forms of collective life and world-making at the expense of others.

Panarchic Government: Care in the Age of the Anthropocene

This analytic grid of a government of things also opens up new avenues for critically engaging capitalist practices. It contributes to a fruitful dialogue between more-than-human accounts and those investigating issues of political economy, bringing together "old materialist" concerns with new materialist commitments. The focus on the environmental not only unsettles the ontological distinction between the technological and the natural, the social and the material; it also undermines the traditional understanding of "the economic" as a sphere different from and external to "the ecological." As Maan Barua and other scholars have stressed, "the economic is configured by, and dependent upon, more-than-human processes and relationships which remake and regenerate the world" (Barua 2019, 664; see also Moore 2015; Battistoni 2017). Taking seriously Foucault's reminder that biopower was an essential element in the development of capitalist economies (Foucault 1978, 140–1), this "relational grammar for anatomizing the nature-capital dynamic" (Barua 2019, 650) examines the different ways in which capitalist accumulation is always already contingent on the control of nonhuman forces and properties. It brings into relief the way in which the category of the economic (itself not unitary or stable) is constituted by nonhuman life and labor from the outset, conceiving of "nature" not just as an already given stage upon which economic practices are performed but a key player in them.

Marxist feminists (e.g., Mies 1999; Federici 2004; Bhattacharya 2017) and researchers in feminist STS (e.g., Thompson 2005; Cooper and Waldby 2014) have argued for the need to recognize activities often considered "natural" like child care or female bodily capacities as forms of reproductive or regenerative labor. These efforts to expand the classical category of labor provide important insights to conceive of "natural" production beyond the human sphere. To acknowledge the productive role of nonhuman forces, scholars have recently mobilized the concepts "hybrid labor" (Battistoni 2017, 6) and "nonhuman labour" (Barua 2017, 275) to acknowledge forces and potentialities beyond or outside the human realm (see also Haraway 2008). One important dimension of this productivity concerns "animal work" (Porcher 2015). Barua discerns three different ways in which the doings of animals contribute to capitalist accumulation, distinguishing between metabolic (e.g., modifying, intensifying, and accelerating cellular processes of intensively farmed animal bodies to realize surplus value), ecological (e.g., the pollination work performed by bees) and affective labor (e.g., the show casting of elephants and other iconic animals).[22] These forms of labor do not just introduce recalcitrance and resistance to human action but are an essential element of economic production (Barua 2019, 652–56; see also Collard and Dempsey 2017).

Once again, we need to address a possible misunderstanding. To go beyond the anthropocentric concept of labor to include extra-human productivity does not mean ignoring, negating, or flattening analytic or normative differences between human and nonhuman labor or humans and other "workers" (Battistoni 2017, 22). Quite on the contrary, it is only the diagnosis of *"structural similarities"* (Collard and Dempsey 2017, 81; emphasis in original) between them that makes it possible to analyze the different (hierarchical) ways human and nonhuman work generates capitalist value.[23] To consider nonhuman productivity in categories of labor instead of as (natural) "capital" marks an important step towards a denaturalization of nature. The apparent reduction of nature to economic categories and its ethical "devaluation" by understanding nonhuman activities in terms of "labor" in fact contributes to repoliticizing the question of nature and brings the problem of the normative status of nonhumans to the fore. The relational-materialist account conceives of nature as an always already economic (and therefore political) domain, and provides a useful corrective to a human exceptionalism that ignores the economic salience of nonhuman forces. It acknowledges nonhuman "doings" without romanticizing or assimilating them and "raises questions about appropriate social relationships of compensation, care, and value. It compels us to consider our shared ends and our contributions to them, while recognizing that they will not always align perfectly" (Battistoni 2017, 21; Barua 2019, 664).[24]

However, in the current "environmentalization" of capitalist economies a rather different concept of care is emerging that is nurtured by the logic of resilience and the focus on self-organization. To account for this contemporary constellation, it is useful to revisit and extend Foucault's notion of pastoral power, which originally referred to the caring relationship between shepherd and flock on which the Christian "government of souls" was modeled (see, e.g., Foucault 2007a, 163–90; 1981b). While scholars have mostly employed the concept of the pastorate to analyze different dimensions of subjectivation processes (Novas 2005; Rose 2007, 73–76), it also provides an innovative and so far unexplored conceptual link between modes of directing human and nonhuman life, articulating notions of responsibility and care.[25]

The pastoral imaginary is by no means exhausted by the direct relationship between shepherd and flock but might also inform more impersonal and complex forms of government. Crawford S. Holling, Lance H. Gunderson, and Garry D. Peterson (2002) coined the notion of "panarchy" to advance a general systems theory capable of integrating society, the economy, and the biosphere. Instead of hierarchical organization characterized by a top-down structure, rigid forms of control, and vertical authority within or between social and ecological systems, panarchy evokes cycles of flexible co-ordination, the mobile interconnectivities between different levels and the evolutionary capabilities of complex systems. Relying on earlier work on resilience as a distinctive property of ecosystems (Holling 1973), the notion of panarchy seeks to capture the dynamics of creation and destruction in adaptation cycles. Holling and his co-authors evoke Pan, the ancient Greek god of the wild, shepherds, and flocks, as a way of understanding the patterns of relationality and change in ecological, economic, and social systems. Pan symbolizes a creative force but "could have a destabilizing, creatively destructive role that is reflected in the word *panic*, derived from one facet of his paradoxical personality" (Holling et al. 2002, 74; emphasis in original). The reference to Pan is an essential part of the attempt to develop a transdisciplinary theory, as he is not only half-goat and half-man but also represents the "all-pervasive, spiritual power of nature" (ibid., 74).[26]

The concept of panarchy seeks to grasp the intricate trajectories of human and nonhuman processes by stressing their interactivity, nonlinearity, and complexity as ways of steering adaptive evolution. Panarchic government promises to contribute to an active management of change by fostering resilience and sustainability. It does not aim at a comprehensive control of ecological or social processes; rather, it advocates institutional design and policy responses to improve the system's capacity to allow for disturbances and disruptions. Like the personal relationship of shepherd and flock, panarchy enacts an ethics of care, albeit one that focuses not on "salvation" and "safety" (see Foucault 2007a, 126) but on governing insecurities and promoting resilience. It is no longer the individual sheep and the collective flock who are addressed in pastoral care (see Foucault 2007a, 128; 1981b), but rather complex socio-techno-ecological "flows" affected by climate change, biodiversity loss, water pollution, and intensive agriculture (Gunderson and Holling 2002; Bröckling 2017, 40–42).[27]

Panarchic government differs from situated and embodied "matters of care" (Puig de la Bellacasa 2017; see also Mol and Pols 2010) that envision new speculative techniques and practices of relating to and sustaining "the world." It enacts a distinctive normative grammar that is informed by the logics of resilience and converts ethical responsibility into an evolutionary ability to respond to future catastrophic events, displacing political questions of justice and equality. Instead of changing patterns of production and consumption and improving conditions for "living on a damaged planet" (Tsing et al. 2017), we are invited to adapt to them: "Accepting the imperative to become resilient means sacrificing any political vision of a world in which we might be able to live better lives freer from dangers, looking instead at the future as an endemic terrain of catastrophe that is dangerous and insecure by design" (Evans and Reid 2013, 95).

The panarchic fixation on adaptation techniques and survival strategies in the shadow of catastrophic events urges us to critically examine the now popular label of the Anthropocene (Crutzen and Stoermer 2000). This term, referring to a new epoch in which humankind has arguably become a planetary geo-physical force, is increasingly gaining currency among geologists, earth systems scientists, and scholars from the humanities and social sciences (Johnson et al. 2014; Bonneuil and Fressoz 2016; Conty 2018). The Anthropocene narrative suggests more modest ontologies that depart from modernist appropriations of nature and nurture the prospect of sustainable futures and mutually supporting human-nonhuman exchanges. While many new materialists conceive of this move beyond human-centered trajectories as more or less directly coupled with a "greener" or more democratic politics characterized by horizontal engagements between species, this altered ontological framing might actually intensify and increase the capitalist enrolment of human and nonhuman natures (Swyngedouw and Ernstson 2018, 11; Neyrat 2018). Cherishing symmetrical ontologies and discarding the modernist split between nature and culture does not necessarily guarantee post-capitalist futures, nor does it automatically undermine technocratic visions of ecological management:

> The main point is that the indeterminacy or constant becoming of matter and life, a decentred—post or anti-humanist—account of human agency and the contestation of any fundamental separation between matter and cognition are assumed by post-constructionist scholarship to have "emancipatory" implications, for both human and nonhuman agents. If the building blocks of reality are not fixed—so the argument goes—politics becomes "ontological" and novel opportunities for change open up. Targeted at the dualisms of naïve or Cartesian realism and of culturalism, however, this argument misses or downplays the politics of ontology inbuilt in the neoliberalization of nature, which builds precisely on these tenets. (Pellizzoni 2015, 8)

The "nonhuman turn" (Grusin 2015) and the move beyond anthropocentric ontologies is not at all incompatible with techno-managerial projects and capitalist imaginaries. It is no accident that Paul Crutzen, who coined the term Anthropocene together with Eugene F. Stoermer, has also advocated climate management and geo-engineering strategies (see, e.g., Crutzen 2002, 23; 2006). The new materialist account of matter as intricate and indeterminate, and its focus on multispecies entanglements, are taken up and harnessed in environmental strategies of adaptive control and resilience management. While it nurtures the promise of symmetrical ontological relations, this program does not in itself secure a political transformation that transcends the capitalist matrix. Rather, the call for a new materialist ontology risks contributing to "a renewed and ecologically sensitive 'hyper-reflexive' capitalism that takes seriously both humans' geo-physical force and the material acting of the non-human, while redeeming the sins of the past" (Swyngedouw and Ernstson 2018, 13; Neyrat 2018).²⁸ Contrary to the erroneous conclusion that the insight into the radical relationality of the world always already makes a difference politically, the point is—taking up an old materialist insight—"to change" the world (Marx) by articulating the political anew and enacting it differently.²⁹

Conclusion

Multiple Materialisms

As is well known, the concept of government Foucault developed in his later work put forward a historically informed and comprehensive understanding of government that goes beyond the conventional meaning of the term. While government is mostly understood in terms of political decision-making and administration, Foucault also attends to knowledge practices and forms of subjectivation. He proposes a "very broad meaning" (2000b, 341) of the notion that does not conceive of subjectivation and state-formation as two independent and separate processes but analyzes them from a single analytical perspective. Thus the "genealogy of the modern state" (2007a, 354) is also a "history of the subject" (ibid., 184), since Foucault does not consider the modern state as a centralized structure but rather as "a tricky combination in the same political structures of individualization techniques and of totalization procedures" (2000b, 332).

In this book I have argued for another revision and expansion of the traditional concept of government. The analytical frame of a government of things displaces the preoccupation with the guidance of human individuals and collectives in order to shift attention to the technological infrastructures and vital milieus informing governmental practices. Exposing the limits of anthropocentric modes of thought, it makes visible the contingent boundaries and material circuits of politics. This more-than-human analytics of government conceives of the human subject as a result of practices of co-emergence and co-becoming with nonhumans rather than as something outside or prior to them. It opens up a whole range of new theoretical and empirical questions: How is the political collective composed, and who (or what) is recognized as a political actor (animals, mountains, algorithms, drones, etc.) (Law 1994, 193–4; Asdal et al. 2008, 6)? How is the government of nonhumans articulated with, and how does it condition, the government of humans (Nimmo 2010)? How should we conceive of the "doings" of human and nonhuman bodies, and their eventfulness and indeterminacy, without resorting to concepts like "resistance" or "recalcitrance" that seem to reinscribe passivity or reconsolidate the opposition of activity vs. passivity (Braun and Whatmore 2010, xx–xxii)?

This conceptual move toward a relational materialism not only makes it possible to extend the territory of government and multiplies the elements it consists of, it also attends to the diverse ways in which the boundaries between the human and the nonhuman world are negotiated, enacted, and stabilized. It examines the fundamental divide between the natural on the one hand and the social on the other, between matter and meaning, as a distinctive instrument and effect of governmental rationalities and technologies. Finally, this theoretical stance contributes to exposing and overcoming the limitations and blind spots many accounts in social and political theory face in addressing the increasing uptake of environmental forms of government in contemporary societies. In the prevailing political imaginary, politics is considered to be based on discourse, communication, will, etc. (Asdal et al. 2008; Marres and Lezaun 2011), while technological and ecological matters are conceived "as the passive and stable foundation on which politics takes place" (Barry 2013, 1; see also Braun and Whatmore 2010).[1] Too many political and social theorists still tend to take for granted "the ontological status of the entities in question" (Woolgar and Neyland 2013, 52). In contrast, the analytical frame of a government of things stresses the materiality of politics by articulating the link between the matter of government and the government of matter.

As I have shown, Foucault's notion of the dispositive and his understanding of technology and the milieu provide analytic tools that enable us to critically investigate environmental rationalities and contemporary governmental practices. Synthesizing Foucault's analytics of government with insights from ANT and with feminist and postcolonial STS, the conceptual proposal of a government of things invites us to disentangle the notions of matter and ontology from determinist or essentialist accounts. In doing so, it challenges political imaginations and critical vocabularies by questioning the idea of nature as solid, stable, and static. This analytical grid takes up important insights and theoretical achievements of new materialist scholarship. It shares the interest in reconceptualizing matter and the focus on the interplay of epistemological, ontological, political, and ethical issues, and insists on the limits of anthropocentric modes of thought. However, in putting forward a relational and performative understanding of materialities that consistently rejects scientific foundationalism and closely attends to the political dimensions of ontogenesis, the conceptual lens of a government of things also goes beyond important strands of new materialism.

This book has examined three highly influential, but very different positions within the field of the new materialisms: object-oriented ontology (OOO), vital materialism, and diffractive materialism. As we have seen, OOO eschews any sense of relationality in order to embrace the withdrawn essence of objects. It puts forward a "staunchly non-relational rationalism" (Gamble et al. 2019, 122) that suffers from a renewed form of subjectivism. OOO's ambition is to engage with objects in their own terms in order to address what lies beyond human rationality, cognition, and knowledge. However, OOO's "desire to cultivate theoretical modesty" to sensitize us to the prevailing problems of anthropocentrism and "human hubris" (Bennett 2015a, 232), stands in stark contrast to the self-promoting rhetoric that protagonists of OOO and speculative realism employ. The vocabulary of branding and innovation prevails, advertising "a new era of scholarship" (Morton 2013a, 159) or the emergence of a "new breed of thinker" (Bryant et al. 2011, 3).

What is more, Harman, Morton, and other representatives of OOO (including speculative realism) tend to engage in "incestuous mutual citing" (Taylor 2016, 205), likely with the intention of promoting the impression of a movement with shared principles and objectives beneath the obvious differences and dissonances. This kind of self-referentiality and self-enclosure points to "a somewhat hermetic research environment" (Norris 2013, 195), which tends to ignore or neglect important theoretical debates that help to correct or complement principles and ideas laid down by authors writing under the rubric of OOO. This intentional theoretical isolation mirrors and consolidates the argument for non-relationality. But cherishing isolated objects also has a drawback: it runs the risk of exempting OOO's work from (self-)critique by reiterating and referencing a quite limited canon of texts and thinkers. Its consequences are often tiresome, repetitive, and excessive theoretical claims. The writings of OOO scholars permanently assure readers that they are witnessing the birth of a new philosophy, and one cannot but be surprised by self-confident and imaginative words used to describe the withdrawal of the real.[2]

In contrast to OOO, vital materialism explicitly rejects essentialist concepts of matter. Bennett's concept of a "vibrancy of things" marks an important step towards a more-than-human politics. It makes it possible to account for the "doings" of nonhuman entities and indicates the limits of liberal concepts of politics as it focuses on associations and assemblages instead of individual and isolated actors. However, as I have shown, Bennett's vital materialism is also characterized by serious conceptual ambiguities and analytic shortcomings. To put it in an old-fashioned vocabulary: Bennett endorses an idealist account of materialism, as she seems to envision an indeterminate but nevertheless fundamental link between an affective ethics and an alternative politics. She claims that attending to the vitality of matter will eventually lead to a "greater appreciation of the complex entanglements of humans and nonhumans" (2010a, 112). In this reading the experience of "the indispensable foreignness that we are" (ibid., 113) provides a strong motivational and affective force to change contemporary economic and political practices. It is quite questionable, and indeed highly improbable, that an increasing acknowledgement of "vibrant matter" will result in a different politics, as this perspective gives undue credit to (human) reflexivity and affectivity at the expense of (more-than-human) assemblages. Thus, Bennett restates and revives a dualism she critically exposes in her work as an integral element of the Western political tradition, namely the opposition of "active Mind and inert Matter" (Bennett 2005, 135). Strangely enough, this understanding of political change contradicts central arguments of vital materialism, which conceives of politics as determined by the agentic forces of the (political) assemblage rather than by reflection and rational insight.

Diffractive materialism shares with vital materialism the emphasis on human-nonhuman entanglements. Both stress the significance of relational accounts, albeit in a different sense. Vital materialism still assumes a foundational agency or a liveliness that pertains to matter itself. In contrast, diffractive materialism promotes a more relational understanding of matter that suspends the idea of originary properties or already given agentive propensities. Barad's agential realism, and specifically the concept of the apparatus, makes it possible to investigate how temporalities, spatialities and materialities are mutually constituted (instead of taking them as absolute and external parameters); it also enables a more concrete analysis of the making/marking of human and nonhuman bodies that goes beyond examining the (re)configuration of the boundaries between them.

While agential realism marks an important step toward a relational materialism, it shares with vital materialism and other variants of new materialism two serious problems that curtail or diminish their radical theoretical impact: scientific positivism and political reductionism.

The first problem pertains to the way new materialist scholarship assesses scientific knowledge. It often claims to go beyond "critical" or "extractive" approaches to "engage" with science (Hird 2009; see also Alaimo and Heckman 2008). Instead of criticizing the construction of research objects or employing scientific concepts in order to understand the social fabric, new materialists endorse "dialogue, conversation, and collaboration with science" (Hird 2009, 331), seeking to "work with, rather than against, scientifically generated theories and data" (Wilson in Kirby and Wilson 2011, 233). While it is certainly important to question and actively transgress disciplinary borders and to bring together different forms of scientific knowledge, this more "science-friendly" (Kirby 2017, 11) stance risks ignoring or disarticulating critical epistemological interventions by Foucauldian and STS scholars, especially those of feminist and postcolonial science studies.

Angela Willey has argued that the "celebratory progress narrative of new materialist *engagements*" with science (2017, 135, emphasis in original) mainly serves to correct and replace presumably outdated or obsolete understandings of matter—providing a better (more inclusive, more complex) science. According to this reading, new materialist scholarship aligns itself with those scientific endeavors that seek to leave behind deterministic concepts and theoretical accounts of stable natural laws and straightforward causal ties. However, the concept of an agentic and non-determinist nature "runs the risk of becoming another natural law" (Willey 2016, 1000). The principle of contingency is translated into a new scientific doctrine, and the argument for indeterminacy might result in a novel form of dogma.[3] While the new materialisms challenge conventional stories of scientific progress, they might also contribute to consolidate them as they tend to reclaim and strengthen epistemological authority—instead of problematizing, destabilizing, or subverting it by showing its conditions of emergence and the "regime of valuation" (Murphy 2017, 5–6; 148–9) it enacts.

Willey's critique of the "neopositivist agenda" (2017, 149) of new materialist scholarship might itself be considered as reductionist, as it does not sufficiently take into account the heterogeneity and complexity of this style of thought. Still, her critical intervention points to a problematic tendency that not only subscribes to the familiar tale of scientific advances and the well-known logic of discovery but also risks renewing a rather conventional and entrenched hierarchy of scientific expertise. New materialisms often promote a highly selective and largely unacknowledged understanding of matter. The call to open up scientific borders and to engage in transdisciplinary endeavors is not reciprocal; rather, new materialists expect social scientists and humanities scholars to revise or even break up their research agendas in order to take in matters traditionally located in the natural sciences (Irni 2013, 356). They endorse a particular disciplinary orientation, as they tend to understand materiality in terms more or less defined by the natural sciences. As materiality is conceived of as a phenomenon studied genuinely in the natural sciences, those interested in processes of materialization need—according this reasoning—to engage with biological, chemical, or physical knowledge. Thus, the call to actively enter into a dialogue with science is often conflated with the claim to account for matter per se. Matter thus becomes "natural science matter" (ibid., 351), excluding or marginalizing processes of materialization that relate more closely to social-scientific and humanities research agendas.

This gesture, reclaiming matter while at the same time restricting it to "natural science matter," reiterates a highly charged normative and epistemological hierarchy that privileges some forms of expertise over others. While materialist accounts that engage with the natural sciences are considered to be "new," i.e., interesting, innovative, and pertinent, other materialist forms of knowledge production appear as failing to address what really matters or are labeled as "familiar," "traditional," "old-fashioned," or even "outdated," e.g., those investigating processes of racialization, gender, or class relations (ibid., 354–55). Thus, there is a "politics of materiality" (ibid., 355) at play that tends to define and delimit the understanding of matter, running the risk of reproducing asymmetric power relations between "hard" and "soft" sciences—thereby reinvigorating the profoundly gendered attributes that go along with them (Willey 2017, 139).[4] What is needed is a thoroughly relational account of materiality that does not privilege or single out one particular form of materiality over others but is attentive to multiple materialities.

Rather than engaging in what Foucault once called a "politics of truth" (1996b, 220) that contests "the status of truth and the economic and political role it plays" (Foucault 1984c, 74), new materialisms tend to endorse "an analytics of truth." Instead of subverting and displacing the existing regime of truth and the asymmetries and hierarchies inscribed into it, they commit themselves to "telling the truth about matter" (Willey 2016, 1001). However, as Foucault reminds us, the political question is not to distinguish true from false knowledge but to investigate the role of truth in (in)forming political rule and scientific authority by inquiring "how effects of truth are produced within discourses which in themselves are neither true nor false" (1984c, 60).

The second issue (intimately connected to the first) concerns the trend in new materialisms to endorse a reductionist or underdeveloped understanding of politics. There are different aspects to this problem. Some versions of new materialism shift attention from political questions to ethical concerns, neglecting the role of antagonisms for radical democratic policies (see Hoppe and Lemke 2015; Rekret 2016; Hoppe 2017b). Also, they often attach desired ethical values and positive political effects to the concepts of indeterminacy and contingency. New materialist scholarship is characterized by a striking paradox: On the one hand it stresses non-teleological dynamics and contingent trajectories, but on the other hand it often tends to assume a rather deterministic and derivative link between ontology and politics. In this view, an alternative or radical politics follows more or less directly from a different ontology. However, it is important to take seriously the principle of contingency as there is no necessary causal link between the idea of matter as "agentic" and an emancipatory or radical politics. As Erik Swyngedouw and Henrik Ernstson note, "the immanentist ontology of earth's multifarious acting does not in itself guarantee a political transformation. That requires a re-thinking and re-enacting of the political too" (2018, 12; see also Ellenzweig and Zammito 2017, 9; Willey 2017, 149).

New materialist scholarship promotes a "romanticization of contingency" (Willey 2016, 1000; 2017, 138) that ignores the fact that indeterminacy and plasticity can be articulated within very different and conflicting regimes of practices. Their political meaning is flexible and mobile rather than fixed and stable. It is exactly this consistent normative preference for instability and flux that makes it more difficult for new materialists to account for the stability, solidity, and persistence of recursive patterns, especially when it comes to relations of power and domination.[5] As we saw in Chapter 8, this celebratory stance obscures the manifold ways in which contingency and indeterminacy are enrolled in contemporary environmental strategies.

By contrast, the analytic grid of a government of things proposed in this book takes political ontologies more seriously. It revises and expands the realm of politics by attending to the material networks that allow for definite phenomena to emerge, instead of understanding politics as the direct outcome and immediate effect of epistemo-ontological configurations, as new materialist scholarship tends to assume. The relational-materialist interest in ontological politics differs substantially from an "idealism of relations" (Jochmaring 2016, 100; see Sprenger 2019, 24). While the former inquires into how relations are materially composed, assembled, and co-ordinated, the latter embraces relationality per se. It promotes the neo-cybernetic idea of a general interconnectivity that fixes relata in a universal network that knows no outside or remainder. This environmental understanding of relationality has recently become a "theoretical, economic, and political trap" (Neyrat 2018, 12) or a "straitjacket" (Swyngedouw and Ernstson 2018, 4), as it does not allow us to imagine and enact substantially different pathways and alternative trajectories, given the universality and immediacy of relational networks.[6] Against this totalizing understanding of relationality, Frédéric Neyrat proposes an "*ecology of separation*" (2018, 14; emphasis in original) that consists in a double-edged intervention. On the one hand it defends and affirms relationalities against the persistence of essentialist concepts and the ontological divide between nature and culture, human and nonhuman, while on the other hand it negates and rejects the idea of a universal and absolute interconnectibility, complementing the notion of relationality by "*a counterprinciple of separation*" (ibid., 14; emphasis in original). This strategy performs a gesture of immanent distancing by affirming exteriority and alterity within and beyond a given relational configuration.

As it takes into account the historical and political dimensions of ontologies, the relational materialism proposed in this book is also a material relationalism. It is attentive to "lines of flight" (Deleuze and Guattari 1987, 3) and displaces concepts of a closed or finite relationality by affirming "excess and subtraction [. . .]. It holds that all relational configurations imply a certain separation and distancing, and, thereby, the always immanent possibility of forms of acting that undermine, transform, or supersede existing relations configurations" (Swyngedouw and Ernstson 2018, 6). Mobilizing this idea of material relationality reopens the question of the political. It gives rise to a new "distribution of the sensible" (Rancière 1999; Swyngedouw and Ernstson 2018, 21) that makes possible more just or egalitarian human-nonhuman encounters.

To be sure, the relational materialism suggested here is still a sketch and needs to be developed further. It is not a finite program but rather a permanent provocation (Murphy 2017, 7). The analytic grid of a government of things is a conceptual construction site, not a fully-fledged proposal but something provisional—a tool to think with and an invitation to think otherwise. The analytics of government investigates the material effects of contingent encounters, examining how relations become stable configurations in order to explore the possibility of composing them differently.[7] It cultivates an experimental gesture (see Chapter 7), and is attentive to excessive or interruptive practices that supersede or undermine the stability and integrity of the relational frame. The conceptual proposal of a government of things contributes to liberating the analytical and critical resources of new materialisms for a (more) radical politics and non-capitalist projects. It opens up a political space of contestation, disagreement, and dissent that facilitates the articulation of alternative, and possibly conflicting, trajectories of socio-techno-ecological futures enacting more-than-human democratic practices.

Thus, the diagnosis of contemporary environmental forms of government does not signal an all-pervasive and totalizing form of domination. While the current dispositive of environmentality further promotes the idea of universal adaptability and controllability, it also gives rise to a more comprehensive understanding of ecology beyond neo-cybernetic power. While this idea of a "general ecology" (Hörl 2017) endorses an integral account that also problematizes the modern opposition of nature and culture, the human and the nonhuman, it goes further by dissecting the selectivities inscribed in current environmental forms of government, in particular their connection to capitalist enrolments of human and nonhuman nature. As Erich Hörl puts it, this understanding of a generalized ecology radically rethinks relationality in non-modern and non-theoretical terms by instituting "a non-philosophical politics of relations" (2017, 7). This "neocritical project" (ibid., 5) contributes to an engagement with ontological politics that transgresses current forms of technocratic and capitalist control. In a similar vein, Sara Nelson has noted an ironic tension in the resilience discourse and its focus on adaptability. While its emergence is intimately linked to mechanisms of flexible accumulation and constitutes an integral element of the rise of neoliberalism since the 1970s, resilience theory could be transformed or (re)articulated as a critical tool that undermines neoliberal strategies to promote a socio-ecological common beyond capitalist control. The critical appropriation of the resilience discourse offers crucial "theoretical tools for an anti-capitalist ecological politics, both for understanding the capitalist ecologies it currently informs and for developing an exit strategy from them" (Nelson 2014, 16).[8]

If we want to envision and struggle for (more) "earthly worlding[s]" (Haraway 2016, 97) beyond the contemporary dispositive of environmentality it is important to take up and extend the new materialist concern of linking epistemological and ontological questions, but without reaffirming scientific truth claims and restating the idea that politics follows directly from ontology. The analytical frame of a government of things suggests cultivating the moment of wonder and surprise in materialist theorizing (Stengers 2011; see also Latour 2007)—a moment often lacking in new materialist accounts. This has nothing to do with cherishing the "opacity" of objects or embracing the "vibrancy" of things, but insists on curiosity and openness as political matters.

This "critical materialism" (Lettow 2017, 118; Willey 2017, 149) reaches out to align the new materialist agenda more closely with the rich conceptual and theoretical resources and repertoires of the materialist tradition. While new materialist scholarship has tended to stress historical breaks, it seems more pertinent to relate the current material turn to the concerns of earlier materialist thought and to investigate its potential to revise and broaden critical theory (Lettow 2017). Also, it would be helpful to shift the accent from breaks and ruptures to continuities and possible alliances, stressing the links between materialist accounts and poststructuralist theory (especially ANT and postcolonial and feminist STS). Rather than reading Barad's account of agential realism in critical distance to the work of Foucault and Butler, it could rather be conceived of as extending their insights into "'natural science matter'" (Irni 2013, 349).

Instead of dismissing the new materialist agenda or opposing it to the work of Foucault and STS scholars, this book has aimed at contributing to a broadening of new materialist concerns. *The Government of Things* proposes to connect new materialist ontologies more closely with an analytics of government informed by empirical investigations and a critical interest in political change. It is an invitation to think the material turn with Foucault, attending to the political dimensions of ontologies and productively engaging with the natural sciences without privileging one form of matter or expertise over others. Moreover, it calls for multiple materialisms and argues for extending the materialist agenda, proliferating materialist concerns instead of engaging in a truth game that separates and hierarchizes them. This relational materialism renews and rearticulates the sensibility materialist thinkers always had: to link philosophical questions and theoretical concerns to political projects addressing human and nonhuman suffering, giving voice to this suffering and engaging in a struggle to end it.

ACKNOWLEDGMENTS

The publication of this book is marked by an unsettling conjuncture. While I was about to finish the work on the manuscript in spring 2020, the COVID-19 crisis was just starting to turn into a global catastrophe. The pandemic has seriously affected social life and economic activities around the world and has brought healthcare systems to the brink of collapse. To contain the spread of the virus, governments in many countries have taken unprecedented and massive measures, intervening drastically in the lives of individuals and populations. No one knows how long this crisis will last and how many more lives will be lost, but it is quite certain that its legacy will remain with us for years. Powerfully, it epitomizes the conceptual proposal of a government of things I seek to flesh out in the book. The COVID-19 pandemic historically and systematically transgresses the borders between the human and the nonhuman, defining a common ground where "thing power" meets biopolitics.

While this strange coincidence between the ambitions of the book and the trajectories of the virus is troubling and thought-provoking, the idea for *The Government of Things: Foucault and the New Materialisms* is much older. It originated in my interest in the theoretical and political prospects of new materialist scholarship and was first discussed in an article published in the journal *Theory, Culture & Society* in 2015. Since then I have been trying to give shape to the analytic frame of a government of things aligning elements from Foucault's work on governmentality and biopolitics with important insights from science and technology studies.

While *The Government of Things: Foucault and the New Materialisms* aims to give substance to the project of a relational materialism, it is itself the result of a rich net of material relations. The project set in motion an enormous range of collegial, institutional, and financial resources. The work on the manuscript was funded by an Opus Magnum grant from the Volkswagen Foundation. I am immensely grateful for the financial support and the Foundation's incredible patience, as my work on the manuscript took much longer than I had originally planned.

While it certainly can be stated for most books that they have many contributors, this is particularly true for this one. My argument builds on the work of many scholars—not least those who figure in this book. Without their inspiration and ideas this book would not have materialized. I presented material from this book at a number of conferences, including the Latin American Conference on Biopolitics in Bogotá (2013), the Annual Meeting of the Society for Social Studies of Science in Buenos Aires (2014) and Barcelona (2016), the Annual Conference of the Australasian Society for Continental Philosophy in Sydney (2015), the Annual Conference on the New Materialisms: The Ethics of Decolonizing Nature and Culture in Paris (2017), and the Bi-Annual Conference of the German Sociological Association in Göttingen (2018). I am grateful to all the participants for animated discussions, engaged comments, and helpful criticisms.

Also, I have been fortunate to have had the opportunity to discuss ideas from the book at several seminars and workshops over the past few years. I thank my colleagues and the audiences in the following universities and research institutes for their generous hospitality and productive conversations: the University of Champaign-Urbana (IL), Cambridge University, the University of São Paulo, Yale University, the Institute for Social Research in Frankfurt am Main, the Technical University of Munich, the University of New South Wales in Sydney, the University of Erlangen-Nuremberg, the RWTH Aachen University, Augsburg University, the University of Sydney, the Central European University Budapest, the New School for Social Researchin New York City, Copenhagen University, and the University of Oslo.

A number of people have commented on the general idea of the project and different versions of the manuscript. I am extremely grateful for their close readings and constructive feedback. Josef Barla, Maan Barua, Endre Dányi, Katharina Hoppe, Vicky Kluzik, Ruzana Liburkina, Doris Schweitzer, Eva Šlesingerová, and Frederic Vandenberghe each read earlier drafts of chapters, and their insightful comments helped me to revise and refine the overall argument. Others I would like to thank for inspiration, comments, criticisms, ideas, and additional help that shaped this book include Andreas Folkers, Samantha Frost, Peta Hinton, Tim

Ingold, Reiner Keller, Vicky Kirby, Vanessa Lemm, Gesa Lindemann, Florian Sprenger, and Miguel Vatter. Of course, responsibility for the final version of the book is mine alone.

I am also grateful to Gerard Holden for his careful language editing. Warm thanks as well to my student assistants Franziska von Verschuer and Ira Zöller, who have provided invaluable help and important suggestions at different times when I was working on the book project, and Eva Sänger and Renate Uhrig, who took care of teaching and administrative work respectively while I was supported by the Volkswagen Foundation. At NYU Press, I am indebted to Ilene Kalish for her encouragement and continued support of the book project.

* * *

Chapters 1 and 2 are based on previously published material: "Materialism Without Matter: The Recurrence of Subjectivism in Object-Oriented Ontology," *Distinktion: Journal of Social Theory* 18(2), 2017: 133–152; and "An Alternative Model of Politics? Prospects and Problems of Jane Bennett's Vital Materialism," *Theory, Culture & Society* 35(6), 2018: 31–54. Content from these articles is reprinted by the permission of the publishers.

NOTES

INTRODUCTION

1 On materialist thought in the history of philosophy, see also Wolfe 2016; 2017.
2 Given the theoretical diversity and multiplicity of research interests and disciplinary perspectives assembled under this label, I shall use the plural instead of the singular form.
3 See Karen Barad's often cited statement that "language has been granted too much power. The linguistic turn, the semiotic turn, the interpretative turn, the cultural turn: it seems that at every turn lately every 'thing'—even materiality—is turned into a matter of language or some other form of cultural representation. [. . .] There is an important sense in which the only thing that does not seem to matter anymore is matter" (2003, 801).
  For a brief overview of how the material has been conceptualized in social theory, see Reckwitz 2002.
4 For alternative cartographies of the new materialist landscape, see Coole and Frost 2010b; Dolphjin and van der Tuin 2012; Connolly 2013; Coole 2013, Devellennes and Dillet 2018; Wilson 2018; and Gamble et al. 2019.
5 It comes as no surprise that a book entitled *The New Materialism* was published some fifty years ago. And it is not only the title that sounds familiar, but also the message the book conveys. In this work, US philosopher James K. Feibleman advanced the thesis that relativity theory and quantum mechanics in the first quarter of the twentieth century had revolutionized our understanding of matter. He claimed that philosophy needed to take into account these scientific insights as a way to envision a "material ontology" (1970, 36). While traditional materialism conceived of matter as "hard, round, impenetrable bits of stuff which were ultimately simple and solid" (ibid., 41), he saw a "new materialism" emerging in which matter "has been acknowledged to be a highly dynamic agent capable of sustaining the most complex activities" (ibid., 42).
  For a review of the book comparing it to a more recent understanding of new materialisms, see Lemke 2015a.
6 See, for example, Barad's understanding of critique: "I am not interested in critique. In my opinion, critique is over-rated, over-emphasized, and over-utilized, to the detriment of feminism. As Bruno Latour signals in an article entitled 'Why Has Critique Run Out of Steam? From Matters of Fact to Matters of Concern' (2004a), critique is a tool that keeps getting used out of habit perhaps, but it is no

longer the tool needed for the kinds of situations we now face" (Barad in Dolphijn and van der Tuin 2012, 49).

This dismissive account notwithstanding, Barad's work is sometimes hailed as an upcoming paradigm for feminist theory and a new foundation for critical thought in general (see, e.g., Hekman 2008, 106; Hinton 2013, 186).

7   In this regard, the rather polemical gestures towards poststructuralism and social constructivism expressed by some new materialists miss an important point. They not only tend to caricature these theoretical perspectives by reducing their diversity and complexity (see also Ahmed 2008; Keller 2019), but also display ignorance concerning their historical conditions of emergence and the political and theoretical problems they responded to. Authors writing under the umbrella of the new materialisms sometimes fail to take the persistent stress on relationality and materiality seriously when they present poststructuralism and social constructivism not as "situated knowledges" (Haraway 1991, 183–201) but rather as "failed materialism" (Gamble et al. 2019, 116–18; Dolphjin and van der Tuin 2012, 48) or "merely reactionary" (Dolphjin and van der Tuin 2012, 138). In short, they tend to see them as an insufficiently complex or altogether false knowledge that has to be replaced by something real and true: new materialisms.

8   Ashley Barnwell provides a very instructive—critical—account of the new materialist "fatigue with critique" (2017, 30).

9   For a quite similar distinction between "negative," "vitalist," and "performative" materialisms, see Gamble et al. 2019. While these authors fully endorse the last variant and criticize object-oriented ontology and Bennett's work to "implicate certain objectivist, non-relational and, thus, idealist assumptions or residuals" (Gamble et al. 2019, 112), I will advance a slightly different evaluation. Agential realism indeed presents a more convincing account of matter compared to Harman's and Bennett's work, but it also suffers from unresolved theoretical problems and blind spots that undermine the relational ontology it proposes.

10 See also William E. Connolly's call for a "militant assemblage" in the light of the new materialisms: "The immediate goal would be to press international organisations, localities, states, corporations, banks, labour unions and universities to defeat neoliberalism, to curtail climate change, to reduce regional and national inequalities, and to infuse a vibrant pluralist spirituality into democratic machines that have lost too much of their vitality" (2013, 412).

11 See Foucault's famous quote: "I would like my books to be a kind of *tool-box* which others can rummage through to find a tool which they can use however they wish in their own area. [. . .] I don't write for an audience, I write for users, not readers" (Foucault 1994a, 523–24; emphasis in original).

12 See, for example, Foucault's programmatic statement in the first volume of the *History of Sexuality*: "The purpose of the present study is in fact to show how deployments of power are directly connected to the body—to bodies, functions, physiological processes, sensations, and pleasures; far from the body having to be effaced, what is needed is to make it visible through an analysis in which the bio-

logical and the historical are not consecutive to one another [. . .] but are bound together in an increasingly complex fashion in accordance with the development of the modern technologies of power that take life as their objective. Hence I do not envision a 'history of mentalities' that would take account of bodies only through the manner in which they have been perceived and given meaning and value; but a 'history of bodies' and the manner in which what is most material and most vital in them has been invested" (Foucault 1978, 151–52).

Even before his "genealogical" writings in the 1970s and 1980s, in his work on the archaeology of knowledge and on the notion of discourse Foucault situated himself within the materialist tradition, pursuing the "at first sight paradoxical direction of a materialism of the incorporeal" (Foucault 1981a, 69; see Elden 2016, 21).

13 The notion of the "more-than-human" was coined by Sarah Whatmore (2002, 159; see also Whatmore 1999, 33; Michael 2017, 112–4). See also Irving A. Hallowell's idea of the "other-than-human" (1960).

14 The conceptual proposal of a government of things incorporates a productive ambiguity, as the term can be read simultaneously as a *genitivus objectivus* and as a *genitivus subjectivus*. On the one hand "things" (itself an empirically open and contested category, as we will see later in this book) are conceived of as a governing agency, while on the other hand they represent what is addressed and targeted in practices of government. Thus, subjects and objects of government are not given in advance and exterior to governmental practices but co-emerge within them.

For a first outline of the idea of a government of things in Foucault's work, see Lemke 2015b.

15 Annemarie Mol, in her seminal article, credits Foucault with a "crucial" role in the "intellectual articulations of ontological politics" (1999, 87, note 2).

16 This provisional and tentative approach is very much inspired by Mol's introduction of the concept "ontonorm" and her reflections on its methodological prospects and limitations. Mol stresses the importance of keeping the meaning of the term fluid and flexible: "What does the term *ontonorms* lead you to see in the cases that you study? Where do you hit upon its limits? How might we adapt and play with it? If we sooner or later end up discarding the term *ontonorms* again because it stops being a strange, terse, productive oxymoron, that is fine by me. But this is my request. Please do not define this term. Abstain from all attempts to make it definite. Let's not make a *turn to ontonorms*, but rather keep them fluid, ambivalent, dancing and gerrymandering" (Mol 2013, 390; emphases in original).

1. IMMATERIALISM

1 Harman has described OOO as "the 'object-oriented' wing of the movement" (Harman 2010a, 1).

2 Meillassoux and other speculative realists turn to Hume in order to resolve the issue of correlationism. They argue that Hume's occasionalism makes it possible

to conceive of laws of nature or physical principles as contingent, malleable, and open to change (Norris 2013, 186–91).

For a variety of criticisms of Meillassoux's analysis in *After Finitude*, see Toscano (2011), Hägglund (2011), Johnston (2011), Brown (2011), and Roffe (2012). On problems and prospects of speculative realism in general, see Gratton 2014.

3 See Harman's account of the differences between his philosophical position and Meillassoux's: "Oversimplifying somewhat, we can say that there are two basic principles underlying the Kantian revolution in philosophy. (1) Kant distinguishes between phenomena and noumena. [. . .] (2) For Kant, the human-world relation is philosophically privileged. [. . .] Now, whereas Meillassoux rejects 1 and affirms 2, my own position affirms 1 and rejects 2. That is to say, Meillassoux rejects Kantian finitude in favor of absolute human knowledge, while I reject absolute knowledge and retain Kantian finitude, though broadening this finitude beyond the human realm to include all relations in the cosmos—including inanimate ones" (2012, 184–85). For a detailed account of the differences, see Harman 2011c.

4 For a more general discussion of Heidegger's work, see Harman 2007a.

5 "The science of geology does not exhaust the being of rocks, which always have a surplus of reality deeper than our most complete *knowledge* of rocks—but our practical use of rocks at construction sites and in street brawls also does not exhaust them. Yet this is not the result of some sad limitation on human or animal consciousness. Instead, rocks themselves are not fully deployed or exhausted by *any* of their actions or relations" (Harman 2013, 32; emphases in original).

6 According to Harman, objects do not have "crystal-clear sets of knowable properties"; rather, they are "dark and stormy events locked in a network with other such events" (2007a, 24). This lack of clarity seems to affect the definition Harman proposes as it provokes the question of what "some sort of" means and what counts as "unitary." Who gets to decide what is unitary enough to qualify as an object?

7 In Levi R. Bryant's words, "The democracy of objects is the *ontological* thesis that all objects, as Ian Bogost has so nicely put it, equally exist while they do not exist equally" (Bryant 2011a, 19; emphasis in original).

8 Harman's account of Latour is presented in two books: *Prince of Networks: Bruno Latour and Metaphysics* (2009) and *Bruno Latour: Reassembling the Political* (2014).

9 In the more systems-theoretically-minded language of Morton, "Objects encounter each other as operationally closed systems that can only (mis)translate one another" (Morton 2011a, 165).

10 Morton's position in *Hyperobjects: Philosophy and Ecology after the End of the World* is quite puzzling. After investing a lot of time and energy in arguing for the distinctiveness of hyperobjects, defining five shared properties—viscosity, nonlocality, temporal undulation, phasing, and interobjectivity (see Morton 2013a, 27–95), the book concludes with the strange observation that "every object is a hyperobject" (ibid., 201; Heise 2015, 461).

11 Harman points to similarities and differences between this idea of symbiosis on the one hand and concepts from evolutionary biology and Deleuze's philosophy on the other (see Harman 2016a, 42–51).
12 See, for example, Harman's critique of John Law's and Annemarie Mol's work—two important figures in ANT and material semiotics—as an "extreme form of anti-realism" (2016a, 26; 22–26).
13 However, other proponents of OOO are less hostile to relational accounts. Iain Hamilton Grant insists that the "conditions" upon which a given object's existence depends "do not belong to that object—they are not 'its' conditions, but conditions that 'possibilize' it" (Grant 2011, 43). See also Levi R. Bryant's focus on "difference" as the "minimal criterion for being." Bryant stresses that only "if a difference is made, then the being *is*" (2011b, 269; emphasis in original). As Geoff Pfeifer notes, in this "proto-structuralist fashion" "it is relationality—rather than thought—that defines the being of objects for Bryant" (2012, 469; see also Sheldon 2015, 221).
14 See also Harman's self-positioning statement: "[M]aterialism is a kind of idealism [. . .] I'm an anti-materialist" (Harman in Brassier et al. 2007, 398).
15 Rebekah Sheldon rightly notes that "object-oriented ontology's split object recalls Plato's account of form and his foundational distinction between that which 'always is and has no becoming' and that which 'comes to be and never is' (*Timaeus*, 58C). For Plato of the *Timaeus* and for Harman, the substance of an object never changes, [. . .] while its accidental qualia and exogenous relations are alone capable of becoming and perishing away again" (Sheldon 2015, 207). Sheldon proposes the notion of *chora* to mediate between form and matter—and between OOO's formalism on the one hand and feminist engagements with matter on the other—in explaining persistence and dynamics (ibid., 211–14).
16 See also Morton's reevaluation of Heidegger, revising his earlier criticism: "If anyone gives us a vivid sense of the uncanny strangeness of coexistence, it is Heidegger" (Morton 2013a, 22).
17 Sevket Benhur Oral uses arguments from OOO to demand a reorientation of educational practice, stressing the need to account for the "weirdness of reality" (Oral 2015, 460). Oral emphasizes the significance of aesthetic issues, as they provide a means to resist the totalizing tendency of rationality and cognition. For him, education should provide "proximate encounters with things with the full awareness of their irreducible autonomous inner reality" (ibid., 462). Following this objective, he argues "for aesthetics—and not cognition—to be the starting point and primary focus of education" (ibid., 460). And elsewhere: "Education is not about knowing facts. It is about interacting with facts as textured and layered events that have the power to surprise" (Oral 2014, 119).
18 For a different understanding of speculation that puts the emphasis on its historical and visionary dimensions and plays an essential role in feminist and critical thought, art, and activism, see Åsberg et al. 2015.
19 Meehan et al. 2013 base their analysis on the television drama *The Wire*, which deals with the drug trade and policing in the city of Baltimore.

20 See Morton's statement that "the aesthetic [. . .] is the secret door through which OOO discovers a theory of what is called 'subject'" (Morton 2011a, 173).
21 Stacy Alaimo also sees Bogost's quest for a posthuman experience ending up by reinstalling "a humanist and masculinist sense of a disembodied subject" (Alaimo 2014, 15). Alaimo quotes Bogost's line of reasoning and the questions he wishes to address with an object-oriented ontology: "What's it like to be a computer or a microprocessor, or a ribbon cable? . . . As operators or engineers, we may be able to describe how such objects and assemblages *work*. But what do they *experience*? What's their proper phenomenology? In short, what is it like to be a thing?" (Bogost 2012, 9–10; emphasis in original, quoted from Alaimo 2014, 15). Alaimo disputes that it is in any way meaningful or useful to imagine what it is like to "be" a cable. Rather, it is the philosophical positioning and the reasoning itself that come to the fore: "I do wonder, however, albeit rather anthropocentrically, what it is like to be a human imagining what it is like to be a thing. In this case, Bogost's speculations on what it means to *be* a particular object emerge from a sovereign, enclosed, rational, speculative, mind. There is no sense of embodied, interactive, intra-active, situated, or scientifically-mediated knowledges here" (Alaimo 2014, 15; emphasis in original).

For a similar argument concerning Meillassoux's proposition of the non-situatedness of thought, see Åsberg et al. 2015, 160–61.
22 See also Brett Bricker's comments on Morton's account of hyperobjects: "[H]uman objects anthropomorphize nonhuman objects by writing human language onto them. We find mountains *thinking*, stones *speaking*, hammers *wanting*, and cigarettes *demanding*, and we act as if we can understand the desires of inanimate objects. Unfortunately, for those invested in OOO, this act of attributing human traits to nonhumans seems to strengthen the primacy of humanness and weaken the democratic coexistence that Morton theorizes" (Bricker 2015, 365; emphases in original; see, e.g., Morton 2013a, 141, 161).
23 In fact, the turn to the isolated object might finally provide insights into its irreducible relationality. For example, Morton contends that hyperobjects like global warming are always already within us, as they are boundless and cannot be separated or contained in a space named "world" or "nature." As they "enter our skin and lungs, threaten our health and survival, and in many ways, *are* us" (Ach 2016, 132; emphasis in original), it is not sufficient to simply state that "[t]hey contacted us" (Morton 2013a, 201).
24 Andrew Cole convincingly argues that Harman and other proponents of OOO endorse a selective and undercomplex reading of Kant as a way of making themselves sound more original. However, Cole's accompanying theoretical claim that OOO must be comprehended as "the metaphysics of capitalism" (Cole 2015, 323) is itself reductionist, as it tends to narrow the intellectual interest in objects and materiality in the humanities and social sciences to a "commodity fetishism in academic form" (ibid.)—ignoring the important theoretical (and political) differences between ANT, vital materialism, and OOO (see also Cole 2013; Galloway 2013).

2. VITAL MATERIALISM

1 In an interview, Bennett names two contemporary political problems that brought her to endorse a vitalist conception of materialism: "The first was the intensification of an alarming trend in American public culture wherein a rise in the invocations by politicians of otherworldly powers (the Judeo-Christian 'Almighty' combating 'forces of evil') was paired with the positioning of violence and torture as legitimate tools of state. [. . . .] The second problem was ecological destruction" (Bennett and Loenhart 2011, 3–4).

2 For an instructive account of the contemporary relevance of vitalism as a concept, see Greco 2021.

3 Bennett opposes the "critical" account of modernity as an epoch marked by the disenchantment of the world by stressing the joy of enchantment as the basis for a successful politics striving for social justice, while at the same time avoiding the impasses of traditional forms of political mobilization. She fears that "acceptance of the disenchantment story, when combined with a sharp sense of the injustice of things by the Left, too often produces an enervating cynicism" (Bennett 2001, 13; 34). According to Bennett, it is the experience of enchantment that provides the motivational and affective prerequisite for political criticism and activism.

4 In fact, given Bennett's own account the chosen vocabulary of "weaken" or "enhance" might be misleading (or analytically "weak") as it suggests a quantitative model of "more or less." The question is rather how the "agency" of nutrients differently impacts on and shapes human wills, intentions and motives.

5 Bennett mentions the concept of the "deodand," which figured in English law from the thirteenth until the mid-nineteenth century and acknowledged the agency of nonhuman entities: "In cases of accidental death or injury to a human, the nonhuman actant, for example, the carving knife that fell into human flesh or the carriage that trampled the leg of a pedestrian—became deodand (literally, 'that which must be given to God'). In recognition of *its* peculiar efficacy [. . .], the deodand [. . .] was surrendered to the crown to be used (or sold) to compensate for the harm done" (Bennett 2010a, 9; emphasis in original; see also Lindemann 2001; Teubner 2006).

6 See also Barry's observation that "human geographers have increasingly argued that they need to attend to what has variously been understood as the liveliness, agency and powers of materials as well as persons. I contend, however, that although this argument is an important one, it does not address the ways in which the existence and the activity of material artefacts have progressively been subject to monitoring, assessment, regulation and management" (Barry 2013, 6).

I will discuss this criticism in more detail in Chapter 7.

7 As Bryan E. Bannon commented in his review of *Vibrant Matter*: "It is unproblematic to assert that all existing bodies are affective and susceptible to affectation, and one need not equate this two-sided capacity with life, even the asubjective life of metal that Bennett describes. If life is a *field of intensities* in the way Bennett

describes, then, far from being a property, it is a particular way of *relating* to the affections that surround an assemblage. Thus, on Bennett's own account, it is possible to assert that matter itself is not alive per se, but that life denotes a particular intricacy of responsiveness within complex alliances between smaller constituent assemblages" (Bannon 2011, 3; emphases in original; see also Hoppe 2017a).

8   Again, Bennett seems to be aware of the problem. In an interview she signals that she is reconsidering the "very provocative" concept of agency. Instead of stressing the "agentic consequences" of things, she is searching for "a better adjective—maybe 'effectivity' consequences" (Bennett in Watson 2013, 149).

See Chapter 7 for a broader discussion of the problem of agency in new materialist scholarship.

9   For a critical discussion of another example Bennett provides—the post-millennial debate in the US over the federal funding of stem cell research (2010a, 84–93)—see Rekret 2016, 237–38.

10 Bennett's vital materialism provokes the following essential question: "Once we start to really lavish attention on things and thing power, on their being made *and* on their being, can we start to highlight the asymmetries in the powers that exist in and alongside assemblages?" (Hinchliffe 2011, 398; emphasis in original; see also Braun 2011).

See also Gay Hawkins' analysis of "plastic materialities," which draws heavily on Bennett's work and her concept of a "force of things." However, beyond the general objective of disturbing an environmental ethics and politics that relies on a material essentialism (e.g., "plastic bags are bad"), her proposal of an "expanded politics of plastic bags" remains obscure (Hawkins 2010, 136).

11 This problem persists in Bennett's recent work, which offers an original engagement with the concept of sympathy in the writings of Walt Whitman (Bennett 2016; see also 2020). According to this interpretation, sympathy is not limited to a moral sentiment but exhibits a material force that draws bodies together. While this instructive account convincingly exposes the limits of many (too) narrow notions of sympathy in political and social theory, it remains one-sided and is unable to fulfill the high (political) expectations Bennett ascribes to it, as it exclusively engages with the affiliative and receptive dimensions of sympathy.

In his comment on Bennett's reading of Whitman's prose and poetry, Romand Coles (2016) suggests a very different and more forceful idea of sympathy. He stresses that articulating feelings and practices of sympathy necessarily involves attacking those who seek to devalue or diminish them. Thus, Coles' interpretation of Whitman's work argues for the necessity of "antagonistic sympathies" to counter Bennett's selective focus on the more productive and positive aspects of sympathy that is ultimately unable to address contemporary political and environmental challenges: "Bennett's ecology of sympathies risks contributing—perhaps unwittingly—to dismissals, deflections, and disavowals of urgently militant forms of struggle" (Coles 2016, 624). Quite contrary to Bennett's ambition to explore sympathy as "an *underdetermined vital force*" (Bennett 2016, 616; emphasis in original) that could be mobilized for political projects addressing inequalities and injustices (see Bennett 2016, 615), in the absence of its antagonistic complement it might "devitalize political vision and powers" (Coles 2016, 624).

3. DIFFRACTIVE MATERIALISM

1 For a more detailed account of representationalism, see Rorty 1980.
2 Barad develops this "new form of realism" (2007, 44, 207) in a critical engagement with Ian Hacking's analysis of the relation between "intervening" and "representing" (Hacking 1983). According to Barad, Hacking successfully exposes important limitations of representationalist accounts. However, she claims that his focus on intervening (instead of representing) ultimately remains within the representationalist frame. In Barad's reading, Hacking still subscribes to the conviction that the "world is composed of individual entities with separately determinate properties," an idea that Barad labels "entity realism" (2007, 55). Going beyond Hacking, her objective is to develop a "nonrepresentationalist realist account of scientific practices" (2007, 56) that not only goes beyond "traditional realism" but also leaves "entity realism" behind.
3 Barad uses the terms "interchangeably" (2007, 81).
4 On diffraction, see Bath et al. 2013; Kaiser and Thiele 2014; Thiele 2014. The term also enables a critical engagement with the concept of intersectionality. Evelien Geerts and Iris van der Tuin (2013) have argued that the debate on intersectionality is often informed by a politics of representation.
5 The concept of intra-action establishes a central point of difference between Barad's account and vital materialism on the one hand and OOO on the other. Concerning the former, it allows Barad to put forward an idea of dynamic relationality different from Bennett's account: "There is a vitality to intra-activity, a liveliness, not in the sense of a new form of vitalism, but rather in terms of a new sense of aliveness" (2007, 234–35). In contrast to OOO, the notion questions the solidity and persistence of individual "objects" in favor of their incessant becomings as "individuals emerge through and as part of their entangled intra-relating" (ibid., ix; for a more substantial comparison see Taylor 2016).

See also Harman's critique of "intra-action" (2016b). Interestingly, Morton suggests a way of mobilizing quantum physics to explore thingness and materiality that is quite different from Barad's relational account. In his view, quantum theory shows that the real consists of isolated units, the quanta (see Morton 2011a, 179–84).
6 For a similar critique of Butler's concept of agency as limited to practices of resignification, see Kerin 1999; Kirby 2006.
7 Butler clarifies her position on the new materialisms in an interview with Vikki Bell, explicitly rejecting the charge that she considers agency an exclusively human capacity: "When we talk about agency, we in fact need to divorce it from the

idea of the subject and allow it to be a complex choreographed scene with many kinds of elements—social, material, human—at work" (Butler in Bell 2010, 151).

8   For Barad, the notion "transhumanism" is also compromised as it "has already been appropriated for unreflective technophilic purposes and suggests a transcendent position" (2007, 428, note 6).

9   Surprisingly, Barad does not—apart from brief references in the footnotes—engage with performative accounts in STS, e.g., the work of Andrew Pickering, Annemarie Mol, John Law and Lucy Suchman (Pinch 2011, 433). See Chapter 7 for a more comprehensive discussion of STS work.

10 Barad uses the terms "object" and "thing" interchangeably without systematically distinguishing between them. Both refer to the idea of stable and fixed ontological entities.

11 The notion of "material-discursive practices" accentuates the inseparability of the discursive and the material. They are not externally related to each other but mutually implicated in the dynamics of intra-activity.

12 An apparatus also needs to be distinguished from an assemblage, which indiscriminately includes nonhumans as well as humans without taking account of the differential boundaries between them (see Barad 2007, 171). As we saw in Chapter 2, the concept of "assemblage" (*agencement*) coined by Deleuze and Guattari occupies a central position in Bennett's vital materialism. It is noteworthy that Barad rarely mentions or refers to Deleuze, although she uses quite frequently and centrally notions like "becoming," "(en)folding," and other terms that can be traced back to his work (with Guattari).

    I will come back to the conceptual distinction between "assemblage" and "apparatus" in the next chapter, when I contrast both of these terms with Foucault's understanding of the dispositive.

13 Barad claims that "the ability to respond to the other [. . .]cannot be restricted to human-human encounters when the very boundaries and constitution of the 'human' are continually being reconfigured and 'our' role in these and other reconfigurings is precisely what 'we' have to face" (2007, 392).

14 As Joseph Rouse (2004) has emphasized, Barad's conception of agential responsibility takes up and continues an older tradition of feminist science studies. This engagement with responsibility seeks to align the notion to understandings of scientific objectivity and conceive of "the *locus* of such responsibility as a prosthetically embodied engagement in material-discursive practices" (Rouse 2004, 154; emphasis in original). Thus, it is not consciousness or humanity that is the agentive force in this ethical inquiry; rather, human and nonhuman bodies are held accountable for the agential cuts they enact. This means agency is neither restricted to humans alone nor is it a property that pertains to some bodies while others are excluded from it. Rather, the "we" is conceived of as a posthuman and open collective (see e.g. Barad 2003, 828) and agency is a "doing" performed by human as well as nonhuman bodies.

    I will come back to Barad's account of ethics in the last part of this chapter.

15 The notion of the "apparatus of bodily production" was introduced by Haraway in her earlier work (1991, 200). See Josef Barla (2019) for an elaboration of the concept.
16 In the light of the profound and pervasive workings of ultrasound technology in reconfigurating bodies, Barad criticizes Foucault's analysis of disciplinary power as limited and outdated, claiming that "while the panopticon may be exemplary of observing technologies in the eighteenth century, ultrasound technology makes for a particularly poignant contemporary apparatus of observation" (2007, 201; for a similar critique see Haraway 1997, 12).
  I will discuss Foucault's concept of technology in Chapter 5.
17 Nete Schwennesen and Lene Koch (2009) take up Barad's insights in order to empirically investigate, in an ethnographic study carried out at an ultrasound clinic in Denmark, how the fetus is configured in prenatal diagnostics and risk assessment. Their ethnographic observations and interviews with couples show how data on fetal life are generated (as visual images and risk figures) and communicated to pregnant persons and their partners to provide them with information for (medical) decision-making (see also Sänger 2020).
18 The usefulness of this perspective is well illustrated by the example of reproductive technologies and obstetric ultrasound. Barad stresses that it is not sufficient to focus on the implications or the impact of these technologies. She points out that the historically and culturally specific materializations of new reproductive technologies enact their own exclusions, inequalities, and asymmetries. The concept of the apparatus makes it possible to go beyond the boundaries of those who are physically engaging with these technologies. The new reproductive technologies not only contribute to producing the fetus as a subject with own rights and needs that could possibly conflict with or contradict the rights and needs of the pregnant person; they also enact and re-configure existing inequalities and asymmetries. Race and class matter in the context of the construction of an "epidemic of infertility" (Barad 2007, 217) which, contrary to public debate and media discourse, disproportionately affects nonwhite and poor women. The new reproductive technologies materialize by specific agential cuts, which exclude some women and couples while simultaneously allowing the production of (more) white babies. Barad concludes that "the new reproductive technologies work to reproduce the fetus and particular race relations marking more women's bodies than just the particular ones that serve as 'maternal environments'" (ibid., 217; see also the concept of "stratified reproduction" put forward by Faye D. Ginsburg and Rayna Rapp [1995]).
19 Fraser refers to Pheng Cheah's warning of "the implausibility of identification as a paradigm of oppression" as "especially salient [. . .] where material marks are constituted through *physical* and not ideational ingestion, not necessarily of the order of the visible, such as the tracings of the digestive tract by inequalities in food production and consumption" (Cheah 1996, 120–21; emphasis in original; see also Jackson 2020).

20 Again, Barad turns to new reproductive technologies to illustrate this point, exploring their potential to go beyond the heterosexual and patriarchal matrix. One example is the use of donors' sperm or IVF by lesbian couples that challenge the heteronormative family model. While these practices might be considered as "subversive acts," Barad insists that they exhibit both destabilizing and reinforcing tendencies and participate in (different) exclusions: "In this case, the destabilizing effects of (mis)appropriations of new reproductive technologies, including challenges to the patriarchal and heteronormative structure, are accompanied by the reinforcement of class asymmetries and the cultural overvaluation of raising children that are genetic offspring" (2007, 220).

See Sarah Dionisius's (2015) review of empirical studies on how lesbian couples using donor insemination to become parents change perceptions and practices of family, parenting, and kinship, and how these practices affect gender relations.

21 This scientism is also quite manifest in Barad's claim to bring together in her account of agential realism the "best" available theories in physics and the social sciences (2007, 24–25).

22 Pinch points to the significance of David Bohm's version of quantum mechanics, which is critical of Bohr (2011, 434–35).

23 Without stating it explicitly in her response to Pinch, Barad seems to revert to the idea that because diffraction can be the object and the method of investigation, a similar claim could be made concerning agential realism in general. In this sense a performative reading of this tension might be available: "While Barad understands scientific practices performatively, her own work seeks to enact what it describes. [. . . .] Quantum physics here works as tool and object of study" (Schrader 2009, 352). Des Fitzgerald and Felicity Callard regard this refusal to "separate the practice of science from the practice of studying science from the outside" (2015, 20) as a particular strength of Barad's agential realism.

24 The tendency to make exaggerated claims about quantum physics as a new basis for epistemology, ontology, and ethics is even more surprising as there are alternative conceptual and theoretical resources available for agential realism. The exclusive reference to Bohr could be easily replaced by turning, for example, to biological theories and concepts (Schweber 2008, 882). In fact, one of the examples Barad uses to explain the inseparability of epistemology, ontology, and ethics in agential realism is informed by biology rather than physics. She refers to certain anatomical and physiological characteristics of the brittlestar, a relative of the starfish, to argue for the entanglements of knowing, being and doing (Barad 2007, 369–84).

Astrid Schrader's impressive study of the "phantomatic ontology" of *pfiesteria piscicida* and the scientific controversies around the existence of microorganisms as a possible "cause" of the deaths of large numbers of fish in the estuaries of the US mid-Atlantic productively makes use of an agential realist framework (Schrader 2010).

25 Sometimes Barad even reverts to a vitalist vocabulary to describe material agency: "Matter feels, converses, suffers, desires, yearns and remembers" (Barad in Dolphijn and van der Tuin 2012, 59).
26 Ahmed's article was followed by several comments and replies (Davis 2009; van der Tuin 2008). See also Peta Hinton's (2013) review of the debate.
27 Sherryl Vint notes that Barad's reading of representationalism is rather reductionist, as it "relies on a rather static word-equals-thing concept" (2008, 316) that tends to ignore the complex understandings of representations in semiotics.
28 As one commentator remarks: "I also would not agree with Barad's assessment of the failure of representational approaches in fundamental physics. The advances at the nuclear and subnuclear levels were due to the possibility of a confluence between ontology and representation" (Schweber 2008, 881).

   However, Barad sometimes acknowledges the significance of a "circumstantial" (or "strategic") account, for example in her discussion of Casper's article. Barad criticizes Casper's principled rejection of fetal agency, asking instead: "Isn't it possible that in certain circumstances there may be a need empirically and strategically to invoke fetal agency to counter the material effects of sexism or other forms of oppression?" (2007, 216)
29 Caroline Braunmühl claims that Barad's stress on the activity of all matter comes at a cost. First, it continues rather than disrupts the devaluation of passivity and "accords with hegemonic, male-supremacist discourse, which feminises that attribute" (2018, 231). Secondly, it tends to negate or understate substantial differences by highlighting sameness and similarity, which might result in "an assimilatory move that may well underestimate power differentials in the rather different senses of 'agency'" (ibid., 236).
30 Barad addresses the problem of justice on several occasions in *Meeting the Universe Halfway*, stating that "the yearning for justice [. . .] is the driving force behind this work" (2007, xi). See also her claim that "questions of space, time, and matter are intimately connected, indeed entangled, with questions of justice" (ibid., 236).

PART II. ELEMENTS OF A MORE-THAN-HUMAN ANALYTICS OF GOVERNMENT

1 See the Introduction to this book.
2 See also Mick Smith's argument that "the implications of biopolitics for ecology and the ecological implications of biopolitics have hardly even been noticed" in the wake of Foucault's work (Smith 2011, xv).

   According to Gesa Lindemann, Foucault remains "naïvely anthropocentric" (2003, 27), as for him the only relevant social bodies are those of human beings (see also Lindemann 2002, 24–5).
3 See Philo 2012 for an extensive engagement with Thrift's critique of Foucault.

4 See Gary Gutting's observation that, "with regard to the well-established natural sciences, Foucault seems content to accept the approach of Bachelard and Canguilhem" (1989, 255; see also 52–54).

5 Foucault's esteem for the natural sciences is very well illustrated in the following interview passage: "[I]f, concerning a science like theoretical physics or organic chemistry, one poses the problem of its relations with the political and economic structures of society, isn't one posing an excessively complicated question? Doesn't this set the threshold of possible explanations impossibly high? But on the other hand, if one takes a form of knowledge (*savoir*) like psychiatry, won't the question be much easier to resolve, since the epistemological profile of psychiatry is a low one and psychiatric practice is linked with a whole range of institutions, economic requirements and political issues of social regulation?" (Foucault 1980a, 109)

However, Foucault seems to have been less convinced by this juxtaposition in his later work, when the notion of *tekhnē* gained more currency. In an interview with Paul Rabinow on architecture, he stated that "if one placed the history of architecture back in this general history of *tekhnē*, in this wide sense of the word, one would have a more interesting guiding concept than by the opposition between the exact sciences and the inexact ones" (Foucault 2000a, 364). I will come back to this point in Chapter 5.

6 Interestingly, Barad's critical reading of Foucault exclusively engages with his work prior to the lectures on governmentality at the Collège de France in the end of the 1970s. She never discusses the theoretical shift that arrives with Foucault's concept of government, which goes beyond his former focus on disciplinary power.

On the emergence of the problem of government in Foucault's work, see Lemke 2019.

4. MATERIAL-DISCURSIVE ENTANGLEMENTS

1 To be sure, there are more-than-human aspects in Foucault's earlier work as well. In his dissertation, *Introduction to Kant's Anthropology* (2008b), submitted in 1961, he stressed how anthropocentrism and humanism have shaped modern thought. While it might be a bit exaggerated to claim that these "views can definitely be considered the opening statements of new materialism" (Dolphijn and van der Tuin 2012, 88; 164–66), both this book and *The Order of Things* (Foucault 1970) emphasize that "man" was a historical figure and a rather recent conceptual invention.

2 Foucault is referring to the book *Le Miroire politique, œuvre non moins utile que necessaire à tout monarches, roys, princes, seigneurs, magistrats, et autres surintendants et gouverneurs de Republicques* (1555). See Foucault 2007a, 112, note 15 for some bibliographical information on the author.

3 Drawing on Martin Heidegger's discussion of the term (1967), Bruno Latour has highlighted the semantic ambiguity of "thing," pointing to older etymologies

in which the term denotes a political assembly, a gathering place, or a space for negotiation: "[I]s this not extraordinary that the banal term we use for designating what is out there, unquestionably, a thing, what lies out of any dispute, out of language, is also the oldest word we all have used to designate the oldest of the sites in which our ancestors did their dealing and tried to settle their disputes? A thing is, in one sense, an object out there and, in another sense, an *issue* very much *in* there, at any rate, a *gathering*" (2004a, 233; emphases in original; see also Latour and Weibel 2005).

4 See also Giorgio Agamben's concept of "bare life" (Agamben 1998).

   In his book *Persons and Things: From the Body's Point of View*, Roberto Esposito goes back to the concept of the person in Roman law, seeking to show that it is grounded in an opposition to things: "[A] thing is a *non*-person and a person is a *non*-thing" (Esposito 2015, 17; emphasis in original). Esposito argues that personhood is intimately linked to the possession of things. According to him, this conceptual framing in Western history since Roman antiquity has allowed a hierarchal distinction not only between humans and nonhumans but also within the human species and within every single individual. It made it possible to deny rights to nonhuman animals, and also to distinguish various levels of personhood down to the status of animality: the line of subordination and exclusion goes from slaves in Roman antiquity to the denomination of Jews as "anti-persons" in Nazi Germany. Also, it enabled the distinction between a rational and an animal part within each individual. This division between persons and things not only produces exclusionary and discriminatory effects on the level of persons but has an equally negative outcome within the realm of "things": "The process of de-personalization of persons is paralleled by that of the derealization of things" (2016, 31). The distinction between persons and things leads to a transformation of things into objects at some person's disposal, commodities defined by exchange value and governed by a logic of equivalence that denies their singularity.

5 All translations from French and German are my own.

   According to Senellart, in his lectures on governmentality Foucault captures very well this transformation from sovereignty to government. However, he cautions that de la Perrière's book is not a particularly well chosen example, since it repeats the traditional idea of a good order of things already formulated by Augustine in the Christian context (Senellart 1995, 43, note 2). In a similar vein, Danica Dupont and Frank Pearce criticize Foucault's interpretation of de la Perrière's work. Rather than pointing to modern politics, they argue, de la Perrière's understanding of government is "more derived from a Renaissance Christian humanist context of cosmic order" (Dupont and Pearce 2001, 135, 135–38). See also Thomas Aquinas' concept of a "government of things" as the ruling of the universe by divine reason (Goerner 1979, 111–12).

6 Joseph Görres declared at the beginning of the nineteenth century that: "If you want to govern humankind, you should govern it as it governs nature: by its own

self." ["*Willst du die Menschheit regieren, . . . so regiere sie, wie sie die Natur regiert, durch sich selbst*"] (quoted by Sellin 1984, 372; emphasis in original).

As Bruce Braun and Sarah J. Whatmore rightly remark, the early political theory of Machiavelli, Hobbes, and Spinoza "understood collectivities [. . .] in decidedly materialist terms, as a question of their ongoing assemblage rather than as primarily theological or philosophical questions" (Braun and Whatmore 2010, xiv).

On Spinoza's concept of government, see Saar 2009.

7 I will come back to this "cybernetic" understanding of government in Chapter 5.

8 See Kafka's (2012, no page number) reconstruction of the debate: Comte "made this argument in the third installment of Saint-Simon's Cathéchisme des industriels. The essay was published in 1822 as the *Plan des travaux scientifiques nécessaires pour réorganiser la société* and then again in 1824 as *Système de politique positive* (it would also be known as the *Opuscule fondamentale*). Saint-Simon wanted to take credit for the publication, which Comte had written at his request, but the younger man insisted on having his name attached to it. The result was a complicated printing history and an even more complicated schism between master and disciple that probably explains why subsequent readers were confused about its authorship."

9 The passage at the beginning of chapter 4 in Book XIX is often mistranslated. Here is the original French version: "Plusieurs choses gouvernent les hommes: le climat, la religion, les lois, les maxims du gouvernement, les exemples des choses passées, les mœurs, les manières" (Montesquieu 2008, 181).

For an analysis of Montesquieu's influence on Comte, see Pickering 1993, 46–48.

10 See Engels' formulation in the *Anti-Dühring*: "The interference of the state power in social relations becomes superfluous in one sphere after another, and then ceases of itself. The government of persons is replaced by the administration of things and the direction of the processes of production. The state is not 'abolished,' *it withers away*" (2000, 355; emphasis in original).

11 Due to this difficult translation process, *dispositif* remained for many scholars in the Anglophone intellectual space an "excessively vague" and "troublesome term" (Dreyfus and Rabinow 1983, 120), while it has attracted a lot of interest among researchers in the French-speaking world (see, e.g., Jacquinot-Delaunay and Monnoyer 1999a; Beuscart and Peerbaye 2006). For a brief conceptual history of the term, see Jacquinot-Delaunay and Monnoyer 1999b; Peeters and Charlier 1999; Abadía 2003; Beuscart and Peerbaye 2006.

Before Foucault took up the notion, it had a central role in the work of Jean-François Lyotard and Jean-Louis Baudry (Lyotard 1973; Baudry 1975). For contemporary uses of the concept in media theory and science and technology studies, see Paech 1997; Gomart and Hennion 1999; Kessler 2003; Thomas 2015; Callon and Muniesa 2003. For an exploration of the different meanings of "disposition" in the history of philosophy and psychology, see Ritter and Pongratz 1972.

12 Agamben (2009, 3–6) traces Foucault's interest in the notion of the dispositive back to *The Archaeology of Knowledge* (Foucault 1972), where the notion of positivity (*positivité*) plays an important role. These two terms share the same etymological source, as they both derive from the Latin *ponere*. Agamben argues that Foucault took up a particular understanding of positivity developed by Jean Hyppolite, one of his teachers, and the interpretation of Hegel he advocated. Hyppolite conceived of the "positivities" in Hegel as the historical horizon that imposes particular rules and constraints on individuals. According to this reading, Foucault was already, in *The Archaeology of Knowledge*, seeking to investigate "concrete modes in which the positivities (or the dispositives) act within the relations, mechanisms, and 'plays' of power" (Agamben 2009, 6, translation modified; see also Pasquinelli 2015, 88, note 7).

13 Foucault distinguishes the "logic of strategy" not only from psychoanalytical accounts but also from a "dialectical logic" by stressing its strong relational understanding of co-existence and difference: "The function of strategic logic is to establish the possible connections between disparate terms which remain disparate. The logic of strategy is the logic of connections between the heterogeneous and not the logic of the homogenization of the contradictory" (2008a, 42).

14 While Agamben (2009, 7) distinguishes between a juridical, a technological and a military use of the term, it seems more pertinent to focus on its ontological dimension instead of a juridical understanding. The important point is not the legal ruling as such but rather the fact that it is announced and thereby brought into being, the enactment of the decision.

Curiously, Agamben's essay *Che cos'è un dispositivo?*, which argues for the etymological and conceptual specificity of *dispositif*, was published in English under the title *What Is an Apparatus?* (Agamben 2009 [2006]; see also Bussolini 2010, 85, note 1).

15 As we will see in Chapter 7, Foucault shares this idea of a heterogeneous and mobile network that links humans and nonhumans, the material and the semiotic with actor-network theory (see, e.g., Law 1987; Callon 1986).

16 On the notion of the aleatory and the idea of an "aleatory materialism," see Althusser 2006 [1994].

See Chapter 6 for an analysis of the government of the aleatory.

In their philosophical theory of causation, Stephen Mumford and Rani Lill Anjum argue that "dispositionality is a primitive, unanalysable modality that is intermediate between pure possibility and necessity" (2011, 193).

17 Claudia Aradau and Rens van Munster analyze the operations of a "dispositif of risk" in the government of terrorism: the dispositive "creates a specific relation to the future, which requires the monitoring of the future, the attempt to calculate what the future can offer and the necessity to control and minimize its potentially harmful effects" (2007, 97–98; Aradau 2010).

18 This double movement is described by Seb Franklin in an abstract for a talk entitled "Forms of Disposal" (2007).
19 Davide Panagia links the concept of the dispositive to a particular reading of Foucault's lectures on Manet (Foucault 2009), arguing that "the distributions of visibilities Foucault enlists in his (and our) viewings become the structuring visual mode that informs both his shift from the language of apparatus to *dispositif* and his formalist readings of modern works of political theory" (Panagia 2019, 717).
20 In this sense, dispositives "encrust themselves and depend for their conditions of exercise on the level of the micro-relations of power. But there are always also movements in the opposite direction" (Foucault 1980b, 199): forms of coordination and expansion of power strategies that are directed "from the top downwards" (ibid., 200).
21 Thomas LaMarre brings Simondon's relational philosophy of technics into a dialogue with Foucault's analytics of power: Simondon "refutes the realism that takes structure or form to be reality; instead he sticks to the realism of relation in order to show not only that the individual is in process but also that stopping or prolonging that process brings into play a dispositif (to use Foucault's term), that is, a set of techniques, an 'apparatus' or 'paradigm,' around which procedures of territorialization, discipline, or control may gather" (2013, 87).

I will discuss the role of technologies in Foucault's work in more detail in Chapter 5.
22 Foucault specified his understanding of strategy in the essay "The Subject and Power," delineating three senses of the word: "(1) to designate the means employed to attain a certain end; [. . .] (2) to designate the way in which a partner in a certain game acts with regard to what he thinks should be the action of the others and what he considers the others think to be his own; [. . .] (3) to designate the procedures used in a situation of confrontation to deprive the opponent of his means of combat and to reduce him to giving up the struggle" (Foucault 2000b, 346).
23 Catherine Millot has suggested this formula in an interview with Foucault (1980b, 202).
24 Noël Nel also stresses the strategic dimension of the dispositive in his analysis of the evolution of French television from the end of the 1960s to the mid-1980s (Nel 1999).
25 In a debate with historians, Foucault insists that the programs he analyzes (e.g., the Panopticon) are not "ideal types" in the Weberian sense. He stresses that programs "never work out as planned. But what I wanted to show is that this difference is not one between the purity of the ideal and the disorderly impurity of the real, but that in fact there are different strategies which are mutually opposed, composed and superposed so as to produce permanent and solid effects which can perfectly well be understood in terms of their rationality, even though they don't conform to the initial programming: this is what gives the resulting

apparatus (*dispositif*) its solidity and suppleness" (Foucault 1991a, 80–81; see also Silva-Castañeda and Trussart 2016).

26 The Foucauldian concept of the dispositive has been used (in the German-speaking social sciences) in qualitative research methodologies to enlarge conventional discourse-analytical approaches by including "*discourses, practices, institutions, objects and subjects*" (Bührmann and Schneider 2008, 68; emphasis in original). The self-declared objective of this "dispositive analysis" is to empirically investigate the networks between knowledge structures, institutional fields, and forms of subjectivation in order to provide a more comprehensive and complex analysis of the social (Bührmann and Schneider 2008; Bührmann 2013; see also Diaz-Bone and Hartz 2017). For an exploration of the analytical potential of dispositive analysis in organizational research, see Raffnsøe et al. 2016.

27 Matteo Pasquinelli argues that Agamben imposes on the Foucauldian notion of dispositive "a Christian lineage that, even from a philological perspective, is not central to it" (2015, 85). Pasquinelli instead traces Foucault's use of the term back to the work of Georges Canguilhem and his understanding of organic and social normativity. Canguilhem seems to have used *dispositif* first in the essay "Machine and Organism" (2008a), originally published in 1952, to discuss Descartes' understanding of a mechanics of power that seeks to replace forms of power relying on personal direction and control (Pasquinelli 2015, 84–85; see also Chapter 5).

28 In their proposal for a "sociology of attachment," Emile Gomart and Antoine Hennion (1999) emphasize the productive dimension of the Foucauldian notion of the dispositive, which makes it possible to circumvent conventional (sociological) dichotomies such as active/passive, free/determined, or subjugated/dominant. It focuses on "the tactics and techniques which make possible the emergence of a subject" (1999, 220) and shifts attention from the concept of agency to analyze "events" and the generation and proliferation of competencies. See Chapter 7 for a more detailed discussion.

29 This interpretation is very much in line with Canguilhem's interpretation of the bourgeoisie as a "normative class" that inaugurated new norms instead of imposing laws and working through repression: "Between 1759, when the word 'normal' appeared, and 1834, when the word 'normalized' appeared, a normative class had won the power to identify [. . .] the function of social norms, whose content it determined, with the use that that class made of them" (Canguilhem 1991).

30 See, for example, the following passage in *Discipline and Punish*: "[T]he sovereign and his force, the social body and the administrative apparatus [*l'appareil*]; mark, sign, trace; ceremony, representation, exercise; the vanquished enemy, the juridical subject in the process of requalification, the individual subjected to immediate coercion; the tortured body, the soul with its manipulated representations, the body subjected to training. We have here the three series of elements that characterize the three mechanisms [*dispositifs*] that face one another in the second half of the eighteenth century" (1979, 131).

31 I will analyze Foucault's understanding of dispositives of security in Chapter 5.
32 Bussolini (2010) has convincingly argued that there are important semantic and conceptual differences, pointing to the Latin derivation of the two terms that still informs their contemporary usages. The etymological source of *appareil* is the Latin word *apparātus*, preparation, from the past participle of *apparāre*, to prepare. It "refers to a preparation or making ready for something: a furnishing, providing, or equipping. [. . .] *Dispositio*, on the other hand, names a regular disposition—an arrangement—and relates to the verb *dispono* and its root *pono* [. . .]. *Dispono* concerns placing here and there, setting in different places, arranging, distributing (regularly), disposing; it also addresses specifically setting in order, arraying, or settling and determining (in military or legal senses). *Pono*, which is intimately related, concerns putting, placing, or setting down (as things in order or troops), or forming or fashioning (as works of art). [. . .] Thus, though apparatus refers to real and movable things, on this reading dispositive has the more robust ontological sensibility as that which creates (possibly) or that which creates an arrangement that gives strategic and decisive import to a state-of-affairs" (Bussolini 2010, 96).
33 The notion of the apparatus is also present in the work of Deleuze und Guattari, especially in their concept of the "apparatus of capture" (*appareil de capture*), which differs from the Althusserian focus on the state (Deleuze and Guattari 1987, 424–73).

Interestingly, Althusser in his essay on ideological state apparatuses also distinguishes between *appareil* and *dispositif*; here, the latter seems to be a subset of the former (e.g., Althusser 1971, 167; Bussolini 2010, 94, note 21). However, in his later work on aleatory materialism he drops the language of the apparatus to turn to the notion of the dispositive instead (see, e.g., Althusser 2006; Panagia 2019, 723, note 27). On the relation between Althusser and Foucault, see Montag 2013, 141–70.
34 Foucault's interest in introducing the notion of government is precisely to disentangle the term from its "rigorous statist meaning" (2007a, 120; see Lemke 2007).
35 The fact that the notion of assemblage also figures in the English translation of Foucault's lectures of 1978 and 1979 at the Collège de France (e.g., 2007a, 296, 315) has prompted some interpreters to note "a fascinating slippage in the language of apparatus/assemblage" (Legg 2011, 129). However, Foucault did not use the term *agencement* but rather employed the French word *ensemble* in these passages.
36 This processual and relational account of dispositives differs crucially from Bennett's restricted understanding of "structure," which precludes the possibility of productive effects and remains bound to an anthropocentric horizon: "a structure can act only negatively, as a constraint on human agency, or passively, as an enabling background or context for it" (Bennett 2010a, 29; see Barnwell 2017, 33).
37 However, the semantic and conceptual difference between "dispositive" and "assemblage" is less clear cut when it comes to the varying definitions of the original terms. Referring to different French dictionaries, Panagia documents how the

meaning of the term *agencement* shifts between a focus on connecting or assembling and interpretations that put the accent on ordering or arranging—the latter being closer to *dispositif*: "The *Dictionnaire de la Langue Française* defines *agencement* as 'Action d'agencer' (the activity of connecting); as well as 'Ajuster, mettre en arrangement' (to adjust, to place in an arrangement); and finally, 'En termes de peinture, arranger des groups, des figures, adjuster les draperies, disposer les accessoires' (in terms of painting, to arrange groups, figures, adjust draperies, and dispose accessories) (*Dictionnaire de La Langue Française*, s.v. 'agencement'). The dictionary of the Académie Française, in contrast, defines 'agencement' as 'Manière d'arranger, de mettre en ordre' (a manner of arranging or placing in order) as well as in architecture, 'dispositions et rapport des différentes parties d'un edifice: l'arrangement, les proportions relatives des divisions d'un plan, d'une façade, d'une décoration' (dispositions and relations of the different parts of an edifice: the arrangement, or the proportions of the relative divisions of a plan, a façade, or a decoration) (*Dictionnaire de l'Académie Française*, s.v. 'agencement')" (Panagia 2019, 716–17, note 7).

My argument here is that most of the literature follows the first line of interpretation ("connection") while neglecting the second ("arrangement").

5. MORE-THAN-SOCIAL CONFIGURATIONS

1 See Steven Dorrestijn's thesis that "the notable relevance of Foucault's work to the philosophy of technology is exactly this approach of revealing the role of (hard) technology for *governing and fashioning human subjects*" (2011, 223; emphasis in original).
2 For a detailed reconstruction of Foucault's different engagements with and usages of "technique" and "technology," see Behrent 2013. Behrent's article also provides instructive historical background on the impact of technology in changing production and consumption schemes in postwar France, and the emergence of a species of "technocrats" who played a central role in shaping the public discourse and policy agenda by setting national economic priorities and determining areas in which French society should be modernized.
3 For notable exceptions see Dorrestijn 2011; Behrent 2013; Matthewman 2013; Lustig 2014.
4 Foucault himself suggests the term "disciplines" to denote "these blocks, in which the deployment of technical capacities, the game of communications, and the relationships of power are adjusted to one another according to considered formulae" (2000b, 339). However, this proposition is not entirely convincing as it tends to confuse a general concept that seeks to capture the coordination of very different technological practices with a specific technology of power.
5 See Andrew Barry's reminder that in "Foucault's account, government is inevitably a technical matter. Practices of government rely on an array of more or less formalized and more or less specialized technical devices from car seat-belts and driving codes to dietary regimes; and from economic instruments to psycho-

therapy. Moreover, government operates both on and across many distinctions which are so critical to our sense of the terrain of politics: public and private; state and market; the realm of culture [...] and the domain of nature [...]. In this way, the study of government [...] opens up a much broader field of politics to inspection" (2001, 5).

6 Mitchell Dean points to the limits of Heidegger's concept of *tekhnē* in contrast to the Foucauldian account, as Heidegger is neither capable of analytically distinguishing between the material, natural, human and technical elements of the standing reserve nor interested in the strategic dimension of technologies (Dean 1996, 57–61; see also Latour 2007, 140–41; Revel 2009).

See Rayner 2007 for a more comprehensive analysis of Foucault's philosophical relationship to Heidegger.

7 On the concept of social constructivism see Hacking 1998; 2000.

8 Foucault does not mention the name of the scholar to whom he refers.

9 Examining Foucault's work against the backdrop of the limits of Marxist theory and politics, Barry Smart identifies, in addition to the waning appeal of Soviet socialism, three theoretical problems: the premise that economic factors prove decisive "in the final instance," insufficient attention to the interrelationship between politics and power, and the claim to scientificity (Smart 1983, 4–31).

Étienne Balibar has stressed that "the whole of Foucault's work can be seen in terms of a genuine struggle with Marx, and [...] this can be viewed as one of the driving forces of his productiveness" (Balibar 1992, 39).

10 Bruno Latour has also written of human individual and firearm in combination as more than the sum of their parts, describing the result as a "gun-citizen" (Latour 1994, 32; see Matthewman 2013, 286).

11 See also the lectures at the Collège de France 1981/82, where Foucault defines *tekhnē* as "an art, a reflected system of practices referring to general principles, notions, and concepts" (2005, 249).

12 Noortje Marres and Javier Lezaun have argued that Foucault's analytics of governmental technologies is limited to a "'sub-political' understanding of the 'politics of things'" (2011, 494). In this account, "hard" technologies and material objects provide a distinctive order of things that operate on subjects to structure possible action and exert specific forms of constraint. While Foucault attends to the question of how objects, artifacts and socio-material architectures are invested with moral and political capacities, he is reluctant to investigate "how objects acquire 'powers of engagement' and how those powers of engagement are articulated, discussed and contested in the public domain" (Marres and Lezaun 2011, 495; see also Marres 2009).

13 On the different metaphorical understandings of the state as a machine, see Stollberg-Rilinger 1986; Mayr 1986; Agar 2003; Koschorke et al. 2007.

14 For a classical case study focusing on the controversy between the political theorist Thomas Hobbes and the natural philosopher Robert Boyle in the seventeenth

century about the status of experimental science and the existence of a vacuum, see Shapin and Schaffer 1985.
15 While Foucault often used physical terms to stress the singularity and distinctiveness of the analytics of power (e.g., its description as a "microphysics of power"), he sometimes seems to be surprised by the simultaneous emergence of important findings and conceptual innovations in physics and the invention of a new form of government in the seventeenth and eighteenth centuries. For example, he describes Leibniz as "the general theorist of force as much from the historical-political point of view as from the point of view of physical science. Why is it like this? What is this contemporaneousness? I confess I know absolutely nothing about it, but I think that the problem inevitably arises insofar as Leibniz is proof that the homogeneity of the two processes was not entirely foreign to the thought of the time" (Foucault 2007a, 296; see Seibel 2016, 54).
16 "For what is the *Heart*, but a *Spring*; and the *Nerves*, but so many *Strings*; and the *Joynts*, but so many *Wheeles*, giving motion to the whole Body [. . .]?" (Hobbes 1962 [1651], 1; emphases in original)
17 Johann Heinrich Gottlob von Justi characterizes the sovereign as the "first mainspring" that "puts everything in motion" ("erste Triebfeder [. . .], die alles in Bewegung setzet") (Justi 1970 [1764], 87).

See also Justi's claim that a well-organized state "should operate like a machine where all wheels and mechanisms fit accurately into each other" ("Ein wohl eingerichteter Staat muß vollkommen einer Maschine ähnlich seyn, wo alle Räder und Triebwerke auf das genaueste in einander passen") (ibid., 86).
18 For a more detailed account, see Türk et al. 2002, 94–100.
19 See Mayr (1969, 123–24) for a short historical account of how the concepts of "regulator," "governor," and "moderator" were used in English and French during the latter part of the eighteenth and the early nineteenth century (see also Canguilhem 2012).

Benjamin Seibel (2016, 49) points out that Foucault's understanding of government grasps a technological problematization aiming at the systematic and permanent regulation of human behavior—in contrast to the idea of steering. While "steering" is mostly reserved to singular events of directing to achieve certain ends, "regulation" refers to a more general establishment and stabilization of multiple systems that allow governing to take a more comprehensive and anticipatory mode.

My argument in this and the following section is indebted to Seibel's "cybernetic" reading of Foucault's work.
20 For a comprehensive history of feedback systems, see Mayr 1970.
21 Adam Smith's *Inquiry into the Nature and Causes of the Wealth of Nations* (1937 [1776]) discusses a variety of social feedback mechanisms, but most famous is the general system of supply and demand. While Smith seems never to have extensively commented on technical feedback devices in his writings, there are still "numerous conceivable points of contact between Adam Smith and the mechani-

cal feedback systems of this time" (Mayr 1971a, 21). One interesting biographical fact is that Smith and Watt were friends, and Smith paid regular visits to Watt's workshop. Also, it needs to be noted that Watt took out no patent on the steam-engine governor he designed, possibly taking the feedback systems that had been around for a long time not as original inventions but rather as commonplace devices (Mayr 1971a, 16–18).

22 In this chapter and the next one I focus on Foucault's account of classical liberalism, as developed in his lectures at the Collège de France in 1978 and 1979. The third part of this book also engages with Foucault's analysis of neoliberal government, especially with his concept of "environmentality" (see Chapter 8).

23 On this problematic and paradoxical relationship within liberal governmentality between the incessant production of freedom and the danger of its destruction, see Dillon and Reid 2009; Lemke 2014.

24 For a political history of statistics, see Desrosières 2002.

25 While the Panopticon is often characterized as typical of the disciplinary mode of power, it is first and foremost a technological machine that no longer relies on external steering but governs processes of self-government: "it arranges things in such a way that the exercise of power is not added on from the outside, like a rigid, heavy constraint, to the functions it invests, but is so subtly present in them as to increase their efficiency by itself increasing its own points of contact. The panoptic mechanism is not simply a hinge, a point of exchange between a mechanism of power and a function; it is a way of making power relations function in a function, and of making a function function through these power relations" (Foucault 1979, 206–7).

26 For an instructive historical contextualization of the paper and a discussion of how Maxwell's interest in governors relates to his scientific work in general, see Mayr 1971b.

27 It is quite ironic that Maxwell's text actually mentions "M. Foucault"—crediting him with designing an arrangement within which "the force acting on the centrifugal piece is the weight of the balls acting downward, and an upward force produced by weights acting on a combination of levers and tending to raise the balls" (Maxwell 1868, 273). Maxwell was certainly referring to Léon Foucault, who wrote extensively on speed regulation (see Mayr 1971b, 428).

28 See, for example, Alfred Russell Wallace's comparison of Watt's feedback mechanism with the workings of natural selection: "The action of this principle is exactly like that of the centrifugal governor of the steam engine, which checks and corrects any irregularities almost before they become evident; and in like manner no unbalanced deficiency in the animal kingdom can ever reach any conspicuous magnitude, because it would make itself felt at the very first step, by rendering existence difficult and extinction almost sure soon to follow" (Wallace 2008 [1871], 291).

29 On the history of the term and the meaning of *kubernētēs*, see Lang 1970, 23–69.

30 Foucault's interest in linking medicine, self-formation, and politics to cybernetics was not limited to the lectures at the Collège de France. As an early text shows,

Foucault seemed to have been familiar with cybernetic ideas and vocabulary by the 1960s. "Message ou bruit?" ("Message or noise?"), published in 1966, discusses medical practice in terms such as code, message, and noise. The text ends with the following words: "We might ask the question if the theory of medical practice has to be rethought not in positivistic terms but in those that are currently developed in practices like the analysis of language or the handling of information" (Foucault 1994f, 560).

See also Friedrich Kittler's (somewhat exaggerated) claim: "The shorter or more occasional his texts, the more Foucault navigated from the safe banks of his libraries to the open sea of media technologies until on the transatlantic horizon of all their theories Wiener's and Shannon's mathematical concept of message emerged" (Kittler 1999, 8).

31 See also Andrew Pickering's account of cybernetics and its relational ontology (2010).

6. BEYOND ANTHROPOCENTRIC FRAMINGS

1 See below for a discussion of this text.
2 Elden (2017, 18) suggests that Foucault gave up the original idea of pursuing a genealogy of modern genetics because he became convinced that Jacob's book already contained important results for such an endeavor. See the quote from *The Order of Discourse*: "This is the work that has just been done by François Jacob with a brilliance and an erudition which could not be equalled" (Foucault 1981a, 73).
3 For a historical account of how genetics and biochemistry were transformed into molecular biology, based on cybernetics and information theory as new paradigms, see Rheinberger 1997; Kay 2000; see also Keller 1995 on informational metaphors of life.
4 This performative account of life entails a critique of neo-Darwinist accounts of evolution and also goes beyond interactionalist paradigms (e.g. the nature-nurture problematic). It is well articulated in developmental systems theory (Oyama 2000; Oyama et al. 2001; Fausto-Sterling 2003) and in the anthropological work of Tim Ingold (2004; Ingold and Pálsson 2013). Ingold stresses that for information theorists like Norbert Wiener, John von Neumann, and Claude Shannon, "information" has "no semantic value whatever; it does not *mean* anything. Information, for them, meant simply those differences, in the input to a system, that make a difference in terms of outcome. This point, however, was entirely lost on the molecular biologists who, having realised that the DNA molecule qualified as a form of digital information in the technical, information-theoretic sense, immediately jumped to the conclusion that it could therefore be treated as a *code* with a specific semantic content" (Ingold 2004, 214; emphases in original).
5 For a different project of an affirmative biopolitics that charts the possibility of aligning it to left politics, see Hannah 2011; see also Esposito 2008; Tierney 2016.

6 For an extensive history of the concept of milieu, see Spitzer 1942; Canguilhem 2008b, 98–120; see also Feuerhahn 2017; Sprenger 2019.

"Milieu" is often translated as "environment" in English editions of Foucault's texts (see, e.g., Foucault 2000c, 150; 2003, 245). Brian Massumi, in his foreword to the translation of *Mille Plateaux* by Deleuze and Guattari, points to the difference between the French and the English understanding of milieu. Whereas in English the term refers only to the natural environment in or on which organisms live, milieu in French also suggests "medium" and "middle" (1987, xvii).

7 See Lamarck's understanding of the milieu as "the habits, mode of life and all the other influences of the environment which have in course of time built up the shape of the body and of the parts of animals" (Lamarck 2011 [1809], 127).

8 On the notion of composition, see Haraway 2016.

Deleuze and Guattari's conceptualization of the milieu draws on Jakob von Uexküll's (2010 [1934]) work on the relationship between the animal and its "umwelt." Following von Uexküll's idea of the milieu as pure relationality, Deleuze and Guattari distinguish between internal, external, intermediate, and associated milieus and stress their fundamentally relational, flexible, and mobile character (Deleuze and Guattari 1987, 51–57; see Altamirano 2014; de Vries 2013).

The genealogy of the milieu can be further extended as the term migrated from biology to sociology in the second half of the nineteenth century, in the work of Comte and Durkheim (Cheung 2014, 249–77; Wessely and Huber 2017). More recently, the notion has been used to go beyond economic (and class-centered) determinants of social inequality and difference to take into account cultural milieus characterized by similar interests and life styles (see, e.g., Bourdieu 1987).

9 This does not mean, however, that the new forms of controlling circulations developed in the eighteenth century only focus on urban spaces. On the contrary, the "problematic of the *économistes* reintroduces agriculture as a fundamental element of rational governmentality. The land now appears alongside, and at least as much as and more than the town, as the privileged object of governmental intervention" (Foucault 2007a, 342; see also Moulton and Popke 2017).

For a detailed account of how the idea of a social environment informed urban planning, architectural design, health policies, and welfare administration in France in the nineteenth and twentieth centuries, see Rabinow 1989.

10 Chandra Mukerji has proposed a distinction between two distinct forms of power: strategics and logistics. While the former operates by political domination and legitimated forms of rule, the latter focuses on the "environment (context, situation, location) in which human action and cognition take place" (2010, 40). It mobilizes the material world in order to shape "the conditions of possibility for collective life. A material regime cultivated this way favors some groups over others, but governs impersonally through an order of things" (ibid., 404; see also

Mezzadra and Neilson 2019). See also Michael Mann's concept of "infrastructural powers" (Mann 1984).

11 Already in the *History of Madness*, published in 1961, Foucault made reference to the notion of the milieu (see Foucault 2007a, 27, note 37).

12 Elden describes Foucault's involvement in a collaborative research project led by Lion Murard and François Fourquet entitled *Les équipements du pouvoir* ("Equipments of power"). The outcome was a book to which Foucault (as well as Deleuze and Guattari) contributed in discussion sections: "The 'equipments' of power analyzed in this book are the three items in the subtitle: towns, territories and 'utilities'—*équipements collectifs*. By these Fourquet and Murard mean something akin to public amenities or the infrastructure of society. These are tools that are utilized collectively—roads, transportation and communication networks, and the more static apparatus of towns. Circulation necessarily plays a crucial role, with the flux and flow of people, goods, and capital as money" (Elden 2017, 169).

13 For an illustration, see Langdon Winner's classic example of a series of expressways erected in Long Island, New York, by the architect Robert Moses in the first half of the twentieth century (Winner 1980). His analysis starts with the thesis that the overpasses were deliberately constructed so low to the ground that public buses couldn't pass under them. According to Winner, the design of the bridges had the social and political effect of preventing poor and black people from easily accessing the beachfront playgrounds. While they had to rely on public transport, the middle and upper classes could go there by car (for a different account of the relations between political relations and technological developments, see Hughes 1983).

14 For a short history of the concept of population, see Foucault 2007a, 81–82, note 13. Michelle Murphy (2017) provides an instructive account of the "economization of life" by focusing on the relations between "economy" and "population" in the twentieth century. While Murphy regards Foucault's work on governmentality and biopolitics as "crucial inspirations" (2017, 149, note 17) for her own work she does not explicitly engage with Foucault's analysis of the figure of population in the governmentality lectures.

15 Ute Tellmann has noted that Foucault largely ignored the fundamental epistemic break that Thomas Robert Malthus' *Essay on the Principle of Population* (Malthus 1986 [1798]) marked within the genealogy of liberal governmentality. Malthus postulates a systematic disequilibrium between the growth of the population and the means of subsistence, resulting in a permanent threat of scarcity. This catastrophic scenario is linked to a colonial hierarchy that differentiates between a presentist and dangerous "savage life" and a future-oriented economic "civilized life" (Tellmann 2013; see also Dean 1991; Stoler 1995; Bohlender 2007).

16 This understanding of a "vital power" that draws on and exploits processes of life is very different from Bennett's notion of "thing power" (Bennett 2010a), since it reformulates the concepts of agency and vitality in political terms—instead of conceiving it as an inherent property of things.

17 On Canguilhem's account of the history of biology, see Macherey 1992; Rabinow and Caduff 2006; Elden 2019.
18 Mark B. Salter has argued that the Foucauldian notion of circulation developed in his lectures at the Collège de France of 1978 exposes serious shortcomings and analytic problems of the "new mobilities turn" (see, e.g., Sheller and Urry 2006; Cresswell 2011; D'Andrea, Luigina, and Breda 2011). It "liberates the mobilities turn from its methodological dependence on movement and its liberal bias towards interpreting freedom as movement. Orienting mobilities research around circulation accounts for processes of control hidden or minimized by the relational mobility/immobility paradigm: circuits that isolate particular individuals or populations without rendering them immobile" (Salter 2013, 16; see also O'Grady 2014).
19 See Lemke 2011b for a more extensive argument concerning this theoretical shift.
20 For bibliographical information and the debate on the contested identity of the author, see Foucault 2007a, 27, note 39; see also Cole 2000, 31–40. Foucault mentions Moheau already in *The History of Sexuality, Volume I* (1978, 140).
21 See the longer passage from Moheau's book quoted by Foucault (in the light of the the current climate crisis): "It is up to the government to change the air temperature and to improve the climate; a direction given to stagnant water, forests planted or burned down, mountains destroyed by time or by the continual cultivation of their surface, create a new soil and a new climate" (Foucault 2007a, 22).
22 Valerie A. Olson's empirical study of space biomedicine illustrates this point well. Combining Canguilhem's account of the milieu with Foucault's understanding of biopolitics, she suggests the concept "ecobiopolitics" to capture the relation between biological features and environmental conditions (e.g., in a spacecraft or on a planet). Analyzing the co-constitution of "humans" and "environments," Olson describes "how astronauts are managed at a fundamentally environmental rather than biological level, how their biologically pathological responses are made 'normal' in ways that politically and socially normalize outer space milieus as well, and how humans are viewed as calculable 'at-risk systems' to be made predictable and manageable on equivalent terms with the technological and environmental systems in which they are situated" (2010, 188–9).
23 The philosopher of biology John Dupré has suggested that "functional biological wholes, the entities that we primarily think of as organisms, are in fact cooperating assemblies of a wide variety of lineage-forming entities" (2012, 126). Dupré rejects the assumption that all cells in an organism belong to the same species. On the contrary, according to this account "living things" (ibid., 126) are "extremely diverse and opportunistic compilations of elements from many distinct sources" (ibid.). Dupré argues for a redefinition of organisms as "cooperating assemblies." In this perspective, human life only exists as the effect of symbiotic systems linking "human" and "nonhuman" life: "A functioning human organism is a symbiotic system containing a multitude of microbial cells—bacteria, archaea, and fungi—

without which the whole would be seriously dysfunctional and ultimately nonviable. Most of these reside in the gut, but they are also found on the skin, and in all body cavities. In fact, about 90 per cent of the cells that make up the human body belong to such microbial symbionts and, owing to their great diversity, they contribute something like 99 per cent of the genes in the human body" (ibid., 125; see also Margulis 1998; Haraway 2008, 3–4; Bennett 2010a, 113).

24 For a useful exploration of how the question of biopolitics has been taken up in (social) geography, with particular reference to nonhuman life, see Rutherford and Rutherford 2013. Chloë Taylor provides an insightful account of how Critical Animal Studies scholars have drawn on Foucault's work to analyze relations between humans and nonhuman animals in agriculture (2013; see also Thierman 2010).

25 Elizabeth Povinelli has extended this critique of the classical notion of biopolitics (Povinelli 2016; Povinelli et al. 2017). According to Povinelli, "biopolitics" still subscribes to an ontology that privileges forms of life, disregarding and neglecting mechanisms of power different from those focusing on the government of and through life in Western politics. She proposes the term "geontopower" to capture the more fundamental division between the active and the inert, life and nonlife that shapes and informs contemporary forms of liberal government (see also TallBear 2017).

26 Nealon's archaeology of biopolitics turns to Foucault's writings of the 1960s, especially *The History of Madness* (2006b) and *The Order of Things* (1970). Concerning the first book, Nealon argues that Foucault demonstrates that madness was understood as a form of secret or hidden animality within mankind. According to Foucault, "the animal realm [. . .] serves to reveal the dark rage and sterile folly that lurks in the heart of mankind" (2006b, 19; see also Foucault 2006b, 147–48). Nealon also proposes an original reading of *The Order of Things*, arguing that Foucault employed the interpretative frame of biopolitics in this book even before he made explicit use of the notion ten years later. Following Nealon, Foucault's archaeology of the human sciences demonstrates that the shift from natural history to biology resulted in a constellation where "animals begin to take priority over plants as the privileged form or figure of life itself" (Nealon 2016, 143). According to this line of interpretation, in Foucault's account—what Nealon terms "biopower 1.0." (ibid., 144)—"animal life is not in fact jettisoned or abjected at the dawn of humanist biopower in the nineteenth century, but instead [. . .] *animality is fully incorporated into biopower as the template for life itself*" (ibid. 144, emphasis in original; Nealon 2015).

PART III. TOWARD A RELATIONAL MATERIALISM

1 For a similar conceptual revision, see Michelle Murphy's proposal to replace the concern for population with the term "distributed reproduction." This theoretical move shifts the accent "from the question of how much and which *bodies* get to reproduce to what *distributions* of life chances and what kinds of infrastructures

get reproduced. [. . .] It stretches beyond bodies, individuals, or heterosexuality into the more-than-human, more-than-biotic relations that have been recomposed in the aftermath of capitalism, the nation-state, and macroeconomy" (2017, 141–43; emphases in original).

2 Nikolas Rose, Pat O'Malley, and Mariana Valverde Rose identified "three principal points of convergence" between ANT and studies of governmentality: a preference for detailed empirical studies, a methodological commitment to investigating "how" (instead of "why" and "in whose interest") questions and a theoretical antihumanism. However, these authors also noted that so far governmentality studies had "not explicitly take[n] up Latour's and Callon's call to consider the agency of things" (2006, 93).

Rose and Peter Miller have sought to link their work on governmentality with ANT, and especially with Callon's and Latour's sociology of translation (Rose 1999, 49; Miller and Rose 2008, 33–34; see also Barry 2001).

3 Yuk Hui (2015a) offers a different and more delimited understanding of relational materialism. Drawing on François Lyotard's idea of the "immaterial," he outlines a concept of "relational materiality" that goes beyond positions which either suggest that relations are immaterial or endorse a substantialist analysis of materiality. According to Hui, this account makes it possible to critically investigate how digital technologies render material—visible and measurable—all sorts of relations.

4 This new dispositive differs from classical biopolitical technologies, but is close to what Deleuze once described as mechanisms of control that operate through "a universal modulation" (1992b, 7).

On the concept of modulation in Deleuze's and Simondon's work, see Hui 2015b.

7. ALIGNING SCIENCE AND TECHNOLOGY STUDIES AND AN ANALYTICS OF GOVERNMENT

1 Andrew Feenberg has noted an essential transformation in recent STS work: "Its latent political critique has become explicit in recent years as STS has responded to the rise of technical politics by broadening its concerns and reaching a wider audience both within and outside the academy" (2017, 4; see also Lezaun 2017).

Mark B. Brown (2015) distinguishes between five distinctive conceptions of politics in STS, which serve different empirical and normative purposes. See also Law and Singleton 2013; Dányi 2018.

2 Richie Nimmo has put forward a similar proposition, arguing for combining ANT and Foucault's work on governmentality to achieve at a "symmetrical governmentality" that attends to "the *government of nonhumans*, that is, to the political technologies through which heterogeneous entities are managed and how these articulate with, underpin and condition the government of humans in historically specific ways" (2008, 78; emphasis in original). While his main interest is in integrating an analytics of government "within a broadly ANT approach" (ibid., 91), my proposition starts from the opposite angle with the goal of introducing insights from STS into a comprehensive understanding of government.

3   Interestingly, the notion of problematization has a crucial role in Foucault's work as well as in ANT (see Foucault 1984a; Callon 1986).
4   Steve Woolgar and Javier Lezaun regard the ontological turn in anthropology as another striking example of this form of reification. They refer, for example, to the work of Viveiros de Castro (2004) on the existence of "Amerindian ontologies" and observe a "tendency to turn the alterities generated through ethnographic inquiry into a form of definitive difference" (Woolgar and Lezaun 2015, 466; note 2). This ontological commitment is also critically assessed by other commentators as "multiplicity realism" (Zuiderent-Jerak 2015) or "ontological anthropology" (Bessire and Bond 2014).

   For an instructive reading of the different meanings of the "turn" metaphor, see Vasileva 2015.
5   Michael Lynch proposes the concept of "ontography," by which he understands "historical and ethnographic investigations of particular world-making and world-sustaining practices that do not begin by assuming a general picture of the world" (2013, 444). See also Mol's term "praxiography" (Mol 2002, 31).
6   Foucault's "historical nominalism" (2008a, 318) breaks with classical nominalism by taking up and radicalizing insights from French epistemology and the work of Paul Veyne (see, e.g., Veyne 1997). On the difference between the two forms of nominalism, see Pfaller (1997, 178–83). Ian Hacking proposes a similar distinction between "static" and "dynamic" nominalism (1986; 2002).

   See Braun (2015, 7) on the parallels between Foucault's historical nominalism and Althusser's "aleatory materialism" (2006).
7   On this point see the statement by Veyne: "The whole difficulty arises from the illusion that allows us to 'reify' objectivizations as if they were natural objects. We mistake the end result for a goal; we take the place where a projectile happens to land as its intentionally chosen target. Instead of grasping the problem at its true center, which is the practice, we start from the periphery, which is the object, in such a way that successive practices resemble reactions to a single object, whether 'material' or rational, that is taken as the starting point, as a given. [. . .] [W]e take the points of impact of successive practices to be preexisting objects that these practices were aiming for: their targets. Madness and the common good throughout the ages have been targeted differently by successive societies whose 'attitudes' were not the same, so that they touched the target at different points" (1997, 161).
8   See also Hacking's work, which takes up Foucault's insight that modes of knowing and categorizing are also practical forms of world-making. His understanding of "historical ontology" (Hacking 2002) engages with classifications of human kinds and how they not only produce representations of individuals and collectives but also affect the people themselves. Historical ontology then seeks to disclose the conditions of emergence and intelligibility of certain ontological fields and entities: "To take Hacking's best-known example, the category 'multiple personality disorder', along with diagnostic tools, the training of various kinds of therapists,

and popular representations, contributes to there being people with multiple personality disorder" (Sismondo 2015, 443).

9   This is of course an originary and essential part of ANT's sensibility (see Callon and Latour 1981); however, it often remains unclear how the insight that "the macro becomes macro only to the extent that it is *done* as macro" (Law and Singleton 2013, 493; emphasis in original) is concretely spelled out in the empirical work.

10 The strong emphasis on the "nontological" also seeks to respond to the growing concern in STS that the notion of ontology might (no longer) be useful as something to "turn to" if we want to address the composition of the real. See the reservations expressed (also in the light of the rise of the new materialisms) in Mol 2013, 380–81; Woolgar and Lezaun 2015, 465; Aspers 2015. As Woolgar and Lezaun put it: "'ontology' reduces the diversity and dimensionality of the practical undertakings that create our worlds. It offers a totalizing answer to the question of the whatness of things—things *are* (or are not)—before the question has been fully parsed" (2015, 465; emphasis in original).

11 Susanne Lettow shows that the opposition between "ontological materialism" and "praxeological materialism" has played a central role within the tradition of historical materialism. While Engels, Lenin, and Bloch provide versions of the former, the latter concept takes Marx's critique of anthropology and his focus on praxis as its starting point. This praxeological version of historical materialism was especially important in the early Frankfurt School of Critical Theory, which also stressed the situatedness of knowledge production (Lettow 2017, 112–16).

There are substantial overlaps between STS and the critique of technocratic reason in this tradition of Critical Theory (Feenberg 2017). See also Foucault's comments on the parallels between his work and the project of the Frankfurt School (Foucault 1988a, 26–27, 1988b, 104; 1991b, 115–29; 1998a, 469).

12 For a different suggestion on how to map accounts of agency in ANT, see Michael 2017, 67–72.

13 Lettow notes that the question of agency is at the same time omnipresent and absent in new materialist literature. On the one hand, new materialisms suggest a universalization and democratization of agency to include formerly excluded entities; on the other hand "agency" is rendered difficult to grasp, as it becomes detached from socio-material relations and is "transferred to anonymous, metahistorical forces like matter or life" (Lettow 2017, 111).

See Povinelli's critique of this gesture of extension: "Rather than dissolving the human-centered theory of *logos* and *demos*, nonlife entities, extensions, and assemblages are *welcomed* into the language and the habitus of the *demos*. The generosity of *extending* our form of semiosis to all forms of existence forecloses the possibility of them provincializing us" (Povinelli in Povinelli et al. 2017, 180; emphases in original; 2015).

14 Talal Asad has criticized the preoccupation with agency in contemporary social science as a backdrop of the current political and social constellation: "Agency has

become a catch word. In a way, this intoxication with 'agency' is the product of liberal individualism. The ability of individuals to fashion themselves, to change their lives, is given ideological priority over the relations within which they themselves are actually formed, situated, and sustained" (Asad 1996, no page number; see Meißner 2013, 166–67).

For a case study investigating the problem of (political) agency see Abrahamsson and Endre Dányi's empirical analysis of a hunger strike that took place in Brussels in 2012. The authors seek to "avoid the historical and theoretical baggage of 'agency'" (2019, 895). Their study shows that "passivity," "silence" and "weakness" do not necessarily signal a lack of agency but might give rise to a different mode of doing politics, inviting us to rethink the liberal grammar of democratic participation.

15 See Chapter 2 for a more detailed discussion of the argument Abrahamsson et al. (2015) put forward against Bennett's claim that omega-3, a particular kind of fatty acid, itself reduces aggression in humans.

16 Angela Willey discusses the work of Elizabeth Grosz, Elizabeth Wilson, Diana Coole, and Samantha Frost and argues that new materialist accounts tend to ignore postcolonial and feminist STS work: "New materialist storytelling narrates human-centric materialism not as an imperialist scientific project but rather as a set of universal tenets only recently displaced by new conceptualizations of the natural in Science and/or critical theory. Neither the posthumanist/queer ecological challenge to anthropocentrism nor the new materialist challenge to the life/nonlife binary can be pulled apart from this epistemological insight of postcolonial feminist science studies. The projects are coimplicated. And this is the opportunity that new materialism has not yet seized and is, in a sense, its great betrayal of feminist and postcolonial critique" (Willey 2016, 1005).

While this assessment might be a bit unfair given the fruitful exchange between feminist and postcolonial theory on the one hand and new materialist concerns on the other, it still captures a troubling tendency. Wilson herself observes that material feminist work often uncritically embraces the results of neurobiological research: "I am increasingly concerned that there is now a tendency to [. . .] side with scientific data in a very literal kind of way. There is a growing credulousness in the humanities about data put in front of us by scientific investigation. I have found this to be particularly evident in the neuro-humanities literatures, which take up certain claims about human and animal neurological function as gospel" (Wilson in Kirby and Wilson 2011, 233; see also Willey 2016; 2017).

17 Des Fitzgerald and Felicity Callard (2015) engage especially with Catherine Malabou's and Brian Massumi's work (Malabou 2008; 2012; Massumi 1996).

Braun criticizes the general "lack of reflexivity" (2015, 3) in new materialist scholarship by engaging with John Protevi's book *Life, War, Earth* (2013), in which findings from neurology, biology, and meteorology seem to confirm Deleuze's philosophical claims and vice versa.

See also the critique by Abrahamsson et al., who note that "much of the literature in new materialism relies on or borrows insights from experimental sciences without acknowledging their situatedness or querying their methods and the transportability of their results" (2015, 5, note 5).

18 See Ben Anderson's reminder that "relational thinking that emphasizes dynamism, complexity or instability has a long history in the US military" (Anderson 2010, 229; see also Edwards 1996).

19 See Chapter 8 for a more detailed analysis.

Similar reservations apply to the notion of plasticity, which is often considered the antidote to forms of essentialism or determinism (especially in the bio- and neurosciences). The term suggests malleability, openness and inclusiveness, promising to transcend naturalist or biologist accounts of the body and personhood, stressing relationality instead of gene- or neurocentric concepts and binary understandings of nature and culture. However, notions of biological plasticity might also be used to reinforce or renew classist or racist markers. In fact, the idea of plasticity was instrumental and essential in the histories of scientific racism and eugenics, and it is flexible enough to happily coexist with contemporary understandings of rigid determination and clear-cut conceptual borders (J. Brown 2015; Willey 2017, 137–39; Schuller and Gill-Peterson 2020; see also Pitts-Taylor 2010 for an analysis of how neoliberal government has relied on and contributed to discourses of plasticity).

20 In STS, there are some prominent proposals to account for the dynamism of these practical associations and arrangements. Charis Thompson (2005) has suggested the notion of choreography as a way of analyzing how heterogeneous entities are linked together. Andrew Pickering's idea of the "dance of agency" (1995) captures the reciprocal emergence of human and nonhuman agency in scientific practices. Haraway has proposed the notion of compost as a way of arranging "unexpected collaborations and combinations" (2016, 4).

21 Casper Bruun Jensen (2015) explores the different meanings of "political materials" in STS and also in anthropology, infrastructure studies, and environmental history.

22 The term "posthumanism" is evoked in very heterogeneous perspectives and positions. For an instructive mapping of the posthumanist landscape, see Castree and Nash 2004; for a classic exploration, see Hayles 1999.

23 This idea takes up Gayatri Spivak's important call for the paradoxical position of a "strategic essentialism" to make possible the formation of collective practices for achieving a set of political ends (Spivak 1988).

24 See the caveat expressed by Bonnie Washick and Elizabeth Wingrove: "[T]he scholarly appeal of a posthumanist ontology makes a great deal of sense. But insofar as that appeal includes its minimal attention to systematically reproduced constraints, its self-evident ethics and its narrowing of collective action to the 'always already' networked dependencies through which we live, the scholarly

imaginary sparked by new materialist ontologies runs the risk of producing a politics that does not really matter" (2015, 77).
25 N. Katherine Hayles argues that new materialists tend to provide very selective and incomplete accounts of material agency. They tend to "encourage overtly general analyses, in which crucial distinctions between kinds of material agency are not acknowledged, presumably because to include them would compromise the decentering project. To reason so confuses decentering the human with its total erasure, an unrealistic and ultimately self-defeating enterprise, considering that the success of the decentering project depends precisely on persuading humans of its efficacy" (2017, 66).
26 Jouni Häkli argues that "posthumanist notions of civil society and citizenship remain normatively hollow, tied to the work of human signification, and as such, parasitic on precisely that kind of humanist conception they set out to transgress" (2018, 7).
27 Note the double meaning of *expérience* in French as both "experiment" and "experience." Timothy O'Leary stresses the "ambiguity within Foucault's use of the term. On the one hand [. . .] experience is the general, dominant form in which being is given to an historical period as something that can be thought. On the other hand, experience is something that is capable of tearing us away from ourselves and changing the way that we think and act" (2008, 14). See also Foucault's statement: "Every time I have tried to do a piece of theoretical work it has been on the basis of elements of my own experience: always in connection with processes I saw unfolding around me" (2000e, 458). On this autobiographical dimension of Foucault's theoretical work, see Eribon (1994).
28 See Foucault's comment on the concept of ethos in his work: "This philosophical ethos may be characterized as a *limit-attitude*. We are not talking about a gesture of rejection. We have to move beyond the outside-inside alternative; we have to be at the frontiers. Criticism indeed consists of analyzing and reflecting upon limits" (1984b, 45; emphasis in original).
29 Martin Jay points to the etymological roots of experience: "The English word is understood to be derived most directly from the Latin *experientia*, which denoted 'trial, proof, or experiment.' [. . .] Insofar as 'to try' (*expereri*) contains the same root as *periculum*, or 'danger,' there is also a covert association between experience and peril, which suggests that it comes from having survived risks and learned something from the encounter (*ex* meaning a coming forth from)" (2005, 10).
30 On the notion of tinkering in STS, see, e.g., Mol et al. 2010.
31 For a discussion of contemporary debates around "experimentation" in geography and sociology, see Gross et al. 2005, Last 2012, Bogusz 2017.

In a historical perspective, Matthias Gross and Wolfgang Krohn (2010) explore how early twentieth-century American sociologists, especially those associated with the Chicago School, emphasized the central role of the notion of experiment. This sociological understanding of experimentation was not modeled on the natural sciences and went beyond the realm of the laboratory and the idea of testing scientific hypotheses in controlled research settings. Rather, they conceived of society as a self-experimental terrain that develops modes of coping with the uncertainties and contingencies of the modern world—an idea of experimentation that deliberately incorporated natural as well as social elements.

32 Lehman and Nelson's vision of experimentation draws on forms of multispecies companionship and practices of rewilding described by Jamie Lorimer and Clemens Driessen (2013; 2014; see also Lorimer and Driessen 2016). These authors refer to "wild experiments" as a new mode of political ecology that exceeds the conservationist paradigm. It attends to emergent events and fosters rather than prevents ecosystem change, involving incessant negotiations between humans and nonhumans. According to Lorimer and Driessen this move to experimental engagements is significant well beyond the domain of conservation policies, as it might provide a template for living well in more-than-human worlds.

33 Fitzgerald and Callard define "experimental entanglements" as "modest, often awkward, typically unequal encounters that work to mobilize specific and often serendipitous moments of potential novelty in and outside the laboratory" (2015, 18). They are based on the premise "that it is 'discipline' that needs explanation, not promiscuity. What might be imagined as a securely 'cultural' or 'social' knowledge is a *product* of collaboration with the biological (and other) sciences: it is not a precursor to that collaboration" (ibid., 23); emphasis in original.

Andrew Barry and Georgina Born (2013) have distinguished three distinctive modes of interdisciplinarity based on empirical investigations of different kinds of collaboration between the natural or technical, on the one hand, and social realms on the other (see also Fitzgerald and Callard 2015; Niewöhner 2015; Neimanis et al. 2015, 86–90; Marguin et al. 2019).

34 Lezaun notes that "the growing implication of ANT scholars in social movements, design work or artistic performance, part of broader orientation in science studies towards collaborative forms of practice, is pushing the theory in more experimental, less categorical directions" (Lezaun 2017, 328).

35 See John Law and Karel Williams (2014) for an understanding of government as "an experimental practice" and their concept of the "learning state" that tests hypotheses about the external world.

36 To illustrate this point, McInerney refers to healthcare policies in the US and the growing medical and social importance of new devices to track diet and monitor exercise. Rather than fostering and promoting forms of participation, these material objects may have an inverse effect: "[T]he private nature of health and healthcare information may limit political participation to the individual level, preventing or undermining the collective behaviors necessary to influence political systems writ large. In other words, by engaging actors on such a personal interactive level, material objects may contribute to the individualization of politics and the maintenance of dominant political systems" (2014, 717).

8. ENVIRONMENTALITY

1 The term proposed by Graham Burchell, the translator of *The Birth of Biopolitics*, is "environmentalism" (Foucault 2008a, 261). Jennifer Gabrys rightly notes that the original French notion *"environnementalité"* (Foucault 2004, 266) more easily connects to the topic of governmentality and does not involve the risk that it will be confused with social movements or political organizations that address environmental issues (Gabrys 2014, 35, note 2; see also Lorimer 2017, 16).
2 As Hörl notes, the term "environmentality" already figured in English translations of Heidegger's *Sein und Zeit* (1962 [1927]), referring to his notion of *"Umweltlichkeit."* However, Foucault's use of the term differs substantively from Heidegger's analysis of worldhood (Hörl 2018, 158).
3 See the work of Humberto Maturana and Francisco Varela for an exploration of the concept of autopoiesis and its crucial role in biology (e.g., Maturana and Varela 1980).
4 On the origins of the resilience discourse and the central role of Holling's concept of the adaptive cycle, see Walker and Cooper 2011; Nelson 2014, 2–7; Folkers 2018, 181–86.

Erik Swyngedouw and Hendrik Ernstson note that Holling was not the first or only author who, at the beginning of the 1970s, proposed a non-equilibrium theory in ecology. They especially point to the early work of Richard Lewontin (1969), who further developed this approach within a Marxist framework, while Holling advanced a managerial account of resilience compatible with neoliberal capitalism (2018, 24).

See also the Gaia hypothesis put forward by James E. Lovelock and Lynn Margulis (1974), which understands the earth as an integrated, complex, and unpredictable dynamic system (for a critical discussion of their work, see Cooper 2008, 34–36).
5 Holling defines resilience as "a measure of the persistence of systems and of their ability to absorb chance and disturbance and still maintain the same relationships between populations or state variables" (1973, 14).
6 See, for example, Morgan Robertson's analysis of how markets for ecosystem services have been created in many regions of the United States since the early 1990s (2006; see also Robertson 2012). For thorough examinations of the ecosystems service economy within the neoliberal agenda since the 1970s, see Dempsey and Robertson 2012; Nelson 2015.
7 See Gabrys 2014 and Folkers 2017 for similar arguments.
8 In Hörl's view, Luhmann's systems-theoretical distinction between system and environment provides a form of thought symptomatic of the new ecological rationality (Hörl 2017, 6; see also Walker and Cooper 2011, 157; Robertson 2006; Clarke 2014).

See Sprenger 2019 for a comprehensive history of the design of artificial environments and their biopolitical dimensions.

9   One example of this shift is the rise of (environmental) epigenetics as a new field of knowledge and intervention in the postgenomic age, destabilizing and subverting the traditional polarity between nature and nurture. It conceives of the body as open to environmental processes, investigating how socio-economic status, exercise habits, diet regimes or traumatic experiences engage with biological processes at the molecular level (Müller et al. 2017; see also Meloni 2014).

    For a comprehensive analysis of environmental sensing technologies, see Gabrys 2016.

10 Drawing on the work of the geologist Peter Haff (2014a; 2014b), Hörl designates the new dispositive of environmentality as a "technosphere" that supplements previous stages of geological history from the lithosphere via the atmosphere and hydrosphere to the biosphere. It critically responds to the label of the Anthropocene, emphasizing the fact that technology is becoming a geological force. Referring to Simondon and Canguilhem, Hörl concludes that "technology in the technosphere becomes the milieu of milieux, a kind of meta- or hypermilieu" (Hörl 2017, 11).

11 See also Ben Anderson's brief analysis of the scent marketing company ScentAir UK, which offers various flavor options to manage consumer experiences and shape affects in diverse business settings. Embedded in air conditioning systems, the scents delivered combine new forms of value creation with "environmental design. Rather than entraining bodily capacities or regulating populations, the milieu of action is the object and target of an intervention" (2014, 31; 25–31).

    However, environmental strategies are not limited to controlling human behavior and affects. In extending Foucault's analysis of biopolitics and pastoral power, Maan Barua has suggested the term "atmospheric politics" (2020) to account for a different set of strategies that mobilize nonhuman labor to generate surplus value. Rather than operating on animals' bodies and populations this form of politics targets an "animal's milieu" (Barua 2020, 15), intervening in physiological processes, communication patterns, architectural designs, and affective propensities. Focusing on the giant panda, Barua shows how interventions that modulate the milieu of panda life-worlds in zoos make it possible to valorize the affective and reproductive labor these iconic animals perform in captivity (on the notion of lively capital, see Haraway 2008; Sunder Rajan 2012).

12 Today, algorithms play a central role for governing "ambividuals" and managing circulations in a vast variety of domains, from search engines, personalized online advertising, educational evaluations, the operation of markets, and the design of political campaigns to the management of investment decisions and social services. They have become powerful operators and decision-making tools, but the calculative practices that allow for searching, ranking, and recommending remain opaque (Ziewitz 2016; Bucher 2018; see also Rouvroy and Berns 2013).

    See also Alexander Galloway's understanding of "protocol" as a new formal and decentralized form of technological control governing how things are done: "a distributed management system that allows control to exist within a heterogeneous material milieu" (Galloway 2004, 8).

13. Lorimer argues that this understanding of "environmental biopower" (2017, 35) resonates with the concepts of "microbiopolitics" (Paxson 2008) or "symbiopolitics" (Helmreich 2009), which refer to the government of microbial relations.

14. Braun's analysis focuses on the Rising Currents exhibition hosted by the Museum of Modern Art (MOMA) in New York City in 2010.

15. Jozef Keulartz briefly discusses flood management in the Netherlands and the decision of the Dutch government in the 1990s to abandon the traditional water policy of dike reinforcement, which had been an essential element of flood control in the past: "Instead of restricting rivers to straightjackets of dikes, the new policy of flood risk reduction is aimed at creating more room for the river and to restore the self-regulating capacities of water systems by allowing dynamic processes to run their course again" (2012, 59).

16. Collier's and Lakoff's analysis focuses on the development in the USA. On the genealogy of vital systems security in Canada and Germany, see Boyle and Speed 2018 and Folkers 2018, respectively.

17. A major event in the rise of vital systems security was the foundation of the Federal Emergency Management Agency (FEMA) in 1979. This integrated "federal emergency management and civic defense functions under the rubric of all-hazards planning. All-hazards planning assumed that [. . .] many kinds of catastrophes could be treated in the same way: earthquakes, floods, major industrial accidents, and enemy attacks were brought into the same operational space, given certain common characteristics. Needs such as early warning, the coordination of response by multiple agencies, public communication to assuage panic, and the efficient implementation of recovery processes were shared across these various sorts of disasters. Thus all-hazards planning focused not on assessing specific threats but on building capabilities that could function across multiple threat domains" (Lakoff and Collier 2010, 258).

18. Lakoff and Collier refer to a list of sectors considered as "critical infrastructures and key resources" in the National Infrastructure Protection Program published by the Department of Homeland Security in 2006 (Department of Homeland Security 2006; Lakoff and Collier 2010, 247).

19. On the politics of infrastructure, see Anand et al. 2018; Hetherington 2019.

20. Note that Foucault employs a very broad understanding of "killing" that includes "every form of indirect murder: the fact of exposing someone to death, increasing the risk of death for some people, or, quite simply, political death, expulsion, rejection, and so on" (2003, 256).

21. Erik Swyngedouw and Henrik Ernstson have suggested that (auto-)immunological responses might even increase and gain significance in socio-technoecological constellations that embrace and endorse nonhuman forces: "Rather than weakening the immunitary logic, these more-than-human disruptions are exactly the kind of anxiety-filled disruptive events that immunitarian responses

feed off. So, while nature's heterogeneous acting might interrupt the smooth functioning of anthropocentric ontologies and human-nonhuman imbroglios—from hurricanes re-ordering people and things to nuclear reactors blowing up, GMOs rekindling DNA, or new virus strains emerging—it is also this excessive performativity that nurtures concerns with risks and immunization. Simply put, rather than undermining an immunitary logic, they could strengthen modes, mechanisms, and subjectivities of auto-immunization" (2018, 22).

22 This list may not be conclusive, as there are probably dimensions of "animal work" not captured by this typology. For example, it does not include the "technical labor" some organisms perform. As Melinda Cooper has pointed out, the emergence of biotechnologies and the use of recombinant DNA could be understood as a series of steps that successfully put bacteria to work. The bacteria act as vectors or carriers for DNA fragments, making it possible for biologists to move sequences of genetic information from one organism to another, transgressing the barriers of species: "recombinant DNA constitutes the first attempt to mobilize the specific reproductive processes of bacteria as a way of generating new life forms" (2008, 33).

23 However, it is important to keep in mind the theoretical and political risks of such an inclusive conceptual proposal. Rosemary-Claire Collard and Jessica Dempsey caution that "any analysis that seeks to think about exploitation across the human-nonhuman boundary must be undertaken carefully because human exploitation so often rests to some degree on a dehumanization or animalization of the exploited humans. The risk of such analysis is that it reifies or reinforces racialized or misogynist comparisons" (2017, 80, note 2).

24 See J. K. Gibson-Graham and Gerda Roelvink (2010) for a sketch of an already emerging more-than-human "economic ethics" in the light of climate change, encompassing a collective of human and nonhuman entities and giving rise to new economic practices: "Each of these practices is involved in building a community economy, in which sustenance and interdependence are key values and ethical negotiations center on the interrelated issues of necessity, surplus, consumption and commons" (2010, 343; see also Bingham 2006).

25 One area of research linked to the question of pastoral power concerns the interactional frames and mutual patterns organizing breeding and herding practices on the one hand and social structures and political regimes on the other. Already in the 1960s, the French anthropologist André-Georges Haudricourt (1969) claimed that human relations with nonhuman animals and plants are not just a mirror or projection of social relations between humans (see, e.g., Chatwin 1989, arguing for a structural link between pastoralism and military organization); rather, both are informed by a common logic or a shared regime of practices. Haudricourt distinguishes between two "extreme types" (1969, 164) of this structural connection. First, an "'indirect, negative' action" without physical contact between the domesticated entity and humans (e.g., the cultivation of yams in New Caledonia). These practices seek to eliminate obstacles and to respect the nature

of the domesticated entity. Second, a "'direct, positive' action" exemplified by sheep-breeding in the Mediterranean region, which requires permanent contact with and care by the herder. Haudricourt argues that there is a risk of this direct control resulting in a structural dependence and an "over-domestication of the sheep—the tamed animal having lost its powers of self-defense and instinctive behaviour" (1969, 164; Descola 2013, 18; Sautchuk 2016).

See also Friedrich Nietzsche's critique of modern societies as characterized by herding (human) animals (see Bröckling 2017, 27–34; Lemm 2009) and Sarah Franklin's analysis of pastoralism in the genealogy of capitalism (Franklin 2007, 46–72).

26 See J. Donald Hughes' reconstruction of the etymological origins of the name Pan and the mythological trajectories of the god: "[M]odern scholarship has demonstrated that the true derivation of the name *Pan* is from *paōn*, 'the nurturer,' 'he who feeds the herds' of sheep and goats, and therefore 'shepherd'" (1986, 8).

For an overview of the research literature using the concept of panarchy for case studies and empirical investigations, see Allen et al. 2014.

27 Günther Anders has suggested a very different understanding of pastoral guidance that focuses neither on forms of subjectivation nor on the co-evolution between human and nonhuman nature. Anders criticizes the Heideggerian idea of man as the "shepherd of being" as anthropocentric (see, e.g., Anders 1980, 129; 461, note 20). As an alternative, he proposes the notion of "object shepherd" (*Objekthirte*), as humans are no longer in charge of the technological apparatus but rather an adjunct to it (ibid., 95; see also 30). According to Anders, technologies have become "the subject of history" (ibid., 279; van Dijk 2000, 101–3).

28 See, for example, the ecomodernist conviction that nature as an independent entity and autonomous force has ceased to exist and has been replaced by human deliberation and design. In this view, the Anthropocene does not signal a moment of crisis and danger but rather affirms human responsibility of planetary stewardship: "A good, or at least a better, Anthropocene is within our grasp. Creating that future will mean going beyond fears of transgressing natural limits and nostalgic hopes of returning to some pastoral or pristine era. Most of all, we must not see the Anthropocene as a crisis, but as the beginning of a new geological epoch ripe with human-directed opportunity" (Ellis 2011, 41–2; for critical analyses of the ecomodernist discourse see Hamilton 2015; Neyrat 2018).

29 See Marx's Theses on Feuerbach: "The philosophers have only interpreted the world, in various ways; the point is to change it."

CONCLUSION

1 As Andrew Barry points out, this political imaginary also leads to an impoverished idea of democracy: "But while radical democratic theorists point to the centrality of dissensus in political life, they say little about the existence and the importance of materials and objects, which frequently come to animate public knowledge controversies. Such controversies revolve around disagreements not

just about the rights and interests of human actors and the identities of social groups [. . .], but also about the causes of climate change, the safety of genetically modified organisms, the origins of diseases, the risks of floods and the consequences of nuclear accident" (2013, 8; see also Latour 2004c).

2 See, for example, Harman's claim that his concepts are not "a taxonomy of entities, but are four structures of reality in general, found everywhere and at all times" (2011a, 96), or the seemingly non-ironic title chosen by Morton for one of his articles: "Here Comes Everything: The Promise of Object Oriented Ontology" (Morton 2011a).

Harman is undoubtedly right when he observes that "OOO and Foucault have little to do with each other" (2018, 210).

3 Braun has noted an "irony" at play in new materialist scholarship: "[E]ven as many new materialists propose an ontology that is non-deterministic and non-teleological they often deploy a very different epistemological position when it comes to the emergence of their ideas, which are viewed as universal rather than particular, and necessary rather than contingent: the world is marked by indeterminacy and contingency, except when it comes to theories of indeterminacy and contingency!" (2015, 4-5)

4 See Willey's observation: "[W]hen we narrate a reconsideration of 'nature' as a re/turn to 'science,' feminized nature is not re-valued as a source of knowledge, so much as science's mastery and authority to name it is reconsolidated" (2017, 146).

Classic accounts of the relation between science and gender include Keller 1985; Harding 1986; Schiebinger 1989.

5 Instead of systematically prioritizing flux and indeterminacy, it is important to empirically distinguish between different kinds of agencies, allowing for moments of balance, cohesion, predictability, and endurance and even for "material forces whose actions are deterministic" (Hayles 2017, 81) to enter the picture—a possibility that many new materialists appear to rule out from the start or are reluctant to admit.

6 See Haraway's reminder: "Nothing is connected to everything; everything is connected to something" (2016, 31).

7 See Althusser's concept of a "materialism of the encounter" (2006) and his idea of the "becoming-necessary of contingent encounters" (Braun 2015, 7).

8 Stephanie Wakefield and Bruce Braun (2014) turn to Agamben's notion of profanization (Agamben 2009) or destituent power (Agamben 2014) to address the challenge to disarticulate the governmental dimension of dispositives, to render inoperable the disparate elements it consists of, and to dissolve the relations it enacts. Agamben refers to the Latin origin of the term profanization: while "to consecrate" designated the exit of things from the sphere of human law, "to profane" signified, on the contrary, to restore things to the free use of men. Profanization in this sense operates as a kind of counter-dispositive that restores to common use what sacrifice had separated and divided (Agamben 2009, 17–19). However, Agamben's use of the term is clearly limited as it reserved exclusively for human communities.

BIBLIOGRAPHY

Abadía, Oscar Moro. 2003. "¿Qué es un dispositivo?" *Empiria: Revista de Metodología de Ciencias Sociales* 6: 29–46.

Abrahamsson, Sebastian, Filippo Bertoni, Annemarie Mol, and Rebeca Ibánez Martin. 2015. "Living with Omega-3: New Materialism and Enduring Concerns." *Environment and Planning D: Society and Space* 33(1): 4–19.

———, and Endre Dányi. 2019. "Becoming Stronger by Becoming Weaker: The Hunger Strike as a Mode of Doing Politics." *Journal of International Relations and Development* 22(4): 882–98.

Ach, Jada L. 2016. "Review of *Hyperobjects: Philosophy and Ecology after the End of the World*, by Timothy Morton." *Clio: A Journal of Literature, History, and the Philosophy of History* 45(1): 127–32.

Adams, Vincanne, Michelle Murphy, and Adele E. Clarke. 2009. "Anticipation: Technoscience, Life, Affect, Temporality." *Subjectivity: International Journal of Critical Psychology* 28(1): 246–65.

Agamben, Giorgio. 1998. *Homo Sacer: Sovereign Power and Bare Life*. Stanford: Stanford University Press.

———. 2009. "What Is an Apparatus?" In *"What Is an Apparatus?" and Other Essays*, 1–24. Stanford: Stanford University Press.

———. 2014. "What Is a Destituent Power?" *Environment and Planning D: Society and Space* 32(1): 65–74.

Agar, Jon. 2003. *The Government Machine: A Revolutionary History of the Computer*. Cambridge: MIT Press.

Agrawal, Arun. 2005. *Environmentality: Technologies of Government and the Making of Subjects*. Durham: Duke University Press.

Ahmed, Sarah. 2008. "Open Forum Imaginary Prohibitions: Some Preliminary Remarks on the Founding Gestures of the 'New Materialism.'" *European Journal of Women's Studies* 15(1): 23–39.

Alaimo, Stacy. 2014. "Thinking as the Stuff of the World." *O-Zone: A Journal of Object-Oriented Studies* 1(1): 13–21.

——— and Susan J. Hekman. 2008. *Material Feminisms*. Bloomington: Indiana University Press.

Allen, Craig R., David G. Angeler, Ahjond S. Garmestani, Lance H. Gunderson, and C. S. Holling. 2014. "Panarchy: Theory and Application." *Ecosystems* 17: 578–89.

Altamirano, Marco. 2014. "Three Concepts for Crossing the Nature-Artifice Divide." *Foucault Studies* 17: 11–35.

Althusser, Louis. 1971. "Ideology and Ideological State Apparatuses." In *Lenin and Philosophy and Other Essays*, translated by Ben Brewster, 121–76. New York: Monthly Review Press.

———. 2006. *Philosophy of the Encounter: Later Writings, 1978–1987*. London: Verso.

———. 2014. *On the Reproduction of Capitalism: Ideology and Ideological State Apparatuses*. London: Verso.

Ampère, André-Marie. 1834. *Essai sur la philosophie des sciences ou exposition analytique d'une classification naturelle de toutes les connaissances humaines*. Paris: Bachelier.

Anand, Nikhil, Hannah Appel, and Akhil Gupta. 2018. *The Promise of Infrastructure*. Durham: Duke University Press.

Anders, Günther. 1980. *Die Antiquiertheit des Menschen II: Über die Zerstörung des Lebens im Zeitalter der dritten industriellen Revolution*. München: Beck.

Anderson, Ben. 2004. "Time-Stilled Space-Slowed: How Boredom Matters." *Geoforum* 25: 739–54.

———. 2010. "Population and Affective Perception: Biopolitics and Anticipatory Action in US Counterinsurgency Doctrine." *Antipode* 43(2): 205–36.

———. 2011a. "Review of *Vibrant Matter: A Political Ecology of Things*, by Jane Bennett." *Dialogues in Human Geography* 1(3): 393–96.

———. 2011b. "Affect and Biopower: Towards a Politics of Life." *Transactions of the Institute of British Geographers* 37(1): 28–43.

———. 2014. *Encountering Affect: Capacities, Apparatuses, Conditions*. Farnham: Ashgate.

Appadurai, Arjun. 1998. *The Social Life of Things: Commodities in Cultural Perspectives*. Cambridge: Cambridge University Press.

Aradau, Claudia. 2010. "Security That Matters: Critical Infrastructure and Objects of Protection." *Security Dialogue* 41(5): 491–514.

———, and Rens van Munster. 2007. "Governing Terrorism Through Risk: Taking Precautions, (un)Knowing the Future." *European Journal of International Relations* 13(1): 89–115.

Asad, Talal. 1996. "Modern Power and the Reconfiguration of Religious Traditions. Interview by Saba Mahmood." *SEHR* 5(1). www.stanford.edu/group/SHR/5-1/text/asad.html.

Åsberg, Cecilia. 2013. "The Timely Ethics of Posthumanist Gender Studies." *Feministische Studien* 31(1): 7–12.

———. 2018. "Feminist Posthumanities in the Anthropocene: Forays into the Postnatural." *Journal of Posthuman Studies* 1(2): 185–204.

———, Kathrin Thiele, and Iris van der Tuin. 2015. "Speculative before the Turn: Reintroducing Feminist Materialist Performativity." *Cultural Studies Review* 21(2): 145–72.

Asdal, Kristin. 2008. "Enacting Things Through Numbers: Taking Nature into Account/ing." *Geoforum* 39: 123–32.

———, Christian Borch, and Ingunn Moser. 2008. "Editorial: The Technologies of Politics." *Distinktion: Scandinavian Journal of Social Theory* 9(1): 5–10.

———, Tone Druglitro, and Steve Hinchliffe. 2017. *Humans, Animals and Biopolitics: The More-Than-Human Condition*. New York: Routledge.

Aspers, Patrik. 2015. "Performing Ontology." *Social Studies of Science* 45(3): 449–53.
Badmington, Neil. 2004. "Mapping Posthumanism." *Environment and Planning A* 36(8): 1344–51.
Baldwin, Andrew. 2013. "Vital Ecosystem Security: Emergence, Circulation, and the Biopolitical Environmental Citizen." *Geoforum* 45: 52–61.
Balibar, Étienne. 1992. "Foucault and Marx: The Question of Nominalism." In *Michel Foucault, Philosopher*, edited by Timothy J. Armstrong, 38–56. Hemel Hempstead: Harvester Wheatsheaf.
Balke, Friedrich, and Maria Muhle. 2016. "Einführung." In *Räume und Medien des Regierens*, edited by Friedrich Balke and Maria Muhle, 8–23. Paderborn: Wilhelm Fink.
Bannon, Bryan. 2011. "Book Review: *Vibrant Matter: A Political Ecology of Things.*" *Environmental Philosophy* 5(1): 1–4.
Barad, Karen. 1998. "Getting Real: Technoscientific Practices and the Materialization of Reality." *Differences* 10(2): 87–128.
———. 2003. "Posthumanist Performativity: Toward an Understanding of How Matter Comes to Matter." *Signs: Journal of Women in Culture and Society* 28(3): 801–31.
———. 2007. *Meeting the Universe Halfway: Quantum Physics and the Entanglement of Matter and Meaning*. Durham: Duke University Press.
———. 2010. "Quantum Entanglements and Hauntological Relations of Inheritance: Dis/continuities, SpaceTime Enfoldings, and Justice-to-Come." *Derrida Today* 3(2): 240–68.
———. 2011. "Erasers and Erasures: Pinch's Unfortunate 'Uncertainty Principle.'" *Social Studies of Science* 41(3): 443–54.
———. 2012a. "Intra-active Entanglements—An Interview with Karen Barad by Malou Juelskjær and Nete Schwennesen." *Kvinder, Køn og forskning/ Women, Gender and Research* (1–2): 10–24.
———. 2012b. "Interview with Karen Barad." In *New Materialism: Interviews & Cartographies*, edited by Rick Dolphijn and Iris van der Tuin, 48–70. Ann Arbor: Open Humanities Press.
———. 2012c. "On Touching—The Inhuman That Therefore I Am." *Differences* 23(3): 206–23.
———. 2012d. "Nature's Queer Performativity." *Kvinder, Køn & Forskning* 1–2: 25–53.
———. 2014. "Diffracting Diffraction: Cutting Together-Apart." *Parallax* 20(3): 168–87.
Barla, Josef. 2019. *The Techno-Apparatus of Bodily Production: A New Materialist Theory of Technology and the Body*. Bielefeld: transcript.
Barnwell, Ashley. 2017. "Method Matters: The Ethics of Exclusion." In *What If Culture Was Nature All Along?*, edited by Vicki Kirby, 26–47. Edinburgh: Edinburgh University Press.
Barry, Andrew. 2001. *Political Machines: Governing a Technological Society*. London/ New York: The Athlone Press.
———. 2013. *Material Politics: Disputes along the Pipeline*. Malden/Oxford: Wiley Blackwell.

———, Thomas Osborne, and Nikolas Rose, eds. 1996. *Foucault and Political Reason: Liberalism, Neo-liberalism and Rationalities of Government*. London: UCL Press.

———, and Georgina Born. 2013. *Interdisciplinarity: Reconfiguration of the Social and Natural Sciences*. New York: Routledge.

Barua, Maan. 2017. "Nonhuman Labour, Encounter Value, Spectacular Accumulation: The Geographies of a Lively Commodity." *Transactions of the Institute of British Geographers* 42(2): 274–88.

———. 2019. "Animating Capital: Work, Commodities, Circulation." *Progress in Human Geography* 43(4): 650–69.

———. 2020. "Affective Economies and the Atmospheric Politics of Lively Capital." Draft Paper.

Bath, Corinna, Hannah Meißner, Stephan Trinkhaus, and Susanne Völker, eds. 2013. *Geschlechter Interferenzen. Wissensformen—Subjektivierungsweisen—Materialisierungen*. Berlin: LIT Verlag.

Battistoni, Alyssa. 2017. "Bringing in the Work of Nature." *Political Theory* 45(1): 5–31.

Baudry, Jean-Louis. 1975. "Le dispositif: Approches métapsychologiques de l'impression de réalité." *Communications* 23: 56–72.

Behrent, Michael C. 2013. "Foucault and Technology." *History and Technology* 29(1): 54–104.

Bell, Vikki 2010. "New Scenes of Vulnerability, Agency and Plurality: An Interview with Judith Butler." *Theory, Culture & Society* 27(1): 130–52.

Bennett, Jane. 2001. *The Enchantment of Modern Life: Attachments, Crossings and Ethics*. Princeton/Oxford: Princeton University Press.

———. 2004. "The Force of Things: Steps Toward an Ecology of Matter." *Political Theory* 32(3): 347–72.

———. 2005. "In Parliament with Things." In *Radical Democracy: Politics between Abundance and Lack*, edited by Lars Tønder and Lasse Thomassen, 133–48. Manchester/New York: Manchester University Press.

———. 2007. "Edible Matter." *New Left Review* 45: 133–45.

———. 2010a. *Vibrant Matter: A Political Ecology of Things*. Durham: Duke University Press.

———. 2010b. "Thing-Power." In *Political Matter: Technoscience, Democracy and Public Life*, edited by Bruce Braun and Sarah J. Whatmore, 35–62. Minneapolis: University of Minnesota Press.

Bennett, Jane 2011a. "Response to Thomas Princen's Review of Vibrant Matter: A Political Ecology of Things." *Perspectives on Politics* 1: 120.

———. 2011b. "Author Response." *Dialogues in Human Geography* 1(3): 404–6.

———. 2015a. "Systems and Things: On Vital Materialism and Object-Oriented Philosophy." In *The Nonhuman Turn*, edited by Richard Grusin, 223–39. Minneapolis/London: University of Minnesota Press.

———. 2015b. "Ontology, Sensibility and Action." *Contemporary Political Theory* 14(1): 82–89.

———. 2016. "Whitman's Sympathies." *Political Research Quarterly* 69(3): 607–20.

———. 2020. *Influx & Efflux: Writing Up with Walt Whitman*. Durham/London: Duke University Press.

———, and Klaus Loenhart. 2011. "Vibrant Matter, Zero Landscape." *GAM Architecture Magazine* 7: 1–7.

Bennett, Tony, and Patrick Joyce, eds. 2010. *Material Powers: Cultural Studies, History and the Material Turn*. London: Routledge.

Benton, Ted. 1991. "Biology and Social Science: Why the Return of the Repressed Should Be Given a (Cautious) Welcome." *Sociology* 25(1): 1–29.

Bergson, Henri. 1998 [1907]. *Creative Evolution*. New York: Dover.

Bernard, Claude. 1957 [1865]. *An Introduction to the Study of Experimental Medicine*, translated by Henry Copley Greene. New York: Dover Publications.

Bessire, Lucas, and David Bond. 2014. "Ontological Anthropology and the Deferral of Critique." *American Ethnologist* 41(3): 440–56.

Beuret, Nicholas. 2019. "Review of *Experimental Practice*, by Dimitris Papadopoulos". *Sociological Review*, February 26, 2019. www.thesociologicalreview.com/book-review-experimental-practice-by-dimitris-papadopoulos/.

Beuscart, Jean-Samuel, and Ashveen Peerbaye, eds. 2006. "Histoires de dispositifs (Introduction)." *Terrains & travaux* 11(2): 3–15.

Bhandar, Brenna, and Jonathan Goldberg-Hiller, eds. 2015. *Plastic Materialities: Politics, Legality, and Metamorphosis in the Work of Catherine Malabou*. Durham: Duke University Press.

Bhattacharya, Tithi, ed. 2017. *Social Reproduction Theory: Remapping Class, Recentering Oppression*. London: Pluto Press.

Bijker, Wiebe, Thomas P. Hughes, and Trevor Pinch. 1987. *The Social Construction of Technological Systems: New Directions in the Sociology and History of Technology*. Cambridge: MIT Press.

Binding, Karl, and Alfred Hoche. 1975 [1920]. *The Release of the Destruction of Life Devoid of Value, Its Measure and Its Form*, with comments by Robert L. Sassone. Santa Ana, CA: Life Quality.

Bingham, Nick. 2006. "Bees, Butterflies, and Bacteria: Biotechnology and the Politics of Nonhuman Friendship." *Environment and Planning A* 38(3): 483–98.

Blaser, Mario. 2016. "Is Another Cosmopolitics Possible?" *Cultural Anthropology* 31(4): 545–70.

Bluwstein, Jevgeniy. 2017. "Creating Ecotourism Territories: Environmentalities in Tanzania's Community-Based Conservation." *Geoforum* 83: 101–13.

Bodén, Linnea, Hillevi Lenz Taguchi, Emilie Moberg, and Carol A. Taylor. 2019. "Relational Materialism." In *Oxford Research Encyclopedias: Education*. New York: Oxford University Press.

Bogost, Ian. 2012. *Alien Phenomenology: What It's Like to Be a Thing*. Minneapolis: University of Minnesota Press.

Bogusz, Tanja. 2017. "Kritik, Engagement oder Experimentalismus? STS als pragmatistische Soziologie kritischer Öffentlichkeiten." In *Pragmatismus und Theorie sozialer*

Praktiken. Vom Nutzen einer Theoriedifferenz, edited by Hella Dietz, Frithjof Nungesser and Andreas Pettenkofer, 283–300. Frankfurt am Main/New York: Campus.

Bohlender, Matthias. 2007. *Metamorphosen des liberalen Regierungsdenkens. Politische Ökonomie, Polizei und Pauperismus*. Weilerswist: Velbrück.

Bolt, Barbara, and Estelle Barrett, eds. 2013. *Carnal Knowledge: Towards a New Materialism Through the Arts*. London: IB Tauris.

Bonneuil, Christophe, and Jean-Baptiste Fressoz. 2016. *The Shock of the Anthropocene: The Earth, History and Us*. London/New York: Verso.

Booth, Rob. 2015. "Review of *The Nonhuman Turn* edited by Richard Grusin." In *Anthropocene Review*. Accessed 12 December 2016. www.theanthropocenereview.com.

Bourdieu, Pierre. 1987. *Distinction: A Social Critique of the Judgement of Taste*. Cambridge: Harvard University Press.

Boyle, Philip J., and Shannon T. Speed. 2018. "From Protection to Coordinated Preparedness: A Genealogy of Critical Infrastructure in Canada." *Security Dialogue* 49: 217–31.

Braidotti, Rosi. 2002. *Metamorphoses: Towards a Materialist Theory of Becoming*. Cambridge: Polity Press.

———. 2006. *Transpositions: On Nomadic Ethics*. Cambridge: Polity Press.

———. 2013. *The Posthuman*. Cambridge: Polity Press.

———. 2018. "A Theoretical Framework for the Critical Posthumanities." *Theory, Culture & Society* 36 (6): 31–61.

Brassier, Ray, Iain Hamilton Grant, Graham Harman, and Quentin Meillassoux. 2007. "Speculative Realism." In *Collapse: Philosophical Research and Development*, Vol. 3, edited by Robin Mackay, 307–450. Falmouth: Urbanomic.

Braun, Bruce. 2004. "Modalities of Posthumanism." *Environment and Planning A* 36(8): 1352–55.

———. 2008. "Environmental Issues: Inventive Life." *Progress in Human Geography* 32(5): 667–79.

———. 2011. "Review of *Vibrant Matter: A Political Ecology of Things*, by Jane Bennett." *Dialogues in Human Geography* 1(3): 390–93.

———. 2014. "A New Urban Dispositif? Governing Life in the Age of Climate Change." *Environment and Planning D: Society and Space* 32: 49–64.

———. 2015. "The 2013 Antipode RGS-IBG Lecture. New Materialism and Neoliberal Natures." *Antipode* 47(1): 1–14.

———, and Sarah J. Whatmore. 2010. "The Stuff of Politics: An Introduction." In *Political Matter: Technoscience, Democracy, and Public Life*, edited by Bruce Braun and Sarah J. Whatmore, ix–xl. Minneapolis: University of Minnesota Press.

Braun, Hermann. 1982. "Materialismus-Idealismus." In *Geschichtliche Grundbegriffe. Historisches Lexikon zur politisch-sozialen Sprache in Deutschland*, edited by Otto Brunner, Werner Conze and Reinhart Koselleck, 977–1019. Stuttgart: Klett-Cotta.

Braunmühl, Caroline. 2018. "Beyond Hierarchical Oppositions: A Feminist Critique of Karen Barad's Agential Realism." *Feminist Theory* 19(2): 223–40.

Brauns, Jörg. 2003. "Schauplätze." *Untersuchungen zur Theorie und Geschichte der Dispositive visueller Medien*. Dissertation, Bauhaus-Universität Weimar, https://e-pub.uni-weimar.de/opus4/files/75/Brauns.pdf.
Braverman, Irus. 2014. "Governing the Wild: Databases, Algorithms, and Population Models as Biopolitics." *Surveillance & Society* 12(1): 15–37.
———. 2017. "Anticipating Endangerment: The Biopolitics of Threatened Species Lists." *BioSocieties* 12(1): 132–57.
Bricker, Brett. 2015. "Hyperobjects: Philosophy and Ecology after the End of the World." *Philosophy and Rhetoric* 48(3): 359–65.
Bröckling, Ulrich. 2017. "Von Hirten, Herden und dem Gott Pan. Figurationen pastoraler Macht." In *Gute Hirten führen sanft: über Menschenregierungskünste*, 15–44. Berlin: Suhrkamp.
Brown, Jayna. 2015. "Being Cellular: Race, the Inhuman, and the Plasticity of Life." *GLQ: A Journal of Lesbian and Gay Studies* 21(2–3): 321–41.
Brown, Mark B. 2015. "Politicizing Science: Conceptions of Politics in Science and Technology Studies." *Social Studies for Science* 45(1): 3–30.
Brown, Nathan. 2011. "The Speculative and the Specific: On Hallward and Meillassoux." In *The Speculative Turn: Continental Materialism and Realism*, edited by Levi Bryant, Nick Srnicek and Graham Harman, 142–63. Melbourne: re:press.
Bruining, Dennis. 2013. "A Somatechnics of Moralism: New Materialism or Material Foundationalism." *Somatechnics* 3(1): 149–68.
Brunner, Eric J., Peter J. S. Jones, Sharon Friel, and Mel Bartley. 2009. "Fish, Human Health and Marine Ecosystem Health: Policies in Collision." *International Journal of Epidemiology* 38(1): 93–100.
Bryant, Levi R. 2011a. *The Democracy of Objects*. Ann Arbor: Open Humanities Press.
———. 2011b. "The Ontic Principle: Outline of an Object-Oriented Ontology." In *The Speculative Turn: Continental Materialism and Realism*, edited by Levi Bryant, Nick Srnicek, and Graham Harman, 261–78. Melbourne: re:press.
———, Nick Srnicek, and Graham Harman, eds. 2011. *The Speculative Turn: Continental Materialism and Realism*. Melbourne: re:press.
Bucher, Taina. 2018. *If . . . Then: Algorithmic Power and Politics*. New York: Oxford University Press.
Bührmann, Andrea D. 2013. "Vom 'Discursive Turn' zumum 'Dispositive Turn'? Folgerungen, Herausforderungen und Perspektiven für die Forschungspraxis." In *Verortungen des Dispositiv-Begriffs: Analytische Einsätze zu Raum, Bildung, Politik*, edited by Johanna Caborn Wengler, Britta Hoffarth and Łukasz Kumięga, 20–34. Wiesbaden: Springer VS.
———, and Werner Schneider. 2008. *Vom Diskurs zum Dispositiv. Eine Einführung in die Dispositivanalyse*. Bielefeld: transcript.
Burchell, Graham. 2006. "Translator's Note." In *Michel Foucault, Psychiatric Power: Lectures at the Collège de France 1973-1974*, edited by Jacques Lagrange, xxiii–xxiv. Hampshire/New York: Palgrave Macmillan. .

———, Colin Gordon, and Peter Miller, eds. 1991. *The Foucault Effect: Studies in Governmentality*. Hemel Hempstead: Harvester Wheatsheaf.
Bussolini, Jeffrey. 2010. "What Is a Dispositive?" *Foucault Studies* 10: 85–107.
Butler, Judith. 1990. *Gender Trouble: Feminism and the Subversion of Identity*. New York: Routledge.
———. 1993. *Bodies that Matter: On the Discursive Limits of Sex*. New York/London: Routledge.
Callon, Michel. 1986. "Some Elements of a Sociology of Translation: The Domestication of the Scallops and the Fishermen of St. Brieuc Bay." In *Power, Action & Belief: A New Sociology of Knowledge?*, edited by John Law, 196–233. London: Routledge & Kegan Paul.
———, and Bruno Latour. 1981. "Unscrewing the Big Leviathan: How Actors Macrostructure Reality and How Sociologists Help Them to Do So." In *Advances in Social Theory and Methodology: Toward an Integration of Micro and Macro-Sociologies*, edited by Aaron V. Cicourel and Karin Knorr-Cetina, 277–303. Boston: Routledge & Kegan Paul.
———, and Fabian Muniesa. 2003. "Les marchés économiques comme dispositifs collectifs de calcul." *Réseaux* 6(122): 189–233.
———, Yuval Millo, and Fabian Muniesa, eds. 2007. *Market Devices*. Oxford: Blackwell.
———, Pierre Lascoumes, and Yannick Barthe. 2009. *Acting in an Uncertain World: An Essay on Technical Democracy*. Cambridge: MIT Press.
Campbell, Norah, Stephen Dunne, and Paul Ennis. 2019. "Graham Harman, Immaterialism: Objects and Social Theory." *Theory, Culture & Society* 36(3): 121–37.
———. 1991. *The Normal and the Pathological*. New York: Zone Books.
———. 2008a. "Machine and Organism: The Living and its Milieu." In *Knowledge of Life*, 75–120. New York: Fordham University Press.
———. 2008b. *Knowledge of Life*. New York: Fordham University Press.
———. 2012. "The Problem of Regulation in the Organism and in Society." In *Writings on Medicine*, 67–78. New York: Fordham University Press.
Cannon, Walter B. 1963. *The Wisdom of the Body: How the Human Body Reacts to Disturbance and Danger and Maintains the Stability Essential to Life*. New York: Norton.
Casper, Monica J. 1994. "Reframing and Grounding Nonhuman Agency: What Makes a Fetus an Agent?" *American Behavioral Scientist* 37(6): 839–56.
———. 1998. *The Making of the Unborn Patient: A Social Anatomy of Fetal Surgery*. New Brunswick: Rutgers University Press.
Castree, Noel. 2003. "A Post-Environmental Ethics?" *Ethics, Place & Environment* 6(1): 3–12.
———, and Catherine Nash. 2004. "Mapping Posthumanism: An Exchange." *Environment and Planning A* 36(8): 1341–63.
———, and Catherine Nash. 2006. "Posthuman Geographies." *Social & Cultural Geography* 7(4): 501–4.

Cavanagh, Connor J. 2014. "Biopolitics, Environmental Change, and Development Studies." *Forum for Development Studies* 41(2): 273–94.
Chandler, Katherine. 2011. "Political Environments." *Qui parle* 19(2): 299–308.
Chatwin, Bruce. 1989. "Nomad Invasions." In *What Am I Doing Here?*, 216–29. London: Picador.
Cheah, Pheng. 1996. "Mattering. Judith Butler: *Bodies That Matter*. Elizabeth Grosz: *Volatile Bodies*." *diacritics* 26(1): 108–39.
Cheung, Tobias. 2014. *Organismen. Agenten zwischen Innen- und Außenwelten 1780–1860*. Bielefeld: transcript.
Chrulew, Matthew. 2011. "Managing Love and Death at the Zoo: The Biopolitics of Endangered Species Preservation." *Australian Humanities Review* 50: 137–57.
Clarke, Bruce. 2014. *Neocybernetics and Narrative*. Minneapolis: University of Minnesota Press.
Cohen, Lawrence. 2005. "Operability, Bioavailability, and Exception." In *Global Assemblages: Technology, Politics, and Ethics as Anthropological Problems*, edited by Aihwa Ong, 79–90. Malden: Blackwell.
Cole, Andrew. 2013. "The Call of Things: A Critique of Object-Oriented Ontologies." *Minnesota Review* 80: 106–18.
———. 2015. "Those Obscure Objects of Desire: Andrew Cole on the Uses and Abuses of Object-Oriented Ontology and Speculative Realism." *Artforum* 6: 318–23.
Cole, Joshua. 2000. *The Power of Large Numbers: Population, Politics, and Gender in Nineteenth-Century France*. Ithaca: Cornell University Press.
Colebrook, Claire. 2008. "On Not Becoming Man: The Materialist Politics of Unactualized Potential." In *Material Feminisms*, edited by Stacy Alaimo and Susan Hekman, 52–84. Bloomington/Indianapolis: Indiana University Press.
Coles, Romand. 2016. "Walt Whitman, Jane Bennett, and the Paradox of Antagonistic Sympathy." *Political Research Quarterly* 69(3): 621–25.
Collard, Rosemary-Claire. 2012. "Cougar-Human Entanglements and the Biopolitical Un/Making of Safe Space." *Environment and Planning D: Society and Space* 30: 23–42.
———, and Jessica Dempsey. 2017. "Capitalist Natures in Five Orientations." *Capitalism Nature Socialism* 28: 78–97.
Collier, Stephen, and Andrew Lakoff. 2008a. "The Vulnerability of Vital Systems: How 'Critical Infrastructure' Became a Security Problem." In *The Politics of Securing the Homeland: Critical Infrastructure, Risk and Securitisation*, edited by Myriam Dunn Cavelty and Kristian Søby Kristensen, 17–39. London: Routledge.
———, and Andrew Lakoff. 2008b. "Distributed Preparedness: The Spatial Logic of Domestic Security in the United States." *Environment and Planning D: Society and Space* 26: 7–28.
———, and Andrew Lakoff. 2015. "Vital Systems Security: Reflexive Biopolitics and the Government of Emergency." *Theory, Culture & Society* 32(2): 19–51.
Collins, Harry M., and Trevor J. Pinch. 1982. *Frames of Meaning: The Social Construction of Extraordinary Science*. London/Boston: Routledge & Kegan Paul.

Comte, Auguste. 1998. *Early Political Writings*, edited by H. S. Jones. Cambridge: Cambridge University Press.

Connolly, William E. 2013. "The 'New Materialism' and the Fragility of Things." *Millennium: Journal of International Studies* 41(3): 399–412.

Conty, Arianne Françoise. 2018. "The Politics of Nature: New Materialist Responses to the Anthropocene." *Theory, Culture & Society* 35(7–8): 73–96.

Coole, Diana. 2013. "Agentic Capacities and Capacious Historical Materialism: Thinking with New Materialisms in the Political Sciences." *Millennium: Journal of International Studies* 41(3): 451–69.

———, and Samantha Frost. 2010a. *New Materialisms: Ontology, Agency and Politics*. Durham: Duke University Press.

———, and Samantha Frost. 2010b. "Introducing the New Materialisms." In *New Materialisms: Ontology, Agency and Politics*, edited by Diana Coole and Samantha Frost, 1–43. Durham: Duke University Press.

Cooper, Melinda. 2008. *Life as Surplus: Biotechnology and Capitalism in the Neoliberal Era*. Seattle: University of Washington Press.

———. 2010. "Turbulent Worlds." *Theory, Culture & Society* 27(2–3): 167–90.

Cortes-Vazquez, Jose A., and Esteban Ruiz-Ballesteros. 2018. "Practising Nature: A Phenomenological Rethinking of Environmentality in Natural Protected Areas in Ecuador and Spain." *Conservation & Society* 16(3): 232–42.

Cresswell, Timm. 2011. "Mobilities I: Catching Up." *Progress in Human Geography* 35(4): 550–58.

Critchley, Simon. 2008. *Unendlich fordernd. Ethik der Verpflichtung, Politik des Widerstands*. Zürich/Berlin: Diaphanes.

Crutzen, Paul. 2002. "Geology of Mankind." *Nature* 415: 23.

———. 2006. "Albedo Enhancement by Stratospheric Sulfur Injections: A Contribution to Resolve a Policy Dilemma? An Editorial Essay." *Climatic Change* 77: 211–20.

———, and Eugene F. Stoermer. 2000. "The 'Anthropocene.'" *Global Change Newsletter* 41: 27–18.

Cudworth, Erika, and Stephen Hobden. 2015. "Liberation for Straw Dogs? Old Materialism, New Materialism, and the Challenge of an Emancipatory Posthumanism." *Globalizations* 12(1): 134–48.

da Costa, Beatriz. 2010. "Reaching the Limit: When Art Becomes Science." In *Tactical Biopolitics: Art, Activism, and Technoscience*, edited by Beatrice da Costa and Kavita Philip, 365–85. Cambridge: MIT Press.

Daly, Herman. 1980. "Introduction to the Steady-State Economy." In *Economics, Ecology, Ethics: Essays Toward a Steady-State Economy*, edited by Herman Daly, 1–31. San Francisco: W. H. Freeman & Co.

D'Andrea, Anthony, Luigina Ciolfi, and Breda Gray. 2011. "Methodological Challenges and Innovations in Mobilities Research." *Mobilities* 6(2): 149–60.

Dányi, Endre. 2018. "Are Parliaments Still Privileged Sites for Studying Politics and Liberal Democracy, and If They Are, at What Price?" In *Routledge Companion to*

Actor-Network-Theory, edited by Andreas Blok, Ignacio Farías and Celia Roberts, 298–305. London: Routledge.

———, and Michaela Spencer. 2020. "Un/Common Grounds: Tracing Politics across Worlds." *Social Studies of Science*. doi:10.1177/0306312720909536.

Davis, Noela. 2009. "New Materialism and Feminism's Anti-Biologism: A Response to Sara Ahmed." *European Journal of Women's Studies* 16(1): 67–80.

de la Cadena, Marisol. 2010. "Indigenous Cosmopolitics in the Andes: Conceptual Reflections Beyond 'Politics.'" *Cultural Anthropology* 25(2): 33470.

de Landa, Manuel. 2000. *A Thousand Years of Nonlinear History*. New York: Swerve.

de Vries, Leonie Ansems. 2013. "Political Life beyond the Biopolitical?" *Theoria* 60(134): 50–68.

Dean, Mitchell. 1991. *The Constitution of Poverty: Toward a Genealogy of Liberal Government*. London: Routledge.

———. 1996. "Putting the Technological into Government." *History of the Human Sciences* 9(3): 47–68.

Deleuze, Gilles. 1992a. "What Is a Dispositive?" In *Foucault: Philosopher*, edited by Timothy J. Armstrong, 159–68. New York: Harvester Wheatsheaf.

———. 1992b. "Postscript on the Societies of Control." *October* 59: 3–7.

———, and Félix Guattari. 1987. *A Thousand Plateaus: Capitalism and Schizophrenia*. Minneapolis: University of Minnesota Press.

Delitz, Heike. 2014. "Gilbert Simondons Ontologie, philosophische Anthropologie und Gesellschaftstheorie: Ein recht verstandener Bergsonismus." In *Philosophische Anthropologie nach 1945*, edited by Guillaume Plas and Gérard Raulet, 277–302. Nordhausen: Traugott Brautz.

Demenchonok, Edwards. 2018. "Michel Foucault's Theory of Practices of the Self and the Quest for a New Philosophical Anthropology." In *Peace, Culture, and Violence*, edited by Fuat Gursozlu, 218–47. Leiden: Brill.

Dempsey, Jessica, and Morgan M. Robertson. 2012. "Ecosystem Services: Tensions, Impurities, and Points of Engagement within Neoliberalism." *Progress in Human Geography* 36(6): 758–79.

Department of Homeland Security. 2006. *National Infrastructure Protection Program*. Washington: Department of Homeland Security.

Descola, Philippe. 2013. "Wahlverwandtschaften. Antrittsvorlesung am Lehrstuhl für die 'Anthropologie der Natur.'" *Mittelweg* 36(5): 4–26.

Despret, Vinciane. 2006. "Sheep Do Have Opinions." In *Making Things Public: Atmospheres of Democracy*, edited by Bruno Latour and Peter Weibel, 360–70. Cambridge: MIT Press.

———. 2008. "The Becomings of Subjectivity in Animal Worlds." *Subjectivity* 23: 123–39.

Desrosières, Alain. 2002. *The Politics of Large Numbers: A History of Statistical Reasoning*. Cambridge: Harvard University Press.

Devellennes, Charles, and Benoît Dillet. 2018. "Questioning New Materialisms: An Introduction." *Theory, Culture & Society* 35(7–8): 5–20.

Diaz-Bone, Rainer, and Ronald Hartz. 2017. *Dispositiv und Ökonomie: Diskurs- und dispositivanalytische Perspektiven auf Märkte und Organisationen*. Wiesbaden: Springer VS.

Dictionnaire Historique de la Langue Française. 2006. Vol. 1, edited by Alain Rey. Paris: Dictionnaires Le Robert.

Dillon, Michael, and Julian Reid. 2009. *The Liberal Way of War: Killing to Make Life Live*. London/New York: Routledge.

Dionisius, Sarah. 2015. "Queer Matters: Family-Building Processes of Lesbian Couples Using Donor Insemination." *Distinktion: Scandinavian Journal of Social Theory* 16(3): 283–301.

Dolphijn, Rick, and Iris van der Tuin. 2012. *New Materialism: Interviews & Cartographies*. Ann Arbor: Open Humanities Press.

Donaldson, Brianne. 2014. "Introduction." In *Beyond the Bifurcation of Nature: A Common World for Animals and the Environment*, edited by Brianne Donaldson, 1–6. Newcastle upon Tyne: Cambridge Scholars Publishing.

Donzelot, Jacques. 1984. *L'invention du social. Essai sur le déclin des passions politiques*. Paris: Seuil.

Dorrestijn, Steven. 2011. "Technical Mediation and Subjectivation: Tracing and Extending Foucault's Philosophy of Technology." *Philosophy & Technology* 25(2): 221–41.

Dotzler, Bernard. 2004. "Der Zusammenhang der Dinge. Regulation und Dämonologie von Watt bis Maxwell." In *Kontingenz und Steuerung. Literatur als Gesellschaftsexperiment 1750–1830*, edited by Torsten Hahn, Erich Kleinschmidt and Nicolas Pethes, 177–89. Würzburg: Königshausen & Neumann.

Dreyfus, Hubert L., and Paul Rabinow. 1983. *Michel Foucault: Beyond Structuralism and Hermeneutics*, 2nd edition. Chicago: University of Chicago Press.

Driesch, Hans. 1908a. *The Science and Philosophy of the Organism: The Gifford Lectures*. Delivered before the University of Aberdeen in the Year 1907. London: Adam and Charles Black.

———. 1908b. *The Science and Philosophy of the Organism: The Gifford Lectures*. Delivered before the University of Aberdeen in the Year 1908. London: Adam and Charles Black.

Dupont, Danica, and Frank Pearce. 2001. "Foucault Contra Foucault: Rereading the 'Governmentality' Papers." *Theoretical Criminology* 5(2): 123–58.

Dupré, John. 2012. *Processes of Life: Essays in the Philosophy of Biology*. Oxford: Oxford University Press.

Edwards, Paul N. 1996. *The Closed World of Computers and the Politics of Discourse in Cold War America*. London: MIT Press.

Elden, Stuart. 2016. *Foucault's Last Decade*. Cambridge: Polity Press.

———. 2017. *Foucault: The Birth of Power*. Cambridge: Polity Press.

———. 2019. *Canguilhem*. Cambridge: Polity Press.

Ellenzweig, Sarah, and John H. Zammito. 2017. "Introduction: New Materialism. Looking Forward, Looking Back." In *The New Politics of Materialism: History, Philosophy, Science*, 1–16. London: Routledge.
Ellis, Erle C. 2011. "The Planet of No Return." *Breakthrough Journal* 2: 37–44.
Engels, Friedrich. 2000. *Herr Eugen Dühring's Revolution in Science (Anti-Dühring)*. London: The Electric Book Company.
Eribon, Didier. 1994. ". . . quelque fragment d'autobiographie." In *Michel Foucault. Les jeux de vérité et du pouvoir*, edited by Alain Brossat, 127–31. Nancy: Presses Universitaires de Nancy.
Esposito, Roberto. 2008. *Bios: Biopolitics and Philosophy*. Minneapolis: University of Minnesota Press.
———. 2011. *Immunitas*. Cambridge: Polity Press.
———. 2015. *Persons and Things: From the Body's Point of View*. Cambridge/Malden: Polity Press.
———. 2016. "Persons and Things." *Paragraph* 39(1): 26–35.
Evans, Brad, and Julian Reid. 2013. "Dangerously Exposed: The Life and Death of the Resilient Subject." *Resilience* 1(2): 83–98.
Ewald, François. 1986. *L'Etat providence*. Paris: Grasset.
Farías, Ignacio, and Laurie Waller. 2016. "A Turn to Nontology? Exploring Environmental Noise in European Cities." Lecture, Technical University Munich, Germany, January 29, 2016.
Fausto-Sterling, Anne. 2003. "The Problem with Sex/Gender and Nature/Nurture." In *Debating Biology: Sociological Reflections on Health, Medicine and Society*, edited by Simon J. Williams, Lynda Birke and Gilian A. Bendelow, 123–32. New York/London: Routledge.
Federici, Silvia. 2004. *Caliban and the Witch: Women, the Body, and Primitive Accumulation*. New York: Autonomedia.
Feenberg, Andrew. 2017. "Critical Theory of Technology and STS." *Thesis Eleven* 138(1): 3–12.
Feibleman, James K. 1970. *The New Materialism*. The Hague: Martinus Nijhoff.
Fernandes, Leela. 1997. *Producing Workers: The Politics of Gender, Class, and Culture in the Calcutta Jute Mills*. Philadelphia: University of Pennsylvania Press.
Feuerhahn, Wolf. 2017. "'Milieu'—Renaissance auf den Schultern von Leo Spitzer und Georges Canguilhem?" In *Milieu: Umgebungen des Lebendingen in der Moderne*, edited by Florian Huber and Christina Wessely, 18–34. Paderborn: Wilhelm Fink.
Fitsch, Hannah. 2014. *. . . dem Gehirn beim Denken zusehen? Sicht- und Sagbarkeiten in der funktionellen Magnetresonanztomographie*. Bielefeld: transcript.
———, and Lukas Engelmann. 2013. "Das Bild als Phänomen. Visuelle Argumentationsweisen und ihre Logiken am Beispiel von Sichtbarmachungen des 'AIDS-Virus' und der funktionellen MRT." In *Visuelles Wissen und Bilder des Sozialen. Aktuelle Entwicklungen in der Soziologie des Visuellen*, edited by Petra Lucht, Lisa-Marian Schmidt and René Tuma, 213–30. Wiesbaden: Springer.

Fitzgerald, Des, and Felicity Callard. 2015. "Social Science and Neuroscience beyond Interdisciplinarity: Experimental Entanglements." *Theory Culture & Society* 32(1): 3–32.

Fletcher, Robert. 2010. "Neoliberal Environmentality: Towards a Poststructuralist Political Ecology of the Conservation Debate." *Conservation & Society* 8(3): 171–81.

———. 2017. "Environmentality Unbound: Multiple Governmentalities in Environmental Politics." *Geoforum* (85): 311–15.

Folkers, Andreas. 2017. "Politik des Lebens jenseits seiner selbst. Für eine ökologische Lebenssoziologie mit Deleuze und Guattari." *Soziale Welt* 68(4): 356–84.

———. 2018. *Das Sicherheitsdispositiv der Resilienz. Katastrophische Risiken und die Biopolitik vitaler Systeme*. Frankfurt am Main: Campus Verlag.

Foucault, Michel. 1970. *The Order of Things: An Archaeology of the Human Sciences*. New York: Pantheon Books.

———. 1972. *The Archaeology of Knowledge*. New York: Pantheon Books.

———. 1978. *The History of Sexuality, Vol. 1. An Introduction*. New York: Pantheon Books.

———. 1979. *Discipline and Punish: The Birth of the Prison*. London: Allen Lane.

———. 1980a. "Truth and Power." (Interview with Alessandro Fontana and Pasquale Pasquino.) In *Power/Knowledge: Selected Interviews and Other Writings 1972–1977*, edited by Colin Gordon, 109–33. New York: Pantheon Books.

———. 1980b. "The Confession of the Flesh." In *Power/Knowledge: Selected Interviews and Other Writings 1972–1977*, edited by Colin Gordon, 194–228. New York: Pantheon Books.

———. 1980c. "Power and Strategies." In *Power/Knowledge: Selected Interviews and Other Writings 1972–1977*, edited by Colin Gordon, 134–45. New York: Pantheon Books.

———. 1981a. "The Order of Discourse." (Inaugural Lecture at the Collège de France, given 2 December 1970.) In *Untying the Text: A Post-Structuralist Reader*, edited by Robert Young, 48–78. Boston/London: Routledge.

———. 1981b. "'Omnes et Singulatim': Towards a Criticism of Political Reason." (Lecture at Stanford University, 10 and 16 October 1979.) In *The Tanner Lectures on Human Values*, edited by Sterlin M. McMurrin, 225–54. Salt Lake City: University of Utah Press.

———. 1984a. "Polemics, Politics and Problematizations: An Interview with Michel Foucault." (Conversation with Paul Rabinow, May 1984.) In *The Foucault Reader*, edited by Paul Rabinow, 381–90. New York: Pantheon.

———. 1984b. "What is Enlightenment?" In *The Foucault Reader*, edited by Paul Rabinow, 32–50. New York: Pantheon.

———.1984c. "Truth and Power." In *The Foucault Reader*, edited by Paul Rabinow, 51–75. New York: Pantheon.

———. 1984d. "Preface to *The History of Sexuality, Vol. II*." (Draft for the foreword to volume 2 of *The History of Sexuality*.) In *The Foucault Reader*, edited by Paul Rabinow 333–39. New York: Pantheon.

———. 1985. "An Interview with Michel Foucault." *History of the Present* 1: 2–3, 14.

———. 1987. *Mental Illness and Psychology*. Berkeley: University of California Press.
———. 1988a. "Critical Theory/Intellectual History." (Conversation with G. Raulet, May 1982.) In *Politics, Philosophy, Culture: Interviews and Other Writings 1977–1984*, edited by Lawrence D. Kritzman, 17–46. New York/London: Routledge.
———. 1988b. "On power." In *Politics, Philosophy, Culture: Interviews and Other Writings 1977–1984*, edited by Lawrence D. Kritzman, 96–109. New York/London: Routledge.
———. 1988c. "Iran: The Spirit of a World without Spirit." In *Politics, Philosophy, Culture: Interviews and Other Writings 1977–1984*, edited by Lawrence D. Kritzman, 211–26. New York/London: Routledge.
———. 1991a. "Questions of Method." In *The Foucault Effect: Studies in Governmentality*, edited by Graham Burchell, Colin Gordon and Peter Miller, 73–86. Hemel Hempstead: Harvester Wheatsheaf.
———. 1991b. *Remarks on Marx: Conversations with Ducio Tromadori*. New York: Semiotexte.
———. 1994a. "Prisons et asiles dans le mécanisme du pouvoir." In *Dits et écrits 1954–1988, par Michel Foucault: Vol. II, 1970–1975*, edited by Daniel Defert and François Ewald, 523–24. Paris: Gallimard.
———. 1994b. "La poussière et le nuage." In *Dits et écrits 1954–1988, par Michel Foucault: Vol. IV 1980–1988*, edited by Daniel Defert and François Ewald, 10–19. Paris: Gallimard.
———. 1994c. "Croître et multiplier." In *Dits et écrits 1954–1988 par Michel Foucault: Vol. II, 1970–1975*, edited by Daniel Defert and François Ewald, 99–104. Paris: Gallimard.
———. 1994d. "Dialogue sur le pouvoir." In *Dits et écrits 1954–1988, par Michel Foucault: Vol. III, 1976–1979*, edited by Daniel Defert and François Ewald, 464–77. Paris: Gallimard.
———. 1994e. "Entretien avec Madeleine Chapsal." In *Dits et écrits 1954–1988, par Michel Foucault: Vol. I, 1954–1969*, edited by Daniel Defert and François Ewald, 513–18. Paris: Gallimard.
———. 1994f. "Message ou bruit?" In *Dits et écrits 1954–1988, par Michel Foucault: Vol. I, 1954–1969*, edited by Daniel Defert and François Ewald, 557–60. Paris: Gallimard.
———. 1994g. "The Art of Telling the Truth." In *Critique and Power: Recasting the Foucault/ Habermas Debate*, edited by Michael Kelly, 139–48. Cambridge: MIT Press.
———. 1994h. "Les mailles du pouvoir." In *Dits et écrits 1954–1988, par Michel Foucault: Vol. IV*, edited by Daniel Defert and François Ewald, 182–201. Paris: Gallimard.
———.1996a. "From Torture to Cellblock." In *Foucault Live: Interviews 1961–1984*, edited by Sylvère Lotringer, 146–49. New York: Semiotext.
———. 1996b. "The End of the Monarchy of Sex." In *Foucault Live: Interviews 1961–1984*, edited by Sylvère Lotringer, 214–25. New York: Semiotext.
———. 1997a. "Technologies of the Self." In *Ethics, Subjectivity and Truth: Essential Works of Michel Foucault, 1954–1984*, Vol. I, edited by Paul Rabinow, 223–51. New York: The New Press.
———. 1997b. "Candidacy Presentation: Collège de France, 1969." In *Ethics, Subjectivity and Truth: Essential Works of Foucault, 1954–1984*, Vol. I, edited by Paul Rabinow, 5–10. New York: The New Press.

———. 1997c. "What is Enlightenment?" In *Ethics, Subjectivity and Truth: Essential Works of Michel Foucault, 1954–1984*, Vol. I, edited by Paul Rabinow, 303–19. New York: New Press.

———. 1997d. "Friendship as a Way of Life." In *Ethics, Subjectivity and Truth: Essential Works of Michel Foucault, 1954–1984*, Vol. I, edited by Paul Rabinow, 135–40. New York: New Press.

———. 1997e. "Sex, Power, and the Politics of Identity." In *Ethics, Subjectivity and Truth: Essential Works of Michel Foucault, 1954–1984*, Vol. I, edited by Paul Rabinow, 163–73. New York: New Press.

———. 1998a. "Life: Experience and Science." In *Aesthetics, Method, and Epistemology: Essential Works of Foucault, 1954–1984*, Vol. II., edited by James D. Faubion, 465–78. New York: The New Press.

———. 1998b. "Different Spaces." In *Aesthetics, Method, and Epistemology: Essential Works of Michel Foucault, 1954–1984*, Vol. II, edited by James D. Faubion, 175–85. New York: The New Press.

———. 1998c. "Nietzsche, Genealogy, History." In *Aesthetics, Method, and Epistemology: Essential Works of Michel Foucault, 1954–1984*, Vol. II, edited by James D. Faubion, 369–91. New York: The New Press.

———. 1998d. "Foucault by Maurice Florence." In *Aesthetics, Method, and Epistemology: Essential Works of Michel Foucault, 1954–1984*, Vol. II, edited by James D. Faubion, 459–63. New York: New Press.

———. 2000a. "Space, Knowledge, and Power." In *Power: Essential Works of Michel Foucault, 1954–1984*, Vol. III, edited by James D. Faubion, 349–64. New York: The New Press.

———. 2000b. "The Subject and Power." In *Power: Essential Works of Michel Foucault, 1954–1984*, Vol. III, edited by James D. Faubion, 326–48. New York: The New Press.

———. 2000c. "The Birth of Social Medicine." In *Power: Essential Works of Michel Foucault, 1954–1984*, Vol. III, edited by James D. Faubion, 134–56. New York: The New Press.

———. 2000d. "The Political Technology of Individuals." In *Power: Essential Works of Michel Foucault, 1954–1984*, Vol. III, edited by James D. Faubion, 403–17. New York: The New Press.

———. 2000e. "So Is It Important to Think?" In *Power: Essential Works of Michel Foucault, 1954–1984*, Vol. III, edited by James D. Faubion, 454–58. New York: The New Press.

———. 2003. *Society Must Be Defended: Lectures at the Collège de France 1975–76*. New York: Picador.

———. 2004. *Naissance de la Biopolitique, Cours au Collège de France, 1978–1979*. Paris: Gallimard/Seuil.

———. 2005. *Hermeneutics of the Subject: Lectures at the Collège de France, 1981–82*. New York: Palgrave Macmillan.

———. 2006a. *Psychiatric Power: Lectures at the Collège de France, 1973–1974*. Hampshire/New York: Palgrave Macmillan.

———. 2006b. *History of Madness*. New York/London: Routledge.

———. 2007a. *Security, Territory, Population: Lectures at the Collège de France, 1977–78*. New York: Palgrave Macmillan

———. 2007b. "The Incorporation of the Hospital into Modern Technology." In *Space, Knowledge, Power. Foucault and Geography*, edited by Jeremy W. Crampton and Stuart Elden, 141–52. Aldershot: Ashgate.

———. 2007c. "The Meshes of Power." In *Space, Knowledge, Power. Foucault and Geography*, edited by Jeremy W. Crampton and Stuart Elden, 153–62. Aldershot: Ashgate.

———. 2008a. *The Birth of Biopolitics: Lectures at the Collège de France, 1978–79*. Basingstoke/New York: Palgrave Macmillan.

———. 2008b. *Introduction to Kant's Anthropology*. Los Angeles: Semiotext(e).

———. 2009. *Manet and the Object of Painting*. London: Tate Publishing.

———. 2014a. *On the Government of the Living: Lectures at the Collège de France 1979–1980*. Basingstoke: Palgrave Macmillan.

———. 2014b. "Bio-history and Bio-politics." *Foucault Studies* 18: 128–30.

Fox, Nick J., and Pam Alldred. 2016. *Sociology and the New Materialism: Theory, Research, Action*. London: Sage.

Franklin, Sarah. 2007. *Dolly Mixtures: The Remaking of Genealogy*. Durham: Duke University Press.

———. 2017. "Staying with the Manifesto: An Interview with Donna Haraway." *Theory, Culture & Society* 34(4): 49–63.

Franklin, Seb. 2007. *Forms of Disposal: Value and the Digital*. Speech, Vanderbilt University, 25 April 2007.

Fraser, Mariam. 2002. "What Is the Matter of Feminist Criticism?" *Economy and Society* 31(4): 606–25.

Friese, Carrie. 2013. *Cloning Wild Life: Zoos, Captivity, and the Future of Endangered Animals*. New York: New York University Press.

Frost, Samantha. 2016. *Biocultural Creatures: Toward a New Theory of the Human*. Durham: Duke University Press.

———. 2018. "Ten Theses on the Subject of Biology and Politics: Conceptual, Methodological, and Biopolitical Considerations." In *The Palgrave Handbook of Biology and Society*, edited by Maurizio Meloni, John Cromby, Des Fitzgerald and Stephanie Lloyd, 897–923. London: Palgrave Macmillan.

Fuller, Matthew. 2005. *Media Ecologies: Materialist Energies in Art and Technoculture*. Cambridge: MIT Press.

Gabbey, Alan. 2002. "Newton, Active Powers, and the Mechanical Philosophy." In *The Cambridge Companion to Newton*, edited by I. Bernard Cohen and George E. Smith, 329–57. Cambridge: Cambridge University Press.

Gabrys, Jennifer. 2014. "Programming Environments: Environmentality and Citizen Sensing in the Smart City." *Environment and Planning D: Society and Space* 32(1): 30–48.

———. 2016. *Program Earth: Environmental Sensing Technology and the Making of a Computational Planet*. Minneapolis: University of Minnesota Press.

Gabrys, Jennifer, and Kathryn Yusoff. 2012. "Arts, Sciences and Climate Change: Practices and Politics at the Threshold." *Science as Culture* 21(1): 1–24.

Galloway, Alexander R. 2004. *Protocol: How Control Exists After Decentralization*. Cambridge/London: MIT Press.

———. 2013. "The Poverty of Philosophy: Realism and Post-Fordism." *Critical Inquiry* 39(2): 347–66.

Gamble, Christopher N., Joshua S. Hanan, and Thomas Nail. 2019. "What is New Materialism?" *Angelaki* 24(6): 111–34.

Garske, Pia. 2014. "What's the Matter? Der Materialitätsbegriff des 'New Materialism' und dessen Konsequenzen für feministisch-politische Handlungsfähigkeit." *Prokla* 44(1): 111–29.

Geerts, Evelien, and Iris van der Tuin. 2013. "From Intersectionality to Interference: Feminist Onto-Epistemological Reflections on the Politics of Representation." *Women's Studies International Forum* 41: 171–78.

Gesch, C. Bernard, Sean M. Hammond, Sarah E. Hampson, Anita Eves, and Martin J. Crowder. 2002. "Influence of Supplementary Vitamins, Minerals and Essential Fatty Acids on the Antisocial Behaviour of Young Adult Prisoners: Randomised, Placebo-Controlled Trial." *The British Journal of Psychiatry* 181(1): 22–28.

Gibson-Graham, J. K., and Gerda Roelvink. 2010. "An Economic Ethics for the Anthropocene." *Antipode* 41: 320–46.

Ginsburg, Fayne D., and Rayna Rapp. 1995. "Introduction." In *Conceiving the New World Order: The Global Politics of Reproduction*, edited by Fayne D. Ginsburg and Rayna Rapp, 1–18. Berkeley: University of California Press.

Goerner, E. A. 1979. "On Thomistic Natural Law: The Bad Man's View of Thomistic Natural Right." *Political Theory* 7(1): 101–22.

Gomart, Emilie, and Antoine Hennion. 1999. "A Sociology of Attachment: Music Amateurs, Drug Users." In *Actor Network Theory and After*, edited by John Law and John Hassard, 220–47. Oxford/Malden: Blackwell.

Gordon, Colin. 1980. "Afterword." In *Power/Knowledge: Selected Interviews and Other Writings 1972–1977*, by Michel Foucault and edited by Colin Gordon, 229–60. New York: Pantheon Books.

Grant, Iain Hamilton. 2011. "Mining Conditions: A Response to Harman." In *The Speculative Turn: Continental Materialism and Realism*, edited by Levi Bryant, Nick Srnicek and Graham Harman, 41–46. Melbourne: re:press.

Gratton, Peter. 2014. *Speculative Realism: Problems and Prospects*. New York: Bloomsbury Academic.

Greco, Monica. 2021. "Vitalism Now–A Problematic." *Theory, Culture & Society* 38(2): 47–69.

Gregson, Nicky. 2011. "Book Review Vibrant Matter: A Political Ecology of Things." *Dialogues in Human Geography* 1(3): 402–04.

―――, Helen Watkins, and Melania Calestani. 2010. "Inextinguishable Fibres: Demolition and the Vital Materialisms of Asbestos." *Environment and Planning A* 42: 1065–83.

Gros, Frédéric. 2005. "Course Context." In *The Hermeneutics of the Subject: Lectures at the Collège de France 1981–1982*, edited by Michel Foucault and Frédéric Gros, 507–50. New York: Palgrave MacMillan.

Gross, Matthias, Holger Hoffman-Riem, and Wolfgang Krohn. 2005. *Realexperimente. Ökologische Gestaltungsprozesse in der Wissensgesellschaft*. Bielefeld: transcript.

Gross, Matthias, and Wolfgang Krohn. 2010. "Society as Experiment: Sociological Foundations for a Self-Experimental Society." *History of Human Sciences* 18(2): 3–86.

Grosz, Elisabeth. 2008. "Darwin and Feminism: Preliminary Investigations for a Possible Alliance." In *Material Feminisms*, edited by Stacy Alaimo and Susan Hekman, 23–51. Bloomington/Indianapolis: Indiana University Press.

Grusin, Richard, ed. 2015. *The Nonhuman Turn*. Minneapolis/London: University of Minnesota Press.

Gunderson, Lance H., and Crawford Stanley Holling, eds. 2002. *Panarchy: Understanding Transformations in Human and Natural Systems*. Washington, DC: Island Press.

Gutting, Garry. 1989. *Michel Foucault's Archaeology of Scientific Reason*. Cambridge/New York: Cambridge University Press.

Habermas, Jürgen. 1970. *Toward a Rational Society: Student Protest, Science, and Politics*. Boston: Beacon Press.

Hacking, Ian. 1983. *Representing and Intervening: Introductory Topics in the Philosophy of Natural Science*. Cambridge: Cambridge University Press.

―――. 1986. "Making Up People." In *Reconstructing Individualism: Autonomy, Individuality, and the Self in Western Thought*, edited by Thomas Heller, Morton Sosna and David E. Wellberry, 222–36. Stanford: Stanford University Press.

―――. 1998. "On Being More Literal about Construction." In *The Politics of Constructionism*, edited by Irving Velody and Robin Williams, 49–68. London: Sage.

―――. 2000. *The Social Construction of What?* Cambridge: Cambridge University Press.

―――. 2002. *Historical Ontology*. Cambridge: Harvard University Press.

Haff, Peter. 2014a. "Humans and Technology in the Anthropocene: Six Rules". *The Anthropocene Review* 1(2): 126–36.

―――. 2014b. Technology as a Geological Phenomenon: Implications for Human Well-Being. In *A Stratigraphical Basis for the Anthropocene*, edited by A. M. Snelling, 301–19. London: Geological Society.

Hägglund, Martin. 2011. "Radical Atheist Materialism: A Critique of Meillassoux." In *The Speculative Turn: Continental Materialism and Realism*, edited by Levi Bryant, Nick Srnicek and Graham Harman, 114–29. Melbourne: re:press.

Häkli, Jouni. 2018. "The Subject of Citizenship—Can There Be a Posthuman Civil Society?" *Political Geography* 67: 166–75.

Hallowell, Irving A. 1960. "Ojibwa Ontology, Behavior and World View." In *Culture in History: Essays in Honor of Paul Radin*, edited by Stanley Diamond, 19–52. New York: Columbia University Press.

Hamilton, Clive. 2015. "The Theodicy of the Good Anthropocene." *Environmental Humanities* 7: 233–38.

Hannah, Matthew G. 2011. "Biopower, Life and Left Politics." *Antipode* 43(4): 1034–55.

Haraway, Donna. 1991. *Simians, Cyborgs, and Women: The Reinvention of Nature*. London/New York: Routledge.

———. 1992. "The Promises of Monsters: A Regenerative Politics for Inappropriate/d Others." In *Cultural Studies*, edited by Lawrence Grossberg, Cary Nelson and Paula Treichler, 295–337. New York: Routledge.

———. 1997. *Modest_Witness@Second_Millennium: FemaleMan Meets Oncomouse*. New York/London: Routledge.

———. 2004. "Ecce Homo, Aint (Ar'n't) I a Woman, and Inappropriate/d Others: The Human in a Post-Humanist Landscape." In *The Haraway Reader*, edited by Donna Haraway, 47–61. New York/London: Routledge.

———. 2008. *When Species Meet*. Minneapolis: University of Minnesota Press.

———. 2012. "Value-Added Dogs and Lively Capital." In *Lively Capital: Biotechnologies, Ethics, and Governance in Global Markets*, edited by Kaushik Sunder Rajan, 93–120. Durham: Duke University Press.

———. 2016. *Staying with the Trouble: Making Kin in the Chthulucene*. Durham: Duke University Press.

Harding, Sandra. 1986. *The Science Question in Feminism*. Ithaca: Cornell University Press.

———, ed. 2004. *The Feminist Standpoint Theory Reader: Intellectual and Political Controversies*. New York: Routledge.

Harman, Graham. 2002. *Tool-Being: Heidegger and the Metaphysics of Objects*. Chicago/La Salle: Open Court.

———. 2005. *Guerrilla Metaphysics: Phenomenology and the Carpentry of Things*. Chicago/La Salle: Open Court.

———. 2007a. *Heidegger Explained: From Phenomenon to Thing*. Chicago: Open Court.

———. 2007b. "Aesthetics as First Philosophy: Levinas and the Non-Human." *Naked Punch* 9: 21–30.

———. 2009. *Prince of Networks: Bruno Latour and Metaphysics*. Melbourne: re:press.

———. 2010a. *Towards Speculative Realism: Essays and Lectures*. Winchester, UK/Washington, DC: Zero Books.

———. 2010b. "Time, Space, Essence, and Eidos: A New Theory of Causation." *Cosmos and History: The Journal of Natural and Social Philosophy* 6(1): 1–17.

———. 2010c. "I Am Also of the Opinion that Materialism Must Be Destroyed." *Environment and Planning D-Society & Space* 28(5): 772–90.

———. 2011a. *The Quadruple Object*. Ropley: Zero Books.

———. 2011b. "Realism without Materialism." *Sub-Stance* 40(2): 52–72.

———. 2011c. *Quentin Meillassoux: Philosophy in the Making*. Edinburgh: Edinburgh University Press.

———. 2011d. "On the Undermining of Objects: Grant, Bruno, and Radical Philosophy." In *The Speculative Turn: Continental Materialism and Realism*, edited by Levi Bryant, Nick Srnicek and Graham Harman, 21–40. Melbourne: re:press.

———. 2011e. "Autonomous Objects." Review of *Vibrant Matter*, by Jane Bennett. *New Formations* 71: 125–30.

———. 2012. "The Well-Wrought Broken Hammer: Object-Oriented Literary Criticism." *New Literary History* 43: 183–203.

———. 2013. *Bells and Whistles: More Speculative Realism*. Winchester: Zero Books.

———. 2014. *Bruno Latour: Reassembling the Political*. London: Pluto Press.

———. 2016a. *Immaterialism: Objects and Social Theory*. Cambridge: Polity Press.

———. 2016b. "Agential and Speculative Realism: Remarks on Barad's Ontology." *Rhizomes: Cultural Studies in Emerging Knowledge* 30. doi.org/10.20415/rhiz/030.e10.

———. 2018. *Object-Oriented Ontology: A New Theory of Everything*. London: Penguin.

Haudricourt, André G. 1969. "Domestication of Animals, Cultivation of Plants and Human Relations." *Social Science Information* 8(3): 163–72.

Hawkins, Gay. 2010. "Plastic Materialities." In *Political Matter: Technoscience, Democracy and Public Life*, edited by Sarah Whatmore and Bruce Braun, 119–36. Minneapolis: University of Minnesota Press.

Hayles, N. Katherine. 1999. *How We Became Posthuman: Virtual Bodies in Cybernetics, Literature, and Informatics*. Chicago: University of Chicago Press.

———. 2017. *Unthought: The Power of the Cognitive Unconscious*. Chicago/London: University of Chicago Press.

Heidegger, Martin. 1962 [1927]. *Being and Time*. Oxford: Blackwell.

———. 1967 [1962]. *What Is a Thing?* Chicago: Henry Regnery Company.

———. 1993 [1954]. "The Question Concerning Technology." In *Basic Writings*, 2nd edition, edited by David Krell, 307–41. London: Routledge.

Heise, Ursula. 2015. "Review of *Hyperobjects: Philosophy and Ecology after the End of the World*, by Timothy Morton." *Critical Inquiry* 41(2): 460–61.

Hekman, Susan. 2008. "Constructing the Ballast: An Ontology for Feminism." In *Material Feminisms*, edited by Stacy Alaimo and Susan Hekman, 85–119. Bloomington: Indiana University Press.

Helmreich, Stefan. 2009. *Alien Ocean: Anthropological Voyages in Microbial Seas*. Berkeley: University of California Press.

———. 2011. "What Was Life? Answers from Three Limit Biologies." *Critical Inquiry* 37(4): 671–96.

Henare, Amiria J. M., Martin Holbraad, and Sari Wastell, eds. 2007. *Thinking Through Things: Theorising Artefacts Ethnographically*. London/New York: Routledge.

Hennessy, Rosemary, and Chrys Ingraham, eds. 1997. *Materialist Feminism: A Reader in Class, Difference, and Women's Lives*. New York: Routledge.

Hetherington, Kregg, ed. 2019. *Infrastructure, Environment, and Life in the Anthropocene*. Durham: Duke University Press.

Hinchliffe, Steve. 2011. "Review of *Vibrant Matter: A Political Ecology of Things*, by Jane Bennett." *Dialogues in Human Geography* 1(3): 396–99.

Hinton, Peta. 2013. "The Quantum Dance and the World's 'Extraordinary Liveliness': Refiguring Corporeal Ethics in Karen Barad's Agential Realism." *Somatechnics* 3(1): 169–89.

Hird, Myra J. 2004. "Feminist Matters: New Materialist Considerations of Sexual Difference." *Feminist Theory* 5: 223–32.

———. 2009. "Feminist Engagements with Matter." *Feminist Studies* 35(2): 329–46.

Hobbes, Thomas. 1962 [1651]. *Leviathan*. London: Dent.

Højgaard, Lis, and Dorte M. Søndergaard. 2011. "Theorizing the Complexities of Discursive and Material Subjectivity: Agential Realism and Poststructural Analyses." *Theory & Psychology* 21(3): 338–54.

Holling, Crawford S. 1973. "Resilience and Stability of Ecological Systems." *Annual Review of Ecology and Systematics* 4(1): 1–23.

———. 2010. "The Resilience of Terrestrial Ecosystems." In *Foundations of Ecological Resilience*, edited by Lance Gunderson, Craig Allen and C. S. Holling, 67–118. Washington, DC: Island Press.

———, Lance H. Gunderson, and Garry D. Peterson. 2002. "Sustainability and Panarchies." In *Panarchy: Understanding Transformations in Human and Natural Systems*, edited by Lance H. Gunderson and Crawford S. Holling, 63–102. Washington, DC: Island Press.

Holloway, Lewis, and Carol Morris. 2012. "Contesting Genetic Knowledge-Practices in Livestock Breeding: Biopower, Biosocial Collectivities, and Heterogeneous Resistances." *Environment and Planning D: Society and Space* 30: 60–77.

Hoppe, Katharina. 2017a. "Eine neue Ontologie des Materiellen? Probleme und Perspektiven neomaterialistischer Feminismen." In *Material Turn. Feministische Perspektiven auf Materialität und Materialismus*, edited by Kathrina Volk, Christine Löw, Imke Leicht and Nadja Meisterhans, 35–50. Leverkusen: Verlag Barbara Budrich.

———. 2017b. "Politik der Antwort. Zum Verhältnis von Politik und Ethik in Neuen Materialismen." *Behemoth. A Journal on Civilisation* 10(1): 10–28.

Hoppe, Katharina, and Thomas Lemke. 2015. "Die Macht der Materie. Grundlagen und Grenzen des agentiellen Realismus von Karen Barad." *Soziale Welt* 66(3): 261–79.

Horkheimer, Max 2012. *Critique of Instrumental Reason*. London/New York: Verso.

Hörl, Erich. 2017. "Introduction to General Ecology: The Ecologization of Thinking." In *General Ecology: The New Ecological Paradigm*, edited by Erich Hörl and James Burton, 1–73. London/New York: Bloomsbury Academic.

———. "The Environmentalitarian Situation: Reflections on the Becoming-Environmental of Thinking, Power, and Capital." *Cultural Politics* 14(2): 153–73.

Hubig, Christoph. 2000. "'Dispositiv' als Kategorie." *Internationale Zeitschrift für Philosophie* (1): 34–47.

Hughes, J. Donald. 1986. "Pan: Environmental Ethics in Classical Polytheism." In *Religion and Environmental Crisis*, edited by Eugene C. Hargrove, 7–24. Athens: University of Georgia Press.
Hughes, Thomas P. 1983. *Networks of Power: Electrification in Western Society, 1880–1930*. Baltimore: Johns Hopkins University Press.
Hui, Yuk. 2015a. "Modulation after Control." *New Formations* 82–83: 74–91.
———. 2015b. "Towards A Relational Materialism." *Digital Culture & Society* 1(1): 131–48.
Husserl, Edmund. 2001 [1900/1901]. *Logical Investigations*. London: Routledge.
Ingold, Tim. 2004. "Beyond Biology and Culture: The Meaning of Evolution in a Relational World." *Social Anthropology* 12(2): 209–21.
Ingold, Tim, and Gísli Pálsson. 2013. *Biosocial Becomings: Integrating Social and Biological Anthropology*. Cambridge: Cambridge University Press.
Irni, Sari. 2013. "The Politics of Materiality: Affective Encounters in a Transdisciplinary Debate." *European Journal of Women's Studies* 20(4): 347–60.
Jackson, Zakiyyah Iman. 2020. *Becoming Human: Matter and Meaning in an Antiblack World*. New York: New York University Press.
Jacob, François. 1973. *The Logic of Life: A History of Heredity*. New York: Pantheon Books.
Jacquinot-Delaunay, Geneviève, and Laurence Monnoyer, eds. 1999a. "Le dispositif—entre usage et concept." *Hermès* 25.
———, and Laurence Monnoyer. 1999b. "Avant-propos." In "Le dispositif—entre usage et concept." *Hermès* 25: 9–14.
Jay, Martin. 2005. *Songs of Experience: Modern American and European Variations on a Universal Theme*. Berkeley: University of California Press.
Jensen, Casper Bruun. 2015. "Experimenting with Political Materials: Environmental Infrastructures and Ontological Transformations." *Distinktion: Scandinavian Journal of Social Theory* 16(1): 17–30.
Jochmaring, Julian. 2016. "Das Unbehagen in der (Medien-)Ökologie." *Internationales Jahrbuch für Medienphilosophie* 2(1): 91–112.
Johnson, Elizabeth, and Harlan Morehouse, eds. 2014. "After the Anthropocene: Politics and Geographic Inquiry for a New Epoch." Special Issue, *Progress in Human Geography* 38(3): 439–56.
Johns-Putra, Adeline. 2013. "Environmental Care Ethics: Notes Toward a New Materialist Critique." *symplokē* 21(1–2): 125–35.
Johnston, Adrian. 2011. "Hume's Revenge: À Dieu, Meillassoux?" In *The Speculative Turn: Continental Materialism and Realism*, edited by Levi Bryant, Nick Srnicek, and Graham Harman, 92–113. Melbourne: re:press.
Kafka, Ben. 2012. "The Administration of Things: A Genealogy." *West 86th*, www.west86th.bgc.bard.edu/articles/the-administration-of-things-a-genealogy/.
Kaiser, Birgit Mara, and Kathrin Thiele. 2014. "Diffraction: Onto-Epistemology, Quantum Physics and the Critical Humanities." *Parallax* 20(3): 165–67.

Kammler, Clemens. 1986. *Michel Foucault: Eine kritische Analyse seines Werks*. Bonn: Bouvier.

Kang, Hyo Yoon, and Sara Kendall. 2020. "Legal Materiality." In *The Handbook of Law and Humanities*, edited by Simon Stern, Maksymilian Del Mar and Bernadette Meyler, 1–20. New York: Oxford University Press.

Kant, Immanuel. 1998. *Critique of Pure Reason*. Cambridge: Cambridge University Press.

Kay, Lily. 2000. *Who Wrote the Book of Life? A History of the Genetic Code*. Stanford: Stanford University Press.

Keller, Evelyn Fox. 1985. *Reflections on Gender and Science*. New Haven/London: Yale University Press.

———. 1995. *Refiguring Life: Metaphors of Twentieth-Century Biology*. New York: Columbia University Press.

Keller, Reiner. 2019. "New Materialism? A View from Sociology of Knowledge." In *Discussing New Materialism: Methodological Implications for the Study of Materialities*, edited by Ulrike Tikvah Kissmann and Joost van Loon, 151–69. Wiesbaden: Springer.

Kelly, Mark G. E. 2013. "Foucault, Subjectivity, and Technologies of the Self." In *A Companion to Foucault*, edited by Christopher Falzon, Timothy O'Leary and Jana Sawicki, 510–25. Malden: Blackwell.

Kerin, Jacinta. 1999. "The Matter at Hand: Butler, Ontology and the Natural Sciences." *Australian Feminist Studies* 14(29): 91–104.

Kessler, Frank. 2003. "La Cinématographie Comme Dispositif (du) Spectaculaire." *Cinémas* 14(1): 21–34.

Keulartz, Jozef. 2012. "The Emergence of Enlightened Anthropocentrism in Ecological Restoration." *Nature and Culture* 7(1): 48–71.

Khan, Gulshan. 2009. "Agency, Nature and Emergent Properties: An Interview with Jane Bennett." *Contemporary Political Theory* 8(1): 90–105.

Kirby, Vicki. 2006. *Judith Butler: Live Theory*. London/New York: Continuum.

———. 2011. *Quantum Anthropologies: Life at Large*. Durham: Duke University Press.

———. 2017. "Matter out of Place: 'New Materialism' in Review." In *What If Culture Was Nature All Along?*, edited by Vicki Kirby, 1–25. Edinburgh: Edinburgh University Press.

———, and Elizabeth A. Wilson. 2011. "Feminist Conversations with Vicki Kirby and Elizabeth A. Wilson." *Feminist Theory* 12(2): 227–34.

Kittler, Friedrich. 1999. "Zum Geleit." In *Botschaften der Macht. Der Foucault-Reader*, edited by Jan Engelmann, 7–9. Stuttgart: DVA.

Klaver, Irene, Josef Keulartz, Henk van den Belt, and Bart Gremmen. 2002. "Born to be Wild: A Pluralistic Ethics Concerning Introduced Large Herbivores in the Netherlands." *Environmental Ethics* 24 (1): 3–23.

Klauser, Francisco, Till Paasche, and Ola Söderström. 2014. "Michel Foucault and the Smart City: Power Dynamics Inherent in Contemporary Governing Through Code." *Environment and Planning D: Society and Space* 32(5): 869–85.

Knorr-Cetina, Karin. 1981. *The Manufacture of Knowledge: An Essay on the Constructivist and Contextual Nature of Science.* Oxford/New York: Pergamon Press.

Koschorke, Albrecht, Susanne Lüdemann, Thomas Frank, and Ethel Matala de Mazza. 2007. *Der fiktive Staat. Konstruktionen des politischen Körpers in der Geschichte Europas.* Frankfurt am Main: Fischer Taschenbuch-Verlag.

Krohn, Wolfgang, and Johannes Weyer. 1994. "Society as a Laboratory: the Social Risks of Experimental Research." *Science and Public Policy* 21(3): 173–83.

Lakoff, Andrew, and Stephen Collier. 2010. "Infrastructure and Event." In *Political Matter: Technoscience, Democracy, and Public Life*, edited by Bruce Braun and Sarah Whatmore, 243–66. Minneapolis: University of Minnesota Press.

Lamarck, Jean Baptiste. 2011 [1809]. *Zoological Philosophy: An Exposition with Regard to the Natural History of Animals.* Cambridge: Cambridge University Press.

LaMarre, Thomas. 2013. "Afterword: Humans and Machines." In *Gilbert Simondon and the Philosophy of the Transindividual*, edited by Muriel Combes, 79–119. Cambridge/London: MIT Press.

Lang, Eberhard. 1970. *Zu einer kybernetischen Staatslehre. Eine Analyse des Staates auf der Grundlage des Regelkreismodells.* Salzburg: Pustet.

Lange, Friedrich Albert. 2010 [1866]. *History of Materialism and Criticism of Its Present Importance.* London/New York: Routledge.

Last, Angela. 2012. "Experimental Geographies." *Geography Compass* 6(12): 706–24.

Latour, Bruno. 1993. *We Have Never Been Modern.* Cambridge: Harvard University Press.

———. 1994. "On Technical Mediation: Philosophy, Sociology, Genealogy." *Common Knowledge* 3(2): 29–64.

———. 1996 "On Actor-Network Theory: A Few Clarifications." *Soziale Welt* 47(4): 369–81.

———. 2004a. "Why Has Critique Run Out of Steam? From Matters of Fact to Matters of Concern." *Critical Inquiry* 30(2): 225–48.

———. 2004b. "Whose Cosmos, Which Cosmopolitics? Comments on the Peace Terms of Ulrich Beck." *Common Knowledge* 10(3): 450–62.

———. 2004c [1999]. *Politics of Nature: How to Bring the Sciences into Democracy.* Cambridge/London: Harvard University Press.

———. 2007. "Can We Get Our Materialism Back, Please?" *Isis* 98(1): 138–42.

———. 2010. "An Attempt at a Compositionist Manifesto." *New Literary History* 41: 471–90.

———, and Peter Weibel. 2005. *Making Things Public: Atmospheres of Democracy.* Cambridge: MIT Press.

Law, John. 1987. "Technology and Heterogeneous Engineering: The Case of Portuguese Expansion." In *The Social Construction of Technological Systems: New Directions in the Sociology and History of Technology*, edited by Wiebe E. Bijer, Thomas P. Hughes and Trevor Pinch, 111–34. Cambridge: MIT Press.

———. 1992. "Notes on the Theory of the Actor-Network: Ordering, Strategy, and Heterogeneity." *Systems Practice* 5: 379–93.

———. 1994. *Organizing Modernity*. Oxford/Cambridge: Blackwell.
———. 1999. "After ANT: Complexity, Naming and Topology." *The Sociological Review* 47(1): 1–14.
———. 2004. *After Method: Mess in Social Science Research*. New York: Routledge.
———. 2009. "Actor Network Theory and Material Semiotics." In *The New Blackwell Companion to Social Theory*, edited by Bryan S. Turner, 141–58. Oxford: Blackwell.
———. 2017. "STS as Method." In *The Handbook of Science and Technology Studies*, edited by Ulrike Felt, Rayvon Fouché, Clark A. Miller and Laurel Smith-Doerr, 31–58. Cambridge: MIT Press.
———, and Annemarie Mol. 1995. "Notes on Materiality and Sociality." *The Sociological Review* 43: 274–94.
———, and Marianne E. Lien. 2012. "Slippery: Field Notes on Empirical Ontology." *Social Studies of Science* 43(3): 363–78.
———, and Vicky Singleton. 2013. "ANT and Politics: Working in and on the World." *Qualitative Sociology* 36(4): 485–502.
Legg, Stephan. 2011. "Assemblage/Apparatus: Using Deleuze and Foucault." *Area* 43(2): 128–33.
Lehman, Jessi, and Sara Nelson. 2014. "Experimental Politics in the Anthropocene." In "After the Anthropocene: Politics and Geographic Inquiry for a New Epoch," edited by Elizabeth Johnson and Harlan Morehouse. Special Issue, *Progress in Human Geography* 38(3): 444–47.
Lemke, Thomas. 2007. "An Indigestible Meal? Foucault, Governmentality and State Theory." *Distinktion: Scandinavian Journal of Social Theory* 8(2): 43–64.
———. 2011a. "Critique and Experience in Foucault." *Theory, Culture & Society* 28(4): 26–48.
———. 2011b. "Beyond Foucault: From Biopolitics to the Government of Life." In *Governmentality: Current Issues and Future Challenges*, edited by Ulrich Bröckling, Susanne Krasmann and Thomas Lemke, 165–84. New York: Routledge.
———. 2011c. *Foucault, Governmentality, and Critique*. Boulder, CO/London: Paradigm Publishers.
———. 2014. "The Risks of Security: Liberalism, Biopolitics and Fear." In *The Government of Life: Foucault, Biopolitics, and Neoliberalism*, edited by Vanessa Lemm and Miguel Vatter, 59–74. New York: Fordham University Press.
———. 2015a. "Varieties of Materialism." *BioSocieties* 10(4): 490–95.
———. 2015b. "New Materialisms: Foucault and the 'Government of Things.'" *Theory, Culture & Society* 32(4): 3–25.
———. 2019. *A Critique of Political Reason: Foucault's Analysis of Modern Governmentality*. London: Verso.
Lemm, Vanessa. 2009. *Nietzsche's Animal Philosophy: Culture, Politics, and the Animality of the Human Being*. New York: Fordham University Press.
Lettow, Susanne. 2017. "Turning the Turn: New Materialism, Historical Materialism and Critical Theory." *Thesis Eleven* 140(1): 106–21.

Levinas, Emmanuel. 1969. *Totality and Infinity: An Essay on Exteriority*. Pittsburgh: Duquesne University Press.
Lewontin, Richard C. 1969. "The Meaning of Stability." *Brookhaven Symposia in Biology* 22: 13–23.
Lezaun, Javier. 2017. "Actor-Network Theory." In *Social Theory Now*, edited by Claudio E. Benzecry, Monika Krause and Isaak Ariail Reed, 305–36. Chicago: University of Chicago Press.
———, Noortje Marres, and Manuel Tironi. 2017. "Experiments in Participation." In *The Handbook of Science and Technology Studies*, edited by Ulrike Felt, Rayvon Fouché, Clark A. Miller and Laurel Smith-Doerr, 195–221. Cambridge, MA/London: MIT Press.
Lindemann, Gesa. 2001. "Die reflexive Anthropologie des Zivilisationsprozesses." *Soziale Welt* 52: 181–98.
———. 2002. *Die Grenzen des Sozialen. Zur sozio-technischen Konstruktion von Leben und Tod in der Intensivmedizin*. München: Wilhelm Fink.
———. 2003. "Prinzipiell sind alle verdächtig. Michel Foucaults Vorlesungen über die Bestrebungen der Psychiatrie, der Justiz das Verbrechen zu entwinden." *Frankfurter Rundschau*, 19 August 2003, 27.
Link, Jürgen. 2008. "Dispositiv." In *Foucault-Handbuch: Leben—Werk—Wirkung*, edited by Clemens Kammler, Rolf Parr and Ulrich Johannes Schneider, 237–41. Stuttgart: Metzler.
Lipp, Benjamin. 2017. "Analytik des Interfacing: Zur Materialität technologischer Verschaltung in prototypischen Milieus robotisierter Pflege." *Behemoth. Journal on Civilisation* 10(1): 107–29.
Lorimer, Jamie. 2017. "Probiotic Environmentalities: Rewilding with Wolves and Worms." *Theory, Culture & Society* 34(4): 27–48.
———, and Clemens Driessen. 2013. "Bovine Biopolitics and the Promise of Monsters in the Rewilding of Heck Cattle." *Geoforum* 48: 249–59.
———, and Clemens Driessen. 2014. "Wild experiments at the Oostvaardersplassen: Rethinking Environmentalism in the Anthropocene." *Transactions of the Institute of British Geographers* 39(2): 169–81.
———, and Clemens Driessen. 2016. "From 'Nazi Cows' to Cosmopolitan 'Ecological Engineers': Specifying Rewilding Through a History of Heck Cattle." *Annals of the American Association of Geographers* 106(3): 631–52.
Lovelock, James E., and Lynn Margulis. 1974. "Atmospheric Homeostasis by and for the Biosphere: The Gaia Hypothesis." *Tellus* 26: 2–10.
Luke, Timothy W. 1995. "On Environmentality: Geo-Power and Eco-Knowledge in the Discourses of Contemporary Environmentalism." *Cultural Critique* 31: 57–81.
Lundborg, Tom, and Nick Vaughan-Williams. 2015. "New Materialisms, Discourse Analysis, and International Relations: A Radical Intertextual Approach." *Review of International Studies* 41(1): 3–25.
Lustig, Nicolas Ferris. 2014. *Rereading Foucault on Technology, Variegation, and Contemporary Power*. Los Angeles: University of California.

Lynch, Michael. 2013. "Ontography: Investigating the Production of Things, Deflating Ontology." *Social Studies of Science* 43(3): 444–62.
Lyotard, Jean-François. 1973. *Des Dispositifs Pulsionnels*. Paris: 10/18.
Macherey, Pierre. 1992. "Towards a Natural History of Norms." In *Michel Foucault: Philosopher*, edited by Timothy J. Armstrong, 176–91. New York/London: Harvester Wheatsheaf.
MacKenzie, Donald A., and Judy Wajcman, eds. 1985. *The Social Shaping of Technology*. Buckingham: Open University Press.
Malabou, Catherine. 2008. *What Should We Do with Our Brain?* New York: Fordham University Press.
——. 2012. *The New Wounded: From Neurosis to Brain Damage*. New York: Fordham University Press.
——. 2016. "One Life Only: Biological Resistance, Political Resistance." *Critical Inquiry* 42(3): 429–38.
Malthus, Thomas Robert. 1986 [1798]. "An Essay on the Principle of Population." In *The Works of Thomas Robert Malthus*, edited by Edward Anthony Wrigley and David Souden. London: William Pickering.
Mann, Michael. 1984. "The Autonomous Power of the State: Its Origins, Mechanisms and Results." *European Journal of Sociology* 25(2): 185–213.
Marcuse, Herbert. 1991. *One-Dimensional Man: Studies in the Ideology of Advanced Industrial Society*. Boston: Beacon Press.
Marguin, Séverine, Henrike Rabe, Wolfgang Schäffner, and Friedrich Schmidgall, eds. 2019. *Experimentieren. Einblicke in Praktiken und Versuchsaufbauten zwischen Wissenschaft und Gestaltung*. Bielefeld: transcript.
Margulis, Lynn. 1998. *Symbiotic Planet: A New Look at Evolution*. New York: Basic Books.
Marlin, Alison. 2014. "Book Review Mundane Governance: Ontology and Accountability by Steve Woolgar and Daniel Neyland." *Science & Technology Studies* 27(3): 115–17.
Marres, Noortje. 2009. "Testing Powers of Engagement: Green Living Experiments, the Ontological Turn and the Undoability of Involvement." *European Journal of Social Theory* 12(1): 117–33.
——. 2012. *Material Participation: Technology, the Environment and Everyday Publics*. New York: Palgrave Macmillan.
——. 2013. "Why Political Ontology Must be Experimentalized: On Eco-show Homes as Devices of Participation." *Social Studies of Science* 43(3): 417–43.
——, and Javier Lezaun. 2011. "Materials and Devices of the Public: An Introduction." *Economy and Society* 40(4): 489–509.
Marso, Lori J. 2011. "Freaks of Nature." *Political Theory* 39(3): 417–28.
Massumi, Brian. 1987. "Translator's Foreword." In *A Thousand Plateaus: Capitalism and Schizophrenia*, by Gilles Deleuze and Felix Guattari, ix–xix. Minneapolis: University of Minnesota Press.
——. 1996. "The Autonomy of Affect." *Cultural Critique* 31: 83–109.
——. 2009. "National Enterprise Emergency: Steps Toward an Ecology of Powers." *Theory, Culture & Society* 26(6): 153–85.

Matthewman, Steve. 2011. *Technology and Social Theory*. London: Red Globe Press.
———. 2013. "Foucault, Technology, and ANT." *Techné: Research in Philosophy and Technology* 17(2): 274–92.
Maturana, Humberto, and Francisco Varela. 1980. *Autopoiesis and Cognition: the Realization of the Living*. Dordecht: D. Reidel Publishing.
Maxwell, James Clerk. 1868. "On Governors." *Proceedings of the Royal Society of London* 16: 270–83.
Mayr, Otto. 1969. *Zur Frühgeschichte der technischen Regelungen*. München: Oldenbourg.
———. 1970. "The Origins of Feedback Control." *Scientific American* 223(4): 111–18.
———. 1971a. "Adam Smith and the Concept of the Feedback System: Economic Thought and Technology in 18th-Century Britain." *Technology and Culture* 12(1): 1–22.
———. 1971b. "Maxwell and the Origins of Cybernetics." *Isis* 62(4): 424–44.
———. 1986. *Authority, Liberty, and Automatic Machinery in Early Modern Europe*. Baltimore/London: Johns Hopkins University Press.
Mbembe, Achille. 2003. "Necropolitics." *Public Culture* 15(1): 11–40.
McInerney, Paul-Brian. 2014. "Book Review*Material Participation: Technology, the Environment and Everyday Publics*, by Noortje Marres." *Contemporary Sociology* 43(5): 716–18.
McKinlay, Alan, and Ken Starke, eds. 1998. *Foucault, Management and Organization Theory: From Panopticon to Technologies of the Self*. London: Sage.
Meadows, Donella H., Dennis L. Meadows, Jørgen Randers, and William W. Behrens III.. 1972. *The Limits to Growth: A Report for the Club of Rome's Project on the Predicament of Mankind*. New York: Universe Books.
Meehan, Katharine, Ian Shaw, Graham Ronald, and Sallie A. Marston. 2013. "Political Geographies of the Object." *Political Geography* 33: 1–10.
Mehrabi, Tara. 2016. *Making Death Matter: A Feminist Technoscience Study of Alzheimer's Sciences*. Linköping: Linköping University.
Meillassoux, Quentin. 2008. *After Finitude: An Essay on the Necessity of Contingency*. London: Continuum.
Meißner, Hanna. 2013. "Feministische Gesellschaftskritik als onto-epistemologisches Projekt." In *Geschlechter Interferenzen. Wissensformen—Subjektivierungsweisen—Materialisierungen*, edited by Corinna Bath, Hanna Meißner, Stephan Trinkaus and Susanne Völker, 163–208. Münster: LIT Verlag.
Meloni, Maurizio. 2014. "Biology Without Biologism: Social Theory in a Postgenomic Age." *Sociology* 48(4): 731–46.
Mezzadra, Sandro, and Brett Neilson. 2019. *The Politics of Operations*. London: Duke University Press.
Michael, Mike. 2017. *Actor-Network Theory: Trials, Trails and Translations*. London: SAGE.
Mies, Maria. 1999. "World Economy, Patriarchy and Accumulation." In *Women in the Third World*, edited by Nelly P. Stromquist, 37–45. New York: Garland.

Miller, Daniel, ed. 2005. *Materiality*. Durham: Duke University Press.
———. 2008. *The Comfort of Things*. Cambridge/Malden: Polity.
Miller, Peter, and Nikolas Rose 2008. *Governing the Present: Administering Economic, Social and Personal Life*. Cambridge: Polity.
Mills, C. Wright. 1972. "The Cultural Apparatus." In *Power, Politics and People*, edited by Irving Louis Horowitz, 405–22. New York: Oxford University Press.
Moheau, Jean-Baptiste. 1994 [1778]. *Recherches et considerations sur la population de la France*, edited by Eric Vilquin. Paris: Inst. National d'Etudes Démographiques.
Mol, Annemarie. 1999. "Ontological Politics. A Word and Some Questions." In *Actor Network Theory and After*, edited by John Law and John Hassard, 74–89. Oxford: Blackwell.
———. 2002. *The Body Multiple*. Durham: Duke University Press.
———. 2013. "Mind Your Plate! The Ontonorms of Dutch Dieting." *Social Studies of Science* 43(3): 379–96.
———, Ingunn Moser, and Jeanette Pols, eds. 2010. *Care in Practice: On Tinkering in Clinics, Homes and Farms*. Bielefeld: transcript.
Montag, Warren. 2013. *Althusser and His Contemporaries: Philosophy's Perpetual War*. Durham: Duke University Press.
Montesquieu, Charles Louis de Sesondat de. 1989. *The Spirit of the Laws*, translated and edited by Anne M. Cohler, Basia Carolyn Miller and Harold Samuel Stone. Cambridge: Cambridge University Press.
———. 2008 [1748]. *De l'esprit des loix*, edited by Catherine de Volpilhac-Auger. Oxford: Voltaire Foundation.
Moore, Jason. 2015. *Capitalism in the Web of Life: Ecology and the Accumulation of Capital*. London/New York: Verso.
Morton, Timothy. 2011a. "Here Comes Everything: The Promise of Object Oriented Ontology." *Qui Parle: Critical Humanities and Social Sciences* 19(2): 163–90.
———. 2011b. "Ecology after Capitalism." *Polygraph* 22: 46–59.
———. 2012. "An Object-Oriented Defense of Poetry." *New Literary History* 43(2): 205–24.
———. 2013a. *Hyperobjects: Philosophy and Ecology after the End of the World*. Minneapolis/ London: University of Minnesota Press.
———. 2013b. "Poisoned Ground." *symplokē* 21(1–2): 37–50.
Moulton, Alex A., and Jeff Popke. 2017. "Greenhouse Governmentality: Protected Agriculture and the Changing Biopolitical Management of Agrarian Life in Jamaica." *Environment and Planning D: Society and Space* 35(4): 714–32.
Muhle, Maria. 2008. *Eine Genealogie der Biopolitik: Zum Begriff des Lebens bei Foucault und Canguilhem*. Bielefeld: transcript.
Mukerji, Chandra. 2010. "The Territorial State as a Figured World of Power: Strategics, Logistic, and Impersonal Rule." *Sociological Theory* 28(4): 403–24.
Müller, Ruth, Clare Hanson, Mark Hanson, Michael Penkler, Georgia Samaras, Luca Chiapperino, John Dupré, Martha Kenney, Christopher Kuzawa, Joanna Latimer, Stephanie Lloyd, Astrid Lunkes, Molly Macdonald, Maurizio Meloni, Brigitte

Nerlich, Francesco Panese, Martyn Pickersgill, Sarah Richardson, Joëlle Rüegg, Sigrid Schmitz, Aleksandra Stelmach, and Paula-Irene Villa.. 2017. "The Biosocial Genome? Interdisciplinary Perspectives on Environmental Epigenetics, Health and Society." *EMBO reports* 18: 1677–82.

Mumford, Stephen, and Rani Lill Anjum. 2011. *Getting Causes from Powers*. Oxford: Oxford University Press.

Murdoch, Jonathan. 2004. "Humanising Posthumanism." *Environment and Planning A* 36(8): 1356–59.

Murphy, Michelle. 2017. *The Economization of Life*. Durham/London: Duke University Press.

Nealon, Jeffrey T. 2015. *Plant Theory: Biopower and Vegetable Life*. Stanford: Stanford University Press.

———. 2016. "The Archeology of Biopower: From Plant to Animal Life in *The Order of Things*." In *Biopower: Foucault and Beyond*, edited by Vernon W. Cisney and Nicolae Morar, 138–57. Chicago/London: Chicago University Press.

Neimanis, Astrida. 2014. "Alongside the Right to Water, a Posthumanist Feminist Imaginary." *Journal of Human Rights and the Environment* 5(1): 5–24.

———, Cecilia Åsberg, and Johan Hedrén. 2015. "Four Problems, Four Directions for Environmental Humanities: Toward Critical Posthumanities for the Anthropocene." *Ethics and the Environment* 20(1): 67–97.

Nel, Noël 1999. "Des dispositifs aux agencements télévisuels 1969-1983." *Hermès, La Revue* 25: 131–41.

Nelson, Sara Holiday. 2014. "Resilience and the Neoliberal Counter-Revolution: From Ecologies of Control to Production of the Common." *Resilience* 2(1): 1–17.

———. 2015. "Beyond *The Limits to Growth*: Ecology and the Neoliberal Counterrevolution." *Antipode* 47(2): 461–80.

Neyrat, Frédéric. 2018. *The Unconstructable Earth: An Ecology of Separation*. New York: Fordham University Press.

Niewöhner, Jörg. 2015. "Epigenetics: Localizing Biology through Co-Laboration." *New Genetics and Society* 34(2): 219–42.

Nimmo, Richie. 2008. "Governing Non-Humans: Knowledge, Sanitation and Discipline in the Late 19th and Early 20th century British Milk Trade." *Distinktion: Scandinavian Journal of Social Theory* 9(1): 77–97.

———. 2010. *Milk, Modernity and the Making of the Human: Purifying the Social*. London/New York: Routledge.

Norris, Christopher. 2013. "Speculative Realism: An Interim Report." In *Philosophy Outside-In: A Critique of Academic Reason*, 181–204. Edinburgh: Edinburgh University Press.

Novas, Carlos. 2005. "Ethics as a Pastoral Practice: Implementing Predictive Genetic Testing in the Medical Genetics Clinic." In *Contesting Moralities: Science, Identity, Conflict*, edited by Nanneke Redclif, 75–89. London: UCL Press.

O'Grady, Nathaniel. 2013. "Adopting the Position of Error: Space and Speculation in the Exploratory Significance of Milieu Formulations." *Environment and Planning D: Society and Space* 31(2): 245–58.

———. 2014. "Securing Circulation Through Mobility: Milieu and Emergency Response in the British Fire and Rescue Service." *Mobilities* 9(4): 512–27.
O'Leary, Timothy. 2008. "Foucault, Experience, Literature." *Foucault Studies* 5: 5–25.
Olson, Valerie A. 2010. "The Ecobiopolitics of Space Biomedicine." *Medical Anthropology* 29(2): 170–93.
Ong, Aihwa, and Stephen J. Collier. 2004. *Global Assemblages: Technology, Politics, and Ethics as Anthropological Problems.* Oxford: Blackwell.
Opitz, Sven. 2016. "Regulating Epidemic Space: The Nomos of Global Circulation." *Journal of International Relations and Development* 19(2): 263–84.
Oral, Sevket Benhur. 2014. "Liberating Facts: Harman's Objects and Wilber's Holons." *Studies in Philosophy and Education* 33(2): 117–34.
———. 2015. "Weird Reality, Aesthetics, and Vitality in Education." *Studies in Philosophy and Education* 34(5): 459–74.
Oyama, Susan. 2000. *The Ontology of Information.* Durham: Duke University Press.
———, Russell Gray, and Paul E. Griffiths. 2001. *Cycles of Contingency: Developmental Systems and Evolution (Life and Mind).* Cambridge, MA/London: MIT Press.
Paech, Joachim. 1997. "Überlegungen zum Dispositiv als Theorie medialer Topik." *Medienwissenschaft* 4: 400–20.
Panagia, Davide. 2019. "On the Political Ontology of the Dispositif." *Critical Inquiry* 45(3): 714–46.
Papadopoulos, Dimitris. 2010. "Alter-ontologies: Towards a Constituent Politics in Technoscience." *Social Studies of Science* 41(2): 177–201.
———. 2018. *Experimental Practice: Technoscience, Alterontologies, and More-Than-Social Movements.* Durham: Duke University Press.
Pasquinelli, Matteo. 2015. "What an Apparatus Is Not: On the Archeology of the Norm in Foucault, Canguilhem, and Goldstein." *Parrhesia* 22: 79–89.
Paxson, Heather. 2008. "Post-Pasteurian Cultures: The Microbiopolitics of Rawmilk Cheese in the United States." *Cultural Anthropology* 23: 15–47.
Peeters, Hugues, and Philippe Charlier. 1999. "Contributions à une théorie du dispositif." *Hermès, La Revue* 25: 15–23.
Pellizzoni, Luigi. 2015. *Ontological Politics in a Disposable World: The New Mastery of Nature.* Farnham: Ashgate.
Pfaller, Robert. 1997. *Althusser—Das Schweigen im Text: Epistemologie, Psychoanalyse und Nominalismus in Louis Althussers Theorie der Lektüre.* München: Fink Verlag.
Pfeifer, Geoff. 2012. "Review of *The Speculative Turn: Continental Materialism and Realism* by Levi Bryant, Nick Srnicek and Graham Harman." *Human Studies* 35(3): 465–69.
Philo, Chris. 2012. "A New Foucault with Lively Implications—or 'the Crawfish Advances Sideways.'" *Transactions of the Institute of British Geographers* 37(4): 496–514.
Pickering, Andrew. 1995. *The Mangle of Practice: Time, Agency, and Science.* Chicago: University of Chicago Press.
———. 2010. *The Cybernetic Brain: Sketches of Another Future.* Chicago: University of Chicago Press.

Pickering, Mary. 1993. *Auguste Comte: An Intellectual Biography*. Vol. 1. Cambridge: Cambridge University Press.
Pierides, Dean, and Dan Woodman. 2012. "Object-oriented Sociology and Organizing in the Face of Emergency: Bruno Latour, Graham Harman and the Material Turn." *The British Journal of Sociology* 63(4): 662–79.
Pinch, Trevor. 2011. "Review Essay: Karen Barad, Quantum Mechanics, and the Paradox of Mutual Exclusivity." *Social Studies of Science* 41(3): 431–41.
Pitts-Taylor, Victoria. 2010. "The Plastic Brain: Neoliberalism and the Neuronal Self." *Health* 14(6): 635–52.
Porcher, Jocelyne. 2015. "Animal Work." In *The Oxford Handbook of Animal Studies*, edited by Linda Kalof, 302–18. Oxford: Oxford University Press.
Post, Werner, and Alfred Schmidt. 1975. *Was ist Materialismus? Zur Einleitung in die Philosophie*. München: Kösel-Verlag.
Povinelli, Elizabeth A. 2015. "The Rhetorics of Recognition in Geontopower." *Philosophy & Rhetoric* 48(4): 428–42.
———. 2016. *Geontologies: A Requiem to Late Liberalism*. Durham/London: Duke University Press.
———, Mathew Coleman, and Kathryn Yusoff. 2017. "An Interview with Elizabeth Povinelli: Geontopower, Biopolitics and the Anthropocene." *Theory, Culture & Society* 34(2–3): 169–85.
Princen, Thomas. 2011. "Book Review *Vibrant Matter: A Political Ecology of Things*, by Jane Bennett." *Perspectives on Politics* 9(1): 118–20.
Protevi, John. 2013. *Life, War, Earth: Deleuze and the Sciences*. Minneapolis: University of Minnesota Press.
Puig de la Bellacasa, María. 2017. *Matters of Care: Speculative Ethics in More Than Human Worlds*. Minneapolis: University of Minnesota Press.
Quélennec, Bruno. 2011. "Review of *Eine Genealogie der Biopolitik*, by Maria Muhle." *Foucault Studies* 11: 222–25.
Rabinow, Paul. 1989. *French Modern: Norms and Forms of the Social Environment*. Chicago/London: University of Chicago Press.
———. 2003. *Anthropos Today: Reflections on Modern Equipment*. Princeton: Princeton University Press.
———, and Carlo Caduff. 2006. "Life—After Canguilhem." *Theory Culture & Society* 23(2–3): 329–30.
———, and Nikolas Rose. 2006. "Biopower Today." *Biosocieties* 1(2): 195–217.
Raffnsøe, Sverre, Marius Gudmand-Høyer, and Morten S. Thaning. 2016. "Foucault's Dispositive: The Perspicacity of Dispositive Analytics in Organizational Research." *Organization* 23(2): 272–98.
Rancière, Jacques. 1999. *Dis-agreement: Politics and Philosophy*. Minneapolis: University of Minnesota Press.
Rayner, Timothy. 2007. *Foucault's Heidegger: Philosophy and Transformative Experience*. London: Continuum.

Reckwitz, Andreas. 2002. "The Status of the 'Material' in Theories of Culture: From 'Social Structure' to 'Artefacts.'" *Journal for the Theory of Social Behaviour* 32(2): 195–217.
Rekret, Paul. 2016. "A Critique of New Materialism: Ethics and Ontology." *Subjectivity* 9(3): 225–45.
———. 2018. "The Head, the Hand, and Matter: New Materialism and the Politics of Knowledge." *Theory, Culture & Society* 35(7–8): 49–72.
Revel, Judith 2009. "Michel Foucault: Penser La Technique. Tracés." *Revue de Sciences humaines* 16: 139–49.
Rheinberger, Hans-Jörg. 1994. "Experimental Systems: Historiality, Narration, and Deconstruction." *Science in Context* 7(1): 65–81.
———. 1997. *Toward a History of Epistemic Things: Synthesizing Proteins in the Test Tube.* Stanford: Stanford University Press.
Richmond, Scott. 2015. "Speculative Realism is Speculative Aesthetics." *Configurations* 23(3): 399–403.
Rieger, Stefan. 2003. *Kybernetische Anthropologie. Eine Geschichte der Virtualität.* Frankfurt am Main: Suhrkamp.
Ritter, Joachim, and Ludwig J. Pongratz. 1972. "Disposition." In *Historisches Wörterbuch der Philosophie*, Vol. 2, edited by Joachim Ritter, 262–66. Darmstadt: Wissenschaftliche Buchgesellschaft.
Robertson, Morgan. 2006. "The Nature that Capital Can See: Science, State, and Market in the Commodification of Ecosystem Services." *Environment and Planning D: Society and Space* 24: 367–87.
———. 2012. "Measurement and Alienation: Making a World of Ecosystem Services." *Transactions of the Institute of British Geographers* 37(3): 386–401.
Roffe, Jon. 2012. "Time and Ground: A Critique of Meillassoux's Speculative Realism." *Angelaki: Journal of the Theoretical Humanities* 17(1): 57–67.
Rooney, David. 1997. "A Contextualising, Socio-technical Definition of Technology. Learning from Ancient Greece and Foucault." *Prometheus: Critical Studies in Innovation* 14(3): 399–407.
Rorty, Richard. 1980. *Philosophy and the Mirror of Nature.* Princeton: Princeton University Press.
Rose, Nikolas. 1999. *Powers of Freedom: Reframing Political Thought.* Cambridge: Cambridge University Press.
———. 2001. "The Politics of Life Itself." *Theory, Culture & Society* 18(6): 1–30.
———. 2007. *The Politics of Life Itself: Biomedicine, Power, and Subjectivity in the Twenty-First Century.* Princeton: Princeton University Press.
———, Pat O'Malley, and Mariana Valverde. 2006. "Governmentality." *Annual Review of Law and Social Science* 2(1): 83–104.
Rouse, Joseph. 1987. *Knowledge and Power: Toward a Political Philosophy of Science.* Ithaca: Cornell University Press.
———. 1993. "Foucault and the Natural Sciences." In *Foucault and the Critique of Institutions*, edited by John D. Caputo and Mark Yount, 137–64. University Park: Pennsylvania UP.

———. 2002. *How Scientific Practices Matter: Reclaiming Philosophical Naturalism*. Chicago/London: University of Chicago Press.

———. 2004. "Barad's Feminist Naturalism." *Hypatia* 19(1): 142–61.

Rouvroy, Antoinette, and Thomas Berns. 2013. "Algorithmic Governmentality and Prospects of Emancipation Disparateness as a Precondition for Individuation Through Relationships?". *Réseaux* 1(177): 163–96.

Ruffié, Jacques. 1976. *De la biologie à la culture*. Paris: Flammarion.

Rutherford, Paul. 1999. "The Entry of Life into History." In *Discourses of the Environment*, edited by Eric Darier, 37–62. Oxford: Blackwell.

———. 2000. *The Problem of Nature in Contemporary Social Theory*. PhD Thesis, Research School of Social Sciences, Australian National University, Canberra.

Rutherford, Stephanie. 2011. *Governing the Wild: Ecotours of Power*. Minneapolis: University of Minnesota Press.

———, and Paul Rutherford. 2013. "Geography and Biopolitics." *Geography Compass* 7(6): 423–34.

Saar, Martin. 2009. "Politik der Natur: Spinozas Begriff der Regierung." *Deutsche Zeitschrift für Philosophie* 57(3): 433–47.

Sänger, Eva. 2020. *Elternwerden zwischen »Babyfernsehen« und medizinischer Überwachung. Eine Ethnografie pränataler Ultraschalluntersuchungen*. Bielefeld: transcript.

Salter, Mark B. 2013. "To Make Move and Let Stop: Mobility and the Assemblage of Circulation." *Mobilities* 8(1): 7–19.

Sarasin, Philipp. 2009. *Darwin und Foucault. Genealogie und Geschichte im Zeitalter der Biologie*. Frankfurt am Main: Suhrkamp.

Sautchuk, Carlos Emanuel. 2016. "Eating (with) Piranhas: Untamed Approaches to Domestication." *Vibrant: Virtual Brazilian Anthropology* 13(2): 38–57.

Sawchuk, Kim. 2001. "The Cultural Apparatus: C. Wright Mills' Unfinished Work." *The American Sociologist* 32(1): 27–49.

Sawicki, Jana. 1991. *Disciplining Foucault: Feminism, Power, and the Body*. London/New York: Routledge.

Schiebinger, Londa. 1989. *The Mind Has No Sex? Women in the Origins of Modern Science*. Cambridge: Harvard University Press.

Schmidt, Jeremy J. 2014. "The Retreating State: Political Geographies of the Object and the Proliferation of Space." *Political Geography* 39: 58–59.

Schmitt, Carl. 2007. *The Concept of the Political*. Chicago: University of Chicago Press.

Schrader, Astrid. 2009. "Diffractive Experiments in the Ethics of Mattering." *Subjectivity* 28(1): 349–53.

———. 2010. "Responding to Pfiesteria piscicida (the Fish Killer): Phantomatic Ontologies, Indeterminacy, and Responsibility in Toxic Microbiology." *Social Studies of Science* 40(2): 275–306.

Schuller, Kyla, and Jules Gill-Peterson. 2020. "Introduction: Race, the State, and the Malleable Body." *Social Text* 38 (2 (143)): 1–17.

Schweber, Silvan S. 2008. "Review of *Meeting the Universe Halfway*, by Karen Barad." *ISIS: Journal of the History of Science in Society* 99(4): 879–82.

Schwennesen, Nete, and Lene Koch. 2009. "Visualizing and Calculating Life: Matters of Fact in the Context of Prenatal Risk Assessment." In *Contested Categories: Life Science in Society*, edited by Susanne Bauer and Ayo Wahlberg, 69–87. Farnham: Ashgate.

Seibel, Benjamin. 2016. *Cybernetic Government. Informationstechnologie und Regierungsrationalität von 1943–1970*. Wiesbaden: Springer.

Sellin, Volker. 1984. "Regierung, Regime, Obrigkeit." In *Geschichtliche Grundbegriffe. Historisches Lexikon zur politisch-sozialen Sprache in Deutschland*, edited by Otto Brunner, Werner Conze and Reinhart Koselleck, 361–421. Stuttgart: Klett-Cotta.

Senellart, Michel. 1995. *Les arts de gouverner: du regimen médiéval au concept de gouvernement*. Paris: Seuil.

Shapin, Steven, and Simon Schaffer. 1985. *Leviathan and the Air-Pump: Hobbes, Boyle and the Experimental Life*. Princeton: Princeton University Press.

Shaviro, Steven 2014. *The Universe of Things: On Speculative Realism*. Minneapolis: University of Minnesota Press.

Sheldon, Rebekah. 2015. "Form / Matter / Chora: Object-Oriented Ontology and Feminist New Materialism." In *The Nonhuman Turn*, edited by Richard Grusin, 193–222. Minneapolis: University of Minnesota Press.

Sheller, Mimi, and John Urry. 2006. "The New Mobilities Paradigm." *Environment and Planning A: Economy and Space* 38(2): 207–26.

Shukin, Nicole. 2009. *Animal Capital: Rendering Life in Biopolitical Times*. Minneapolis: University of Minnesota Press.

Silva-Castañeda, Laura, and Nathalie Trussart. 2016. "Sustainability Standards and Certification: Looking through the Lens of Foucault's Dispositive." *Global Networks* 16(4): 490–510.

Simondon, Gilbert. 2017 [1958]. *On the Mode of Existence of Technical Objects*. Minneapolis: Univocal Publishing.

Sismondo, Sergio. 2015. "Ontological Turns, Turnoffs and Roundabouts." *Social Studies of Science* 45(3): 441–48.

Smart, Barry. 1983. *Foucault, Marxism and Critique*. London/New York: Routledge.

Smith, Adam. 1937 [1776]. *An Inquiry into the Nature and Causes of the Wealth of Nations*, edited by Edwin Cannan. New York: The Modern Library/Random House.

Smith, Mick. 2011. *Against Ecological Sovereignty*. Minneapolis: University of Minnesota Press.

Spitzer, Leo. 1942. "Milieu and Ambiance: An Essay in Historical Semantics." *Philosophy and Phenomenological Research* 3(1–2): 1–42; 169–218.

Spivak, Gayatri Chakravorty. 1988. "Subaltern Studies: Deconstructing Historiography." In *Selected Subaltern Studies*, edited by Ranajit Guha and Gayatri Chakravorty Spivak, 3–32. New York: Oxford University Press.

Sprenger, Florian. 2019. *Epistemologien des Umgebens. Zur Geschichte, Ökologie und Biopolitik künstlicher environments*. Bielefeld: transcript.

Srinivasan, Krithika. 2014. "Caring for the Collective: Biopower and Agential Subjectification in Wildlife Conservation." *Environment and Planning D: Society and Space* 32(3): 501–17.

Stengers, Isabelle. 2005. "The Cosmopolitical Proposal." In *Making Things Public: Atmospheres of Democracy*, edited by Bruno Latour and Peter Weibel, 994–1003. Cambridge: MIT Press.

———. 2010. "Including Nonhumans in Political Theory. Opening Pandora's Box?" In *Political Matter: Technoscience, Democracy, and Public Life*, edited by Bruce Braun and Sarah Whatmore, 3–33. Minneapolis: University of Minnesota Press.

———. 2011. "Wondering about Materialism." In *The Speculative Turn: Continental Realism and Materialism*, edited by Levi Bryant, Nick Srnicek and Graham Harman, 368–80. Melbourne: re:press.

Stoler, Ann Laura. 1995. *Race and the Education of Desire: Foucault's History of Sexuality and the Colonial Order of Things*. Durham/London: Duke University Press.

Stollberg-Rilinger, Barbara. 1986. *Der Staat als Maschine. Zur politischen Metaphorik des absoluten Fürstenstaats*. Berlin: Duncker & Humblot.

Suchman, Lucy. 2007. *Human-Machine Reconfigurations*. Cambridge: Cambridge University Press.

Sunder Rajan, Kaushik. 2006. *Biocapital: The Constitution of Postgenomic Life*. Durham/London: Duke University Press.

———, ed. 2012. *Lively Capital: Biotechnologies, Ethics, and Governance in Global Markets*. Durham/London: Duke University Press.

Swyngedouw, Erik, and Henrik Ernstson. 2018. "Interrupting the Anthropo-ObScene: Immuno-Biopolitics and Depoliticizing Ontologies in the Anthropocene." *Theory, Culture & Society* 35(6): 3–30.

Talcott, Samuel. 2014. "Errant Life, Molecular Biology, and Biopower: Canguilhem, Jacob, and Foucault." *History and Philosophy of the Life Sciences* 36(2): 254–79.

TallBear, Kim. 2017. "Beyond the Life/Not-Life Binary: A Feminist-Indigenous Reading of Cryopreservation, Interspecies Thinking, and the New Materialisms." In *Cryopolitics: Frozen Life in a Melting World*, edited by Johanna Radin and Emma Kowal, 179–202. Cambridge.: MIT Press.

Tanner, Jakob. 1998. "'Weisheit des Körpers' und soziale Homöostase. Physiologie und das Konzept der Selbstregulation." In *Physiologie und industrielle Gesellschaft. Studien zur Verwissenschaftlichung des Körpers im 19. und 20. Jahrhundert*, edited by Philipp. Sarasin and Jakob Tanner, 129–69. Frankfurt am Main: Suhrkamp.

Taylor, Carol A. 2016. "Close Encounters of a Critical Kind: A Diffractive Musing In/Between New Material Feminism and Object-Oriented Ontology." *Cultural Studies* 16(2): 201–12.

Taylor, Chloë. 2013. "Foucault and Critical Animal Studies: Genealogies of Agricultural Power." *Philosophy Compass* 8(6): 539–51.

Tellmann, Ute. 2013. "Catastrophic Populations and the Fear of the Future: Malthus and the Genealogy of Liberal Economy." *Theory, Culture & Society* 30(2): 135–55.

Terranova, Tiziana. 2009. "Another Life: The Nature of Political Economy in Foucault's Genealogy of Biopolitics." *Theory, Culture & Society* 26(6): 234–62.

Teubner, Guenther. 2006. "Rights of Non-Humans? Electronic Agents and Animals as New Actors in Politics and Law." *Journal of Law and Society* 33(4): 497–521.

Thiele, Kathrin. 2014. "Ethos of Diffraction: New Paradigms for a (Post)humanist Ethics." *Parallax* 20(3): 202–16.
Thierman, Stephan. 2010. "Apparatuses of Animality: Foucault Goes to a Slaughterhouse." *Foucault Studies* 9: 89–110.
Thomas, Monique Martinez. 2015. "Dispositive, Intermediality and Society: Tales of the Bed in Contemporary Spain." *SubStance* 44(3): 98–111.
Thompson, Charis. 2005. *Making Parents: The Ontological Choreography of Reproductive Technologies*. Cambridge: MIT Press.
Thrift, Nigel. 2007. "Overcome by Space: Reworking Foucault." In *Space, Knowledge and Power: Foucault and Geography*, edited by Jeremy W. Crampton and Stuart Elden, 53–58. Aldershot: Ashgate.
———. 2008. *Non-Representational Theory: Space, Politics, Affect*. Abingdon/New York: Routledge.
Tierney, Thomas F. 2016. "Toward an Affirmative Biopolitics." *Sociological Theory* 34(4): 358–81.
Tischleder, Babette Bärbel. 2014. *The Literary Life of Things: Case Studies in American Fiction*. Frankfurt am Main/New York: Campus.
Toscano, Alberto 2011. "Against Speculation, or, A Critique of the Critique of Critique: A Remark on Quentin Meillassoux's *After Finitude* (After Colletti)." In *The Speculative Turn: Continental Materialism and Realism*, edited by Levi Bryant, Nick Srnicek, and Graham Harman, 84–91. Melbourne: re:press.
Traweek, Sharon. 1988. *Beamtimes and Lifetimes: The World of High Energy Physicists*. Cambridge/London: Harvard University Press.
Tsing, Anna Lowenhaupt, Heather Swanson, Elaine Gan, and Nils Bubandt, eds. 2017. *Arts of Living on a Damaged Planet: Ghosts and Monsters of the Anthropocene*. Minneapolis/London: University of Minnesota Press.
Türk, Klaus, Thomas Lemke, and Michael Bruch. 2002. *Organisation in der modernenodernen Gesellschaft. Eine historische Einführung*. Wiesbaden: Springer Fachmedien.
Usher, Mark. 2014. "Veins of Concrete, Cities of Flow: Reasserting the Centrality of Circulation in Foucault's Analytics of Government." *Mobilities* 9(4): 550–69.
van der Tuin, Iris. 2008. "Deflationary Logic: Response to Sara Ahmed's 'Imaginary Prohibitions: Some Preliminary Remarks on the Founding Gestures of the 'New Materialism'" *European Journal of Women's Studies* 15(4): 411–16.
———. 2011a. "New Feminist Materialisms." *Women's Studies International Forum* 34: 271–77.
———. 2011b. "'A Different Starting Point, a Different Metaphysics': Reading Bergson and Barad Diffractively." *Hypatia* 26(1): 22–42.
———. 2014. "Diffraction as a Methodology for Feminist Onto-Epistemology: On Encountering Chantal Chawaf and Posthuman Interpellation." *Parallax* 20: 231–44.
van Dijk, Paul. 2000. *Anthropology in the Age of Technology: The Philosophical Contribution of Günther Anders*. Amsterdam: RODOPI.

van Wyk, Alan R. 2012. "What Matters Now?" *Cosmos and History: The Journal of Natural and Social Philosophy* 8(2): 130–36.
Vasileva, Bistra. 2015. "Stuck with/in a 'Turn': Can We Metaphorize Better in Science and Technology Studies?" *Social Studies of Science* 45(3): 454–61.
Veyne, Paul. 1997 [1978]. "Foucault Revolutionizes History." In *Foucault and His Interlocutors*, edited by Arnold Ira Davidson, 146–82. Chicago: University of Chicago Press.
Viveiros de Castro, Eduardo. 2004. "Perspectival Anthropology and the Method of Controlled Equivocation." *Tipití: Journal of the Society for the Anthropology of Lowland South America* 2(1): 3–20.
Vint, Sherryl. 2008. "Entangled Posthumanism." *Science Fiction Studies* 35(2): 313–19.
Vogl, Joseph. 2004. "Regierung und Regelkreis." In *Cybernetics—Kybernetik. The Macy-Conferences 1946–1953, Bd. II: Essays und Dokumente*, edited by Claus Pias, 67–80. Zürich/Berlin: Diaphanes.
von Justi, Johann Heinrich Gottlob. 1970 [1764]. Gesammelte politische und Finanz-Schriften über wichtige Gegenstände der Staatskunst, der Kriegswissenschaften und des Kameral-und Finanzwesens, Vol. 3. Aalen: Scientia-Verlag. von Uexküll, Jakob. 2010 [1934]. *A Foray into the Worlds of Animals and Humans with A Theory of Meaning*. Minneapolis: Minnesota University Press.
Wakefield, Stephanie, and Bruce Braun. 2014. "Governing the Resilient City." *Environment and Planning D: Society and Space* 32(1): 4–11.
Walker, Jeremy, and Melinda Cooper. 2011. "Genealogies of Resilience: From Systems Ecology to the Political Economy of Crisis Adaptation." *Security Dialogue* 42(2): 143–60.
Wallace, Alfred Russel. 2008 [1871]. "On the Tendency of Varieties to Depart Indefinitely from the Original Type." *Journal of Science Education* 13(3): 283–91.
Walters, William. 2012. *Governmentality: Critical Encounters*. London/New York: Routledge.
Washick, Bonnie, and Elizabeth Wingrove. 2015. "Politics that matter: Thinking about Power and Justice with the New Materialists." *Contemporary Political Theory* 14(1): 63–79.
Watson, Janell. 2013. "Eco-sensibilities: An Interview with Jane Bennett." *Minnesota Review* 81: 147–58.
Wessely, Christina, and Florian Huber. 2017. "Milieu. Zirkulationen und Transformationen eines Begriffs." In *Milieu: Umgebungen des Lebendigen in der Moderne*, edited by Christina Wessley and Florian Huber, 7–17. Paderborn: Wilhelm Fink.
Whatmore, Sarah. 1999. "Hybrid Geographies: Rethinking the 'Human' in Human Geography." In *Human Geography Today*, edited by Doreen Massey, John Allen and Phil Sarre, 22–39. Cambridge: Polity Press.
———. 2002. *Hybrid Geographies: Natures Cultures Spaces*. London: Sage Publications.
———. 2004. "Humanism's Excess: Some Thoughts on the 'Post-human/ist' Agenda." *Environment and Planning A* 36(8): 1360–63.
Wiener, Norbert. 1948. *Cybernetics: Or Control and Communication in the Animal and the Machine*. Paris: Hermann.

Wiley, Stephan. 2005. "Spatial Materialism: Grossberg's Deleuzean Cultural Studies." *Cultural Studies* 19(1): 63-99.

Willey, Angela. 2016. "A World of Materialisms: Postcolonial Feminist Science Studies and the New Natural." *Science, Technology & Human Values* 41(6): 991-1014.

———. 2017. "Engendering New Materializations: Feminism, Nature, and the Challenge to Disciplinary Proper Objects." In *The New Politics of Materialism: History, Philosophy, Science*, edited by Sarah Ellenzweig and John H. Zammito, 131-53. London: Routledge.

Williams, Raymond. 1980. *Problems of Materialism: Problems in Materialism and Culture*. London: Verso.

Wilson, Alexander. 2018. "Beyond the Neomaterialist Divide: Negotiating Between Eliminative and Vital Materialism with Integrated Information Theory." *Theory, Culture & Society* 35(7-8): 97-116.

Wilson, Elizabeth A. 2015. *Gut Feminism*. Durham/London: Duke University Press.

Winner, Langdon. 1980. "Modern Technology: Problem or Opportunity?" *Daedalus* 109(1): 121-36.

Witmore, Christopher. 2014. "Archaeology and the New Materialisms." *Journal of Contemporary Archaeology* 1(2): 203-46.

Wolf, Burkhardt. 2008. "Das Schiff, eine Peripetie des Regierens. Nautische Hintergründe von Kybernetik und Gouvernementalität." *MLN* 123(3): 444-68.

Wolfe, Cary. 2013. *Before the Law: Humans and Other Animals in a Biopolitical Frame*. Chicago/London: University of Chicago Press.

Wolfe, Charles T. 2016. *Materialism: A Historico-Philosophical Introduction*. Dordrecht: Springer.

———. 2017. "Materialism Old and New." *Antropología Experimental* 17(13): 215-224.

Woolgar, Steve, and Javier Lezaun. 2013. "The Wrong Bin Bag: A Turn to Ontology in Science and Technology Studies?" *Social inudies of Science* 43(3): 321-40.

———, and Daniel Neyland. 2013. *Mundane Governance: Ontology and Accountability*. Oxford: Oxford University Press.

———, and Javier Lezaun. 2015. "Missing the (Question) Mark? What Is a Turn to Ontology?" *Social Studies of Science* 45(3): 462-67.

Youatt, Rafi. 2008. "Counting Species: Biopower and the Global Biodiversity Census." *Environmental Values* 17(3): 393-417.

Zahavi, Dan. 2016. "The End of What? Phenomenology vs. Speculative Realism." *International Journal of Philosophical Studies* 24(3): 289-309.

Zalloua, Zahi. 2015. "On Meillassoux's 'Transparent Cage': Speculative Realism and Its Discontents." *symplokē* 23(1-2): 393-409.

Zedler, Johann Heinrich. 1739. "Materialismus." In *Grosses vollständiges Universal-Lexicon Aller Wissenschafften und Künste*, Vol. 19, edited by Johann Heinrich Zedler, 2025-39. Halle and Leipzig.

Ziewitz, Malte. 2016. "Governing Algorithms: Myth, Mess, and Methods." *Science, Technology & Human Values* 41(1): 3-16.

Zuiderent-Jerak, Teun. 2015. *Situated Intervention: Sociological Experiments in Health Care*. Cambridge: MIT Press.

INDEX

Abrahamsson, Sebastian, 53–54, 150–51, 238n14, 239n15, 239n17
actant, 43–45, 50–51, 110, 213n5
actor-network theory (ANT), 12, 110, 238n9; Bennett and, 43–46; Critical Theory and, 238n11; Foucault and, 141, 192, 223n15, 236n2; governmentality and, 236n2; Harman and, 26, 43; on human and nonhuman, 149–50; on networks and power, 147; overmining in, 27; problematization and, 237n3
administration of things, 85–89
aesthetics: on epistemology and objects, 33; as first philosophy, 32–35; Harman on ethics and, 32; OOO for, 34; speculative realism for, 34
Agamben, Giorgio: on dispositive, 97–98, 139, 221n4, 223n12, 223n14, 225n27, 248n8
agency, 4, 248n5; for assemblages, 50–51, 53; Bennett on, 214n8; displacement of, 149–54; as distributive, 44; as fetal, 69–70; Foucault and, 225n28; government of things and, 81, 149; human and nonhuman, 43, 130–31; liberal individualism and, 238n13; as material, 241n25; as more than relational, 48–51; neo-animistic ontology and, 42; new materialisms and, 151; OOO and vital materialism on, 150; relational materialism and, 149–54; responsibility and, 52–53; will and, 51, 221n4
agential realism: apparatus and, 58, 64–66, 70–71, 100; of Barad, 8, 13, 15, 57, 62, 73–74, 150; diffraction and, 74; on dualism, 63, 65; on entanglements, 57; epistemology, ontology, ethics of, 15; intra-actions and, 75; matter and, 208n9; methodology of, 15; paradox in, 75; on phenomena, 8; on power, 68–69; problems for, 73–78; quantum physics and, 74, 218n24; relational ontology in, 57; responsibility and, 76; technoscience and, 72–73
agriculture, governmentality and, 232n9
Ahmed, Sara, 75, 219n26
Alaimo, Stacy, 212n21
aleatory, 92, 176; government of, 122, 129–33; materialism, 223n16, 226n33, 237n6
alterontologies, 161
Althusser, Louis, 99, 226n33, 237n6, 248n7
ambividuals, 178–79, 244n12
Ampère, André-Marie, 118–19
Anderson, Ben, 48, 142, 240n18, 244n11
ANT. *See* actor-network theory
Anthropocene, 28, 163, 170, 247n28; care in, 185–90; panarchy and, 18, 185–90, 247n26
anthropocentrism, 2, 7, 35–36, 39, 138; as enlightened, 180; Foucault and, 9–10, 80, 121, 220n1; Haraway on, 121, 124, 158–59; order of things and, 121, 125; posthumanism and, 160, 239n16; probiotic strategies and, 180; problems of, 193; as strategic, 157
anthropology, 2–3, 237n4, 238n11
antihumanism, 63
anti-materialism, 211n14

anti-realism, 211n12
apparatus: agential realism and, 58, 64–66, 70–71, 100; Althusser on state, 99, 226n33, 237n6, 248n7; assemblage vs., 216n12, 226n35; Barad on, 58, 64–66, 72–73, 100; causality and, 65; Deleuze and Guattari on, 226n33; dispositive dissociated from, 98–100, 102; ethics and, 71–72; Foucault and, 82; Haraway on, 217n15; on human and nonhuman, 71; power and, 70–71, 100; as static collection, 99–100; subversive practices and, 72
architecture, 232n9; critical infrastructures and, 181, 233n13; Foucault and, 91, 110, 127, 132–33, 220n5, 228n12, 244n11
assemblages: agency for, 50–51, 53; apparatus vs., 216n12, 226n35; Braun on, 100, 221n6; of car-driver, 177–78; definition of, 101; dispositives and, 100–102, 226n37; food as, 45–46; Foucault and, 82; free will to, 45; human and nonhuman in, 101; responsibility and, 52–53; thing groupings as, 42

Bannon, Bryan E., 213n7
Barad, Karen, 216n13, 216nn9–10; on agential realism, 8, 13, 15, 57, 62, 73–74, 150; on apparatus, 58, 64–66, 72–73, 100; on Bohr, 57–59, 63–65, 73–74, 151, 218n22, 218n24; on Butler, 58, 62–63, 72; on Casper, 69–70, 219n28, 240n21; critique and, 207n6; diffraction and, 60–62, 64, 75, 77, 218n23; ethics and, 66–68, 76–78; on Fernandes, 68–69, 77; Foucault and, 9–10, 100, 217n16, 220n6; on Haraway, 58, 60; on human and nonhuman, 57–58; intra-action by, 61, 68; on justice, 219n30; material foundationalism and, 72–75; on matter and language, 207n3; politics and, 77; posthumanism and, 63; poststructuralism and, 57, 63, 75, 201; on power, 15;

problems for, 73–78; quantum physics and, 3, 15, 57–58, 73–74, 215n5, 218n24; on representationalism, 59–61, 75, 215n2, 219nn27–28; reproductive technologies and, 218n20; scientism and, 73, 218n21; STS and, 72; on transhumanism, 216n8; vital materialism and, 219n25
Barnwell, Ashley, 208n8
Barry, Andrew, 48, 153, 213n6, 227n5, 242n33, 247n1
Barua, Maan, 185–86, 244n11
Behrent, Michael C., 104, 227n2
Bennett, Jane: on agency, 214n8; ANT and, 43–46; on deodand, 221n5; on electricity grid, 41, 44–45, 52; on enchanted materialism, 40–41, 43, 49, 54, 213n3; ethics and, 15, 52; on fatty acid, 41, 45–46, 53–54, 221n4, 239n15; Harman on, 42; on nature paralleling culture, 46; on OOO and matter, 37; politics and, 40, 46, 48–56, 213n1; positive ontology of, 15; problems and theory of, 48–56, 157; on structure, 226n36; thing-power by, 14–15, 40–41, 50, 54, 150, 214n10, 233n16; vibrancy of things and, 8, 13–15, 40–56; on vital materialism, 193–94, 213n1, 214n10
Beuret, Nicholas, 166
bioavailability, 92
biology: Canguilhem on, 134, 234n22; Dupré and, 234n23; Foucault and, 122–23; on genetics, 125–26; Harman on, 211n11; as historical and contingent, 4; machine within machine in, 117–18; milieu and, 128–29; norms and, 134; physics to cybernetics and, 116–20
biopolitics, 219n2, 235n25, 236n4; Foucault and, 120–21, 137, 176, 235n24; Friese on, 139; genetics and, 16, 123–24; government of things and, 138; on humans and nonhumans, 16; liberalism and, 136; milieu and, 16, 122, 137, 140; as

more-than-human, 136–40; Nelson on, 163, 176, 200, 242n32; of population, 133–34; as resilient, 171, 176
biopower, 79, 138, 140, 185, 235n26, 245n13
Bogost, Ian, 21, 36, 210n7, 212n21
Bohm, David, 218n22
Bohr, Niels, 57; Heisenberg and, 58–59; on nature, 58–59; on phenomena, 63–64; quantum physics and, 73–74, 151, 218n22, 218n24
Braun, Bruce, 239n17, 245n14, 248n8; on assemblage, 100, 221n6; on Foucault and history, 237n6; on government, 142; on individual will, 177–78; on matter, 49; on new materialisms, 171, 248n3; on nonhuman, 181; on posthumanism, 155; on urban flood management, 181, 245n15
Braunmühl, Caroline, 219n29
Bricker, Brett, 212n22
Bruining, Dennis, 75
Bryant, Levi R., 211n13
Burchell, Graham, 243n1
Butler, Judith, 58, 62–63, 72, 201, 215n7

Canguilhem, Georges, 127, 225n27, 244n10; on biology, 134, 234n22; on equilibrium, 117–18; genetics and, 122–23, 126; on milieu, 121, 128–29, 234n22; on power, 111
capitalism: cybernetics in, 174; government of things in, 185; on human and nonhuman, 189; labor and, 185–86, 246n22; new materialisms and, 171; political technologies on, 170; resilience in, 174, 187, 200, 243nn4–5
Cartesian-Newtonian ontology, 4, 24, 31, 189
Casper, Monica, 69–70, 219n28, 240n21
Cheah, Pheng, 217n19
circulation: control technology on, 114, 127, 129–33, 135–37, 139, 142, 169, 232n9, 244n12; Foucault and, 234n18; human and nonhuman in, 16; liberalism and, 134–36, 233n12, 234n18; milieu on, 16, 129–32, 135–36; Moheau on, 136–37, 234n21; as vital systems, 179, 182
climate change, 28; experiments for, 163; Marres on, 165–67, 228n12; neoliberalism and, 175–76; OOO and, 35–36, 38–39; vibrancy of things and, 42
clock, 16, 104; as political order, 110–11
Cole, Andrew, 212n24
Coles, Romand, 214n11
Collier, Steven, 181–84, 245n16, 245n18
Comte, Auguste, 87–88, 222nn8–9, 232n8
consciousness: Harman on, 24; Husserl and, 24–25, 38
control, technology of: on circulation, 114, 127, 129–33, 135–37, 139, 142, 169, 232n9, 244n12; self-regulation contrasted with, 114, 131
Cooper, Melinda, 175, 246n22
correlationism, 22–24
critique: Barad and, 207n6; Barnwell on, 208n8; Latour on, 6, 207n6; new materialisms and, 6–7, 200–201; OOO on, 38
Crutzen, Paul, 189
cybernetics: Ampère and, 118–19; in capitalism, 174; communication and, 16; ecology and, 181; Foucault and, 119, 230n30; genetics and, 125; government and, 104, 116–20; governors vs. moderators for, 118; milieu and, 127; neocybernetics and, 17, 120, 171–79; physics to biology and, 116–20; resilience of neo-, 17, 120, 170–71, 174, 177, 198, 200; Wiener and, 118, 124, 230n30, 231n4

Deleuze, Gilles, 211n11, 236n4; apparatus and, 226n33; assemblage and, 100, 216n12; milieu and, 232n6, 232n8; new materialisms and, 239n17; power and, 233n12
deodand, 213n5

diffraction, 73, 215n4; agential realism and, 74; Barad and, 60–62, 64, 75, 77, 218n23; Haraway on, 58, 60; relationalism and, 58, 61; representationalism and, 59–61, 75, 215n2, 219n27

diffractive materialism, 57–78, 193–95

Dionisius, Sarah, 218n20

dispositif (dispositive): Agamben on, 223n14; Foucault and, 11, 16, 82, 90–91, 96, 99, 101, 150, 192, 222n11, 223n14, 224n21, 224n25, 225n27, 226n33, 226n37; French meanings of, 16, 91–92; material-discursive and, 82; translations of, 90, 92, 222n11, 226n32

dispositive, 224n24, 236n4; Agamben on, 97–98, 139, 221n4, 223n12, 223n14, 225n27, 248n8; apparatus dissociated from, 98–100, 102; assemblage and, 100–102, 226n37; characters of, 17; definition of, 89, 92–93; dimensions of, 89–96; as discursive and non-discursive, 97; Foucault and, 14, 82, 89–91, 96–97, 141, 152, 161–62, 224n19, 225n26; Gomart and Hennion on, 149–50, 225n28; human-nonhuman and, 96; for material-discursive entanglements, 80–81; network in ontology, 92; new materialisms and, 14; ontology and, 16, 82, 92, 102; power and, 224n20; for relational materialism, 12; of security, 226n31; of sexuality, 98; strategy and, 16, 82, 94–97, 102; technology and, 16, 82, 93, 102

duomining, 26–27, 38

Dupré, John, 234n23

economy: Barua on, 185–86, 244n11; Gabrys on environment and, 142, 178, 243n1; government and political, 134–35; liberalism on, 175; neoliberalism and, 174–76; omega-3 in global, 54; panarchy and, 18, 185–90, 247n26

electricity grid, 41, 44–45, 52

enchanted materialism, 40–41, 43, 49, 54, 213n3

Engels, Friedrich, 88–89, 222n10, 238n11

environment, 17; Barua on, 185–86, 244n11; body open to, 244n9; cybernetics and, 181; Gabrys on economy and, 142, 178, 243n1; governmentality and, 168, 243n1; human and, 177–78, 244n9; Marres on, 165–67, 228n12; more-than-human and, 157; Neimanis on water and, 159; resilience and, 172, 187; technology and, 177; vibrancy of things and, 47–48

environmentality: Burchell and, 243n1; definition of, 176–77; ecology and, 199–200; Foucault on, 14, 17, 120, 168–70, 178, 230n22, 234n22; government and, 142, 170–71, 192, 199–200; Hörl on, 142, 177, 200, 243n2, 243n8, 244n10; neoliberal modes on, 168–69; politics and, 168–69; as probiotic, 179; vital systems security as, 181–84, 245n17

episteme, 97

epistemology: aesthetics and, 33; of agential realism, 15; of new materialisms, 4

Esposito, Roberto, 139, 221n4

essentialism, 8, 28–32

ethics: accountability in, 53; of agential realism, 15; apparatus and, 71–72; Barad and, 66–68, 76–78; Bennett and, 15, 52; Harman on aesthetics and, 32; Levinas and, 22, 32, 58, 66–67; of new materialisms, 4–5; politics and, 75–76; responsibility and, 52–53, 76

experimental ethos, 161–67, 241nn27–31, 242nn32–35

fatty acid. *See* omega-3 fatty acid

Feibleman, James K., 207n5

feminism: Marxism and, 185; materialism and, 38; STS and, 239n16; technoscience as, 12, 57, 141, 216n14
Fernandes, Leela, 68–69, 77
fetal agency, 69–70
force of things, 41–42, 43–46, 50, 57, 214n10
Foucault, Michel: on government of things, 9, 12–13; on killing, 245n20. *See also specific subjects*
Fraser, Mariam, 71, 217n19
freedom of choice: assemblages and, 45; liberal governmentality and, 135; nonhuman and, 43; security technology and, 114–15
Friese, Carrie, 139

Gabrys, Jennifer, 142, 178–79, 243n1, 244n12
genealogy: of governmentality, 82; of milieu, 123–24, 127–29, 232n6, 232n8
genetics, 138, 231nn3–4; biology on, 125–26; Canguilhem on, 122–23, 126; cybernetics and, 125; Foucault on, 16, 122–26; Jacob and, 122–26, 231n2
Global North vs. Global South, 54
global warming. *See* climate change
Gomart, Emilie, 149–50, 225n28
Görres, Joseph, 221n6
government, 228n13; of aleatory, 122, 129–33; on architecture, 84, 233n13; biopolitics as milieu and, 16, 122, 137, 140; Braun on, 142; cybernetics and, 104, 116–20; definition of, 82, 85; democracy and, 46–48, 247n1; environmentality and, 142, 170–71, 192, 199–200; environmental modes of, 17, 177–78; Foucault on, 81–83, 191, 226n34, 227n5; on materials, 153; as material-semiotic, 17, 141–42; milieu guided by, 142, 169; as more-than-human, 79, 191; old and new types of, 170–71; panarchy in, 18, 185–90, 247n26; political economy and, 134–35; with probiotic strategies, 17–18; Quesnay's Principle and, 85–89; resilience and, 172–73; ship with, 84–85; sovereignty and, 52, 82, 84, 89, 99, 103, 129, 151, 221n5, 225n30, 229n17; STS and, 141–42; as technology, 16, 104–6, 111; on things, 82; transformation in, 17–18; vital systems security and, 181–84, 245n17; Woolgar and Neyland on, 153–54
governmentality: agriculture and, 232n9; ANT and, 236n2; environment and, 168, 243n1; Foucault on, 82, 99, 104–5, 110, 120, 122, 132, 135, 137, 141, 168, 220n6, 221n5, 233n14, 236n2; genealogy of, 82; humans and, 121–22; as liberal, 16, 111–14, 122, 127, 132–37, 230nn22–23, 233nn15–16; milieu and, 122, 127, 132–33, 168–69, 176, 191; studies of, 141, 236n2
government of things, 14, 84, 199, 201; definition of, 13; Foucault on, 9, 12–13, 77–78. *See also specific subjects*
governors, 229n19; cybernetics with moderators and, 118; machine within machine in, 117–18; Maxwell on, 116–18, 230n26; moderators vs., 116–18; on politics, 112–13; steam-engine as, 104, 112, 228n13; Watt inventing, 112, 116, 229n21, 230n28
Grant, Iain Hamilton, 211n13
Guattari, Félix, 42, 132, 233n12; apparatus and, 226n33; assemblage and, 100, 216n12; milieu and, 232n6, 232n8

Hacking, Ian, 215n2
Haraway, Donna, 48, 232n8, 240n20, 248n6; on apparatus, 217n15; Barad on, 58, 60; on diffraction, 58, 60; on Foucault and anthropocentrism, 121, 124, 158–59; on human and nonhuman, 138; on posthumanism, 158–59

Harman, Graham: on aesthetics and ethics, 32; ANT and, 26, 43; antimaterialism by, 211n14; anti-realism by, 211n12; on Bennett, 42; on biology, 211n11; on intra-actions, 215n5; on Latour, 26–27, 29, 210n8; on Meillassoux, 210n3; on objects, 210nn5–6, 211n15; on object weirdness, 21, 33–34; on OOO, 7–8, 13–14, 21–22, 25–39, 209n1, 212n24, 248n2; on relationism, 29–30; science and knowledge, 33–34; on scientific naturalism, 23–24; on symbiosis, 29–30; on vibrant matter, 42
Hawkins, Gay, 214n10
Heidegger, Martin, 22, 33, 38, 243n2, 247n27; Morton on, 211n16; on objects, 24–26; technology and, 106, 228n6; on things, 220n3
Heisenberg, Werner, 58–59
Hennion, Antoine, 149–50, 225n28
Hinchliffe, Steve, 50
Holling, Crawford S., 172–75; panarchy by, 18, 185–90, 247n26; on resilience, 243nn4–5
Hörl, Erich, 142, 177, 200, 243n2, 243n8, 244n10
human, 210n3; agency of, 43, 130–31; as complex material, 47; environment and, 177–78, 244n9; Foucault on nonhuman and, 11; more-than-posthuman on, 158; nature and, 247n28; nonhuman influencing, 41–42; as object, 25, 35, 83; beyond politics of, 55–56; posthumanism and, 17, 155, 157; power and, 156; technology and, 103; as things, 82–84
human and nonhuman: apparatus on, 71; in assemblages, 101; Barad on, 57–58; biopolitics on, 16; breeding practices in, 246n25, 247n27; capitalism on, 189; in circulation, 16; democracy for, 46–48; diffractive materialism on, 194–95; dispositive and, 96; experimentation for, 161; Haraway on, 138; interconnections of, 40, 191; labor by, 185–86, 246n22; materiality for, 47–48; milieu and, 16; new materialisms on, 3; OOO on, 14; panarchy in, 18, 185–90, 247n26; STS on, 240n20; technology and, 103; Woolgar and Neyland on, 153–54
humanism: Foucault and, 220n1; posthumanism vs., 156–58; on technology, 108
Hume, David, 209n2
Husserl, Edmund, 24–26, 38
hyperobjects, 27–28, 212nn22–23

immaterialism, 21, 31
intra-actions, 215n5, 216n11; agential realism and, 75; by Barad, 61, 68; matter and, 68, 74; piezo-electric transducer for, 70

Jacob, François, 122–26, 231n2

Kafka, Ben, 86–89, 222n8
Kant, Immanuel, 22–24, 210n3

Lakoff, Andrew, 181–84, 245n16, 245n18
Lamarck, Jean-Baptiste, 128–29, 232n7
Latour, Bruno: ANT and, 26; on critique, 6, 207n6; Harman on, 26–27, 29, 210n8; posthumanist political theory and, 46–48; Rancière and, 46–48; on things, 220n3, 236n2
Law, John, 12, 211n12
Lettow, Susanne, 238n11, 238n13
Levinas, Emmanuel, 22, 32, 58, 66–67
Lezaun, Javier, 165–66, 228n12, 237n4, 238n10, 242n34
liberal governmentality, 111–12, 230n22; Foucault on, 133, 135, 137, 233n15; freedom and, 135; milieu and, 16, 127, 132–33, 136; on population, 134; self-regulation for, 113–14; technology of security and, 114; vital politics and, 133–36, 233n16

liberalism: agency and individualism in, 238n13; biopolitics and, 136; circulation and, 134–36, 233n12, 234n18; on economy, 175; as living, 122; of natural, 134
Lorimer, Jamie, 179–81, 242n32, 245n13

Marres, Noortje, 165–67, 228n12
Marxism, 2, 247n29; Foucault and, 98, 104, 228n9; on government of things, 88; hyperobjects and, 28; poststructuralism and, 57, 208n7; on power, 108
material agency, 241n25
material-discursive: dispositive for entanglements and, 80–81; Foucault on, 11; on government, 82
material foundationalism, 72–75
materialism: Bennett on politics and, 40, 46, 48–56, 213n1; challenges to, 1–2; Foucault and, 208n12; as immaterialism, 21; new materialisms vs. old, 5–6; ontology and, 238n11; OOO vs., 30–31; vital materialism vs., 43. *See also* new materialisms
materiality, 208n7; for humans and nonhumans, 47–48; plastic, 3; of technology, 107–10
materials: government on, 153; performance of, 153; STS and, 14, 141
material-semiotic, 17, 141–42
matter: agential realism and, 208n9; Barad on, 207n3; Bennett on OOO and, 37; Braun on, 49; Foucault and negation of, 9; history and politics of, 18; intra-action and, 68, 74; life vs., 47; as malleable information, 48; passivity of, 3, 5, 10, 40–42, 48, 57, 62; in relation, 54
Maxwell, James Clerk, 116–18, 230nn26–27
Meehan, Katharine, 35, 211n19
Mehrabi, Tara, 93
Meillassoux, Quentin, 22–24, 209n2, 210n3

milieu: Bennett on, 45; biology and, 128–29; biopolitics as government and, 16, 122, 137, 140; Canguilhem on, 121, 128–29, 234n22; on circulation, 16, 127–32, 135–36; cybernetics and, 127; definition of, 131; Deleuze and, 232n6, 232n8; Foucault on, 11, 14, 16, 94–95, 99, 119–22, 124, 127, 131, 141, 192, 233n11; genealogy of, 123–24, 127–29, 232n6, 232n8; governmentality and, 122, 127, 132–33, 168–69, 176, 191; government of things and, 102, 137, 141; history and, 104–5, 110, 232n6, 233n11; Hörl on, 142, 177, 200, 243n2, 243n8, 244n10; as interactive space, 130–31; Lamarck on, 128–29, 232n7; liberal governmentality and, 16, 127, 132–33, 136; new materialisms on, 14; nonhuman and, 244n11; physics and, 128; population and, 133–34; of power, 178; for relational materialism, 12; security and, 129–30; as spatial-temporal, 131; technology and, 192
Moheau, Jean-Baptiste, 136–37, 234n21
Mol, Annemarie, 211n12; on new materialisms and STS, 148; on ontological politics, 209n15; on ontonorm, 209n16
more-than-human, 194, 246n24; biopolitics, 136–40; environment and, 157; Foucault and, 10–11, 15–16, 121–22, 220n1; government as, 79, 191; government of things as, 103; immunological responses and, 245n21; posthumanism vs., 17, 155; responsibility in, 157–58; Whatmore and, 209n13
more-than-posthuman, 154–60
Morton, Timothy: on Heidegger, 211n16; hyperobjects and, 27–28, 212nn22–23; on objects, 210nn9–10; on object weirdness, 28; on OOO, 212n20; on quantum physics, 215n5

nature: Bohr on, 58–59; culture paralleling, 46; human and, 247n28
Nealon, Jeffrey T., 140, 235n26
Neimanis, Astrida, 159
Nelson, Sara, 163, 176, 200, 242n32
neo-animistic ontology, 42
neocybernetics, 17, 120, 170–71, 174, 177, 198, 200
neoliberalism: for economic complexity, 175–76; economy and, 174–75; ecosystem services and, 243n6; on environmentality, 168–69
new materialisms, 2; types of, 3–4, 14. *See also specific subjects*
Newton, Isaac, 128
Neyland, Daniel, 153–54
Neyrat, Frédéric, 198–99
nominalism, 237nn6–7
nonhuman, 246n23; agency of, 43, 130–31; biopolitics, 16; deodand as, 221n5; Foucault and, 141; freedom of choice and, 43; milieu and, 244n11; OOO on, 14; posthumanist accounts of, 158; probiotic strategies on, 179–80
nontology, 17, 144–49, 238n10
norms: biology and, 134; class and, 225n29; discipline for, 115; new materialisms and, 196; ontonorms and, 148, 209n16

object-oriented ontology (OOO), 192, 211n13, 212n22; for aesthetic questions, 34; on agency, 150; on ANT and knowledge, 27–28; on architecture, 21; Bennett on matter and, 37; climate change and, 35–36, 38–39; Cole on, 212n24; on critique, 38; essentialism and, 8, 28; feminist materialism and, 38; Harman and, 7–8, 13–14, 21–39, 209n1, 212n24, 248n2; on human and nonhuman, 14; for beyond human experience, 21; for immaterialism, 31; for knowledge limits, 33; materialism vs., 30–31; Morton on, 212n20; on real and unreal, 26; on realism and things, 27; on relationism, 30–32; for subjectivity, 36–37; on symbiosis, 29–30; on things, 21–22, 27, 34; on weird objects, 165
objects: aesthetics on epistemology and, 33; Harman on, 210nn5–6, 211n15; Heidegger and, 24–26; humans as, 25, 35, 83; Husserl and, 24–25, 38; Morton on, 210nn9–10; new materialisms on, 8; speculative realists on, 22–24; as thing-itself, 27
object weirdness, 165; Harman on, 21, 33–34; Morton on, 28; subjectivity vs., 36–37
omega-3 fatty acid, 41, 45–46, 53–54, 221n4, 239n15
ontography, 237n5
ontological politics, 17, 97–102, 209n15
ontology: of agential realism, 15; alterontologies as, 161; anthropology and, 237n4; Bennett and positive, 15; Foucault on, 146–47, 160; government of things and, 81, 148; as historical, 237n8; materialism and, 238n11; as neo-animistic, 42; new materialisms and, 4, 148; nontology and, 17, 144–49, 238n10; of politics, 17, 97–102, 209n15; STS and, 147; things and, 144–45
ontonorm, 148, 209n16
OOO. *See* object-oriented ontology
Oral, Sevket Benhur, 211n17
overmining, 26–27

panarchy, 18, 185–90, 247n26
Panopticon, 90, 132, 217n16, 224n25, 230n25
Pfeifer, Geoff, 211n13
Pinch, Trevor, 72–74, 218nn22–23
plasticity, 3, 240n19
political technologies, 108–9; on dysfunctions and emergencies, 170; vital systems security and, 181–84, 245n17

politics, 240n23; Barad and, 77; Bennett and, 40, 46, 48–56, 213n1; clock and order in, 110–11; environmentality on, 168–69; ethics and, 75–76; experimentation in science and, 161–67; Foucault and, 77–78, 81, 164; government of things and, 81, 103–4, 166, 192, 198; governor on, 112–13; healthcare and, 242n36; beyond human, 55–56; liberal government and vital, 133–36, 233n16; metaphors on, 16; new materialisms and, 4–5, 8, 196–98; ontological, 17, 97–102, 209n15; of population, 133; Rancière on, 46–48; relational materialism on, 166; STS and, 14, 236n1, 240n21; vibrancy of things and, 42; vital materialism and environmental, 47–48

population: biopolitics and milieu on, 133–34; as distributed reproduction, 235n1; Foucault and, 233n14; liberal governmentality on, 134; Moheau on, 136–37, 234n21; vital systems security for, 181–84, 245n17

posthumanism, 240n22, 240n24, 241n26; anthropocentrism and, 160, 239n16; Barad and, 63; Braun on, 155; government of things and, 160; Haraway on, 158–59; as historical break, 155, 157; human and, 17, 155, 157; humanism vs., 156–58; more than, 154–60; on nonhuman, 158; political theory of, 46–48; relational materialism on, 158–59

postmodernism, 28, 69

poststructuralism, 21; antihumanism by, 63; Barad and, 57, 63, 75, 201; correlationism and, 23; Marxism and, 57, 208n7

power, 17; agential realism on, 68–69; ANT on, 147; apparatus and, 70–71, 100; Barad on, 15; Bennett on thing-, 14–15, 40–41, 50, 54, 150, 214n10, 233n16; Butler on, 62; Canguilhem on, 111; clock and, 110–11; Deleuze and, 233n12; dispositive and, 224n20; Fernandes and, 68–69, 77; Foucault and, 9–10, 18, 58, 62, 79–80, 101, 227n4, 228n12, 229n15; humans and, 156; Marxism on, 108; milieu of, 178; new materialisms and, 4–5, 9; Panopticon and, 230n25; right vs. physics of, 84; as strategics and logistics, 232n10; technology and, 108

probiotic strategies: anthropocentrism and, 180; definition of, 179; government with, 17–18; Lorimer on, 179–81, 242n32, 245n13; on nonhuman, 179–80; urban flood management and, 181, 245n15

quantum physics: agential realism and, 74, 218n24; Barad and, 3, 15, 57–58, 73–74, 215n5, 218n24; Bohm and, 218n22; Bohr and, 73–74, 151, 218n22, 218n24; Feibleman and, 207n5; Heisenberg and, 58–59; Morton on, 215n5; Pinch on, 72–74, 218nn22–23

Quesnay's Principle, 85–89

race, 125–26
Rancière, Jacques, 46–48
reductionist materialism, 3
reflexivity, 60–61
relationalism: agency as more than, 48–51; diffraction and, 58, 61; essentialism and, 29; military and, 240n18; OOO on, 30–32

relational materialism, 208n7, 236n3; dispositive for, 12; government of things and, 14, 152, 192, 201; Haraway on, 158–59; material relationalism and, 152, 199; on politics, 166; on posthumanism, 158–59; STS and, 17

relational ontology, 57, 208n9
representationalism, 59–61, 75, 215n2, 219nn27–28

reproductive technologies, 218n20, 235n1
resilience: adaptability and, 200; as biopolitics, 171, 176; on capitalism and environment, 174, 187, 200, 243nn4–5; definition of, 243n5; as dominant life paradigm, 171–74, 187–89; environment and, 172, 187; government and, 172–73; of neo-cybernetics, 17, 120, 170–71, 174, 177, 198, 200; new materialisms and, 189–90; technologies of, 179, 183
responsibility: accountability in, 53; agency and assemblages in, 52–53; agential realism and, 76; Anthropocene and, 247n28; ethics and, 52–53, 76; experiments and, 163; as full or partial, 51–52; in more-than-human practices, 157–58
Rheinberger, Hans-Jörg, 162–63, 231n3
Rouse, Joseph, 80, 216n14
Rutherford, Paul, 79–80

Schrader, Astrid, 218n24
science: Foucault and natural, 220nn4–5; Harman on knowledge and, 33–34; naturalism and, 23–24; new materialisms and, 151, 195–96; for security technologies, 115
science and technology studies (STS): analytics of government and, 141–42; Barad and, 72; on becoming, 145; experimentation in, 161–67; Foucault and, 12, 142, 192, 216n9; government of things and, 144, 166–67; on human and nonhuman, 240n20; Lezaun on, 165–66, 228n12, 237n4, 238n10, 242n34; for material practices, 14; materials and, 141; Mol on new materialisms and, 148; new materialisms and, 145, 165; new materialisms and feminist, 239n16; ontology and, 147; politics and, 14, 236n1, 240n21; relational materialism and, 17
scientism, 73, 149–54, 195–96, 218n21

security, technologies of, 104; dispositive and, 226n31; Foucault on, 116, 122, 129–30; freedom and, 114–15; naturalness conceptions in, 115–16; science and expertise for, 115
self-regulation: control technology and, 114, 131; government of things and, 116; for liberal government, 113–14; after perturbations, 172
Senellart, Michel, 84, 221n5
sex: Foucault on, 108; technology of, 98
Sheldon, Rebekah, 38, 211n15
Simondon, Gilbert, 94, 224n21, 236n4, 244n10
Smith, Mick, 219n2
social constructivism, 58–59, 104, 146, 208n7; beyond, 107–10; Foucault and, 9–11, 80, 103, 107–9; technology vs., 16
sovereignty: government and, 52, 82, 84, 89, 99, 103, 129, 151, 221n5, 225n30, 229n17; government of things vs., 89
speculative realism, 22–24; for aesthetic questions, 34; Zahavi on, 37
steam-engine governor, 104, 112, 228n13
strategy: anthropocentrism and, 157; dispositive and, 16, 82, 94–97, 102; Foucault and, 223n13, 224n22, 224n25; power and, 232n10
STS. *See* science and technology studies
subjectivity: Harman on, 24; object weirdness vs., 36–37; OOO for, 36–37; Taylor on, 34, 36
subversive practices, 72
symbiosis, 29–30
sympathy, 214n11

Taylor, Carol, 34, 36
technological determinism: beyond, 107–10; technology and, 16
technology: Bennett on grid and, 41, 44–45, 52; cybernetics from, 104, 116; dispositive and, 16, 82, 93, 102; ecology and, 177; Foucault on, 14, 104–7, 109,

141, 192, 227n4, 227nn1-2, 228n11; government as, 16, 104–7, 111; Heidegger on, 106, 228n6; of human affairs, 103; humanism on, 108; intra-actions and, 70; materiality of, 107–10; power and, 108; for relational materialism, 12; of resilience, 179, 183; of sex, 98; Simondon and, 94, 224n21, 236n4, 244n10; social constructivism vs., 16; technological determinism and, 16. *See also* political technologies
technoscience, 63, 155, 160; agential realism and, 72–73; as feminist, 12, 57, 141, 216n14; as postcolonial, 12, 141
thing-itself, 27; as concrete, 31–32, 34; relationism on, 29
thing-power: Abrahamsson on, 150–51; by Bennett, 14–15, 40–41, 50, 54, 150, 214n10, 233n16
things: as actors, 110; assemblages of, 42; Bennett and vibrancy of, 8, 13–15, 40–56; Comte on, 87–88, 222nn8–9, 232n8; Esposito on, 221n4; force of, 41–42, 43–46, 50, 57, 214n10; Foucault on, 82–83; government of, 89; Heidegger on, 220n3; humans and, 82–84; irreducible dark side of, 27; Latour on, 220n3, 236n2; Moheau on, 136–37, 234n21; new materialisms on, 8; ontology and whatness of, 144–45; OOO on, 21–22, 27, 34; beyond subject and object, 43–44; vital materialism on, 8, 40–56, 194
Thrift, Nigel, 79, 152

ultrasound, 217nn16–18
undermining, 26–27

vibrancy of things: Bennett and, 8, 13–15, 40–56; environment and, 47–48; for greener human culture, 42; on politics, 42
vital materialism, 193; on agency, 150; Barad and, 219n25; criticism of, 48; environmental politics and, 47–48; materialism vs., 43; on things, 8, 40–56, 194. *See also* Bennett, Jane
vital systems security: circulation and, 179, 182; Collier and Lakoff on, 181–84, 245n16, 245n18; government and, 181–84, 245n17

Wakefield, Stephanie, 142, 248n8
Washick, Bonnie, 53, 77, 240n24
Watt, James, 112, 116, 229n21, 230n28
weird objects. *See* object weirdness
Whatmore, Sarah, 49, 155, 209n13, 221n6
Whitman, Walt, 214n11
Wiener, Norbert, 118, 124, 230n30, 231n4
will: agency and, 51, 221n4; by inorganic matter, 44
Willey, Angela, 195–96, 239n16, 248n4
Wingrove, Elizabeth, 53, 77, 240n24
Wolfe, Cary, 138–39
Woolgar, Steve, 153–54, 237n4, 238n10

Zahavi, Dan, 37

ABOUT THE AUTHOR

THOMAS LEMKE is Professor of Sociology with a Focus on Biotechnologies, Nature, and Society in the Faculty of Social Sciences at Goethe University Frankfurt. He is the author of *Foucault's Analysis of Modern Governmentality: A Critique of Political Reason* and of *Biopolitics: An Advanced Introduction*.

www.ingramcontent.com/pod-product-compliance
Lightning Source LLC
Chambersburg PA
CBHW020356080526
44584CB00014B/1041